T0320361

Human Rights and Emerging Technologies

Human Rights and Emerging Technologies

Analysis and Perspectives in Europe

Daniele Ruggiu

Foreword by Roger Brownsword

PAN STANFORD PUBLISHING

Published by

Pan Stanford Publishing Pte. Ltd.
Penthouse Level, Suntec Tower 3
8 Temasek Boulevard
Singapore 038988

Email: editorial@panstanford.com
Web: www.panstanford.com

British Library Cataloguing-in-Publication Data
A catalogue record for this book is available from the British Library.

Human Rights and Emerging Technologies: Analysis and Perspectives in Europe

ISBN 978-981-4774-93-2 (Hardcover)
ISBN 978-0-429-49059-0 (eBook)

To the invisible ones hiding behind the progress

*Quoi qu'elle donne à voir et quelle que soit sa manière,
une photo est toujours invisible: ce n'est pas elle qu'on voit.*

Whatever it grants to vision and whatever its manner,
a photograph is always invisible: it is not it that we see.

Roland Barthes, *La chambre claire: Note sur la photographie,*
1980, p. 18

Contents

Foreword
by Roger Brownsword

Nearly thirty years ago, I became interested in the patentability of the novel processes and products that were then being developed in biotechnology. Knowing very little about either patent law or biotechnology, this was pure serendipity – because, looking back, I regard this as the moment when my more general interest in law and emerging technologies originated. That said, it should be emphasised that, at the time, I would not have characterised my interest in those terms and, indeed, I believe that it was to be some years before 'law and technology' was itself to crystallise as an accepted field of legal inquiry.

The case that caught my attention all those years ago was the application to patent the Harvard Onco-mouse (a mouse that was genetically engineered with a view to assisting researchers who sought a better understanding of the way in which tumours develop). The arguments in support of the patent application were that the claims covered original and innovative processes and products that promised to help with treatments for human cancers. While the US Patent Office had signalled that they treated the mouse as patentable subject matter, at the European Patent Office the application was regarded as much more controversial. Lying at the core of the controversy, Article 53(a) of the European Patent Convention (EPC) 1973 provides that no matter how inventive and original the process or product, the subject matter of the claim should be treated as unpatentable if its commercial exploitation would be contrary to ordre public or morality. Feelings were running high. Patent offices are geared for granting patents and supporting innovation; but, outside the European Patent Office in Munich, there were many people carrying boards that protested that there should be no patents on life.

Inside the Patent Office the examiners were, of course, required to take Article 53(a) seriously. However, there was really no jurisprudence to assist the examiners, and the patent community, generally taking a 'technical' view of the law, professed no expertise in determining what might be contrary to morality. The $1000 question was, whose or which morality was it that was the reference point for the EPC?

In a book that staked out a position that was extremely unpopular with patent lawyers, Deryck Beyleveld and I argued that the relevant morality had to be that given by the European Convention on Human Rights (because, quite apart from any other considerations, when the EPC was opened for signature, the Contracting States were already committed to the ECHR and simply could not set aside their human rights' obligations) (Beyleveld and Brownsword, 1993). At that time, even if this argument were accepted, it was not entirely clear how much purchase the ECHR would have on patentability because the biotechnological applications that were at the heart of this early debate concerned genetically modified plants and non-human animals. However, as the years have gone by, and as developments in human genetics have been brought to patent offices, the potential application of human rights to questions of patentability is more widely appreciated.

This is not to say that the application of human rights is no longer controversial. As the authors of a report prepared by the Rathenau Institute (and written for the Council of Europe's Committee on Bioethics) remark, emerging technologies tend to be 'strongly intertwined' with human rights and human dignity (van Est et al., 2014, p. 10). However, the intertwining has both positive and negative aspects: while these technologies might 'strengthen human rights and human dignity', they might also 'give rise to risks and ethical issues and therefore threaten human rights and human dignity' (Ibid.). So, for example, appeals to human rights (particularly coupled with human dignity) might be made to argue against patentability – especially where researchers have made use of human embryos, or materials derived from human embryos, or where it is not clear that the human sources of research materials have consented to commercial exploitation. Equally, though, appeals to human rights might be made to argue (as Aurora Plomer has done) that 'the global patent system is spectacularly failing to facilitate realization of the human rights ideal of universal access to science' (Plomer, 2015, p. 178).

But, of course, the class of emerging technologies goes way beyond biotechnologies. During the last thirty years, there have been major developments in information and communication technologies, in machine-learning, robotics and artificial intelligence, in nanotechnologies, in neurotechnologies, in synthetic biology, in additive manufacturing tools (such as those used for 3D printing), and in the convergence of these technological streams. Similarly, patentability is not the only legal issue in relation to these technologies. The effect of withholding a patent is to offer no encouragement for investment in a line of research, but it is not a formal prohibition, and neither, of course, is it a formal endorsement or recognition that support should be given.

To the extent that the recognition of human rights gives citizens some security against overbearing State actions, one of the more obvious contexts for appeals to the ECHR is that of criminal justice. Thus far, emergent technologies (such as CCTV surveillance and DNA profiling) have largely been employed to discourage and to identify offenders, but the latest developments in AI might encourage a shift to a more predictive and preventive approach – and it will be a major surprise if an algorithmic approach to criminal justice does not provoke challenges that appeal to the due process provisions of human rights instruments.

Already, though, the relevance of the ECHR is clear. For example, in *R v. Chief Constable of South Yorkshire Police, ex parte LS and Marper*,[1] the UK's legislation that authorised the taking and retention of biosamples, and the making and retention of DNA profiles, was challenged as incompatible with the privacy provisions of Article 8 of the ECHR. Although the domestic courts were generally prepared to accept that the right to informational privacy was engaged, they held that the State could justify its measures by reference to the compelling public interest in the prevention and

---

[1] [2002] EWCA Civ 1275, [2004] UKHL 39.

detection of serious crime. However, on appeal to the Strasbourg Court, the Grand Chamber took a quite different view, holding that the sweeping powers for the taking and retention of DNA profiles of persons suspected but not actually convicted of offences constituted a disproportionate interference with the right to respect for private life and could not be considered to strike a fair balance between the competing public and private interests.[2]

Accordingly, there are grounds for thinking that appeals to human rights will become more frequent and significant as the applications of emerging technologies penetrate ever deeper into our daily lives. Whether the technologies impact on early human life or the end of life, on enhanced human capacities or smart machines, whether the context is criminal justice or health, the communities of the twenty-first century face unprecedented disruption and technological turbulence. If we are to stay focused on our fundamental values, we need to be clear about the bearing of human rights in relation to emergent technologies, and this is exactly what Daniele Ruggiu sets out to do in this book.

Distinctively, Ruggiu treats human rights not only as the benchmark for what the State, science and commerce should not be permitted or encouraged to do in relation to the development, application, and exploitation of emerging technologies, but also as the reference point for what we should be permitted to do and, indeed, for the responsibilities and obligations that the State has to assist us (as bearers of human rights). What, though, are the responsibilities of the State and the research community?

At the heart of Ruggiu's book is the important thesis that responsible research and innovation – understood as research that is ethically acceptable, sustainable, and societally desirable – needs to be fully attuned to respect for human rights. Stated simply, if the governance of research is to channel researchers towards the development of processes and products that are ethically acceptable, sustainable, and societally desirable, it needs to encourage (i) an inclusive, flexible, agile, and reflexive process, and (ii) a normative sensitivity. However, neither inclusive, flexible, agile, and reflexive processes nor a sensitivity to what one ought to do (rather

---

[2] *S and Marper v. United Kingdom* [2008] ECHR 1581.

than simply to what one can do) can, without more, guarantee rational and reasonable outcomes. Even if the process is normatively sensitised, there is, as Ruggiu rightly says, 'no guarantee that the values chosen via the process of negotiation coincide with those at the basis of the system, in particular with fundamental rights' (in Section 3.6). If we are to secure responsible research and innovation, human rights needs to be brought in as the anchoring point for the enterprise.

This, surely, makes very good sense. Human rights instruments treat humans alike in recognising that we all have the same basic needs – needs that derive in part from our biological characteristics and needs that relate to our prospective agency. If researchers compromise the conditions for human *existence* (for example, by taking forward genetic engineering, or nanotechnology, or synthetic biology in the dystopian way that some writers imagine), then they compromise the commons for all human agents. If researchers are complicit in developing surveillance tools that inhibit our agency, then once again the commons is compromised. The surveillance society does not threaten our existence, but without some privacy our agency is liable to be frustrated and inhibited. And, if researchers are complicit in developing technologies that crowd out moral reasoning, then again the commons is compromised. As Ian Kerr put it so eloquently, there is no automating of moral virtue (Ker, 2010, p. 247). To be clear, the preservation and protection of the commons is not a prescription for homogeneity either of individual agents or groups or communities. There is still ample scope for plurality at second, third and fourth base, but without the commons secured at first base, there is no future for human agents or the kind of person or kind of community that they want to be.

It follows that responsible research must not only respect these basic conditions but do so *above all other considerations*. These are cosmopolitan conditions; they represent values and goods that no agent can coherently or reasonably deny; from any regulatory perspective, these are non-negotiable conditions. Thinking about the impact of innovation (both negative and positive) on human rights is a good place for both regulators and researchers to start. Again, to quote Ruggiu: 'Human rights can have the paramount function of strengthening the protection of individual rights in the balancing with other public interests at the national and EU level, avoiding cases of system failure of innovation and the abrupt shift of governance arrangements. Thanks to them and their practice some principles applicable to the technoscientific progress can be promptly identified and proactively used for consistently steering governance of emerging technologies' (Section 3.6).

Thirty years ago, as I have said, 'law and technology' was barely on the radar, and the books that now fill the libraries on this burgeoning field of inquiry were still to be written. In his book, Daniele Ruggiu makes an important contribution to the literature, recounting the ways in which European regulators have sought to engage more effectively and acceptably with emerging technologies but, most importantly, putting human rights front and centre in his narrative.

References

Beyleveld, D., and Brownsword, R. (1993). *Mice, Morality, and Patents (The Oncomouse Application and Article 53(a) of the European Patent Convention)*, with a foreword by Lord Scarman, London: Common Law Institute of Intellectual Property.

Kerr, I. (2010). Digital Locks and the Automation of Virtue, in: Geist, M. (ed.), *From 'Radical Extremism' to 'Balanced Copyright': Canadian Copyright and the Digital Agenda*, Toronto: Irwin Law, pp. 247–304.

Plomer, A. (2015). *Patents, Human Rights and Access to Science*, Cheltenham: Edward Elgar.

van Est, R., Stemerding, D., Rerimassie, V., Schuijff, M., Timmer, J., and Brom, F. (2014). *From Bio to NBIV Convergence: From Medical Practice to Daily Life*, The Hague: Rathenau Instituut.

Table of Legislation

UNITED NATIONS

- Preamble of the *Charter of the United Nations* adopted in S. Francisco on 26 June 1945 (entered into force on 24 October 1945).
- *Universal Declaration on Human Rights* (UDHR) adopted in New York on 10 December 1948 by the General Assembly of the United Nations.
- *Proclamation of Tehran, Final Act of the International Conference on Human Rights*, 22 April to 13 May 1968.
- *Declaration on the Use of the Scientific and Technological Progress in the Interest of the Peace and for the Benefit of the Mankind* adopted with Resolution 3384 (XXX) proclaimed by the General Assembly on 10 December 1975.
- *Convention on the Prohibition of the Development, Production and Stockpiling of Bacteriological (Biological) and Toxin Weapons and on Their Destruction*, entered into force on 26 March 1975.
- *International Covenant on Civil and Political Rights* (ICCPR) adopted in New York on 16 December 1966 (entered into force on 23 March 1976).
- *International Covenant on Economic, Social, and Cultural Rights* (ICESCR) adopted in New York on 16 December 1966 (entered into force on 3 January 1976).
- Resolution 38/111 of the General Assembly on *Implication of scientific & technological developments for human rights*, 16 December 1983.
- Resolution 38/112 and 113 of the General Assembly on *Human Rights and Scientific and Technological Developments*, 16 December 1983.

- Resolution 1986/9 of the Commission on Human Rights, *Use of the Scientific and Technological Developments for the Promotion and the Realization of Human Rights and Fundamental Freedoms*, 27 March 1983.
- *Convention on Biological Diversity* of Rio de Janeiro adopted on 5 June 1992.
- *Kyoto Protocol to the United Nations Framework Convention on Climate Change* adopted in Kyoto, Japan, 11 December 1997. (greenhouse gas reduction)
- The UNECE *Aarhus Convention on Access to Information, Public Participation in Decision-Making and Access to Justice in Environmental Matters*, adopted in Aarhus, Denmark, on 25 June 1998.
- Resolution 1999/63 of the General Assembly on *Human Rights and Bioethics*, 28 April 1999.
- *Cartagena Protocol on Biosafety* on 29 January 2000, entered into force on 11 September 2003. The EU and all EU Member States have ratified the protocol. (Regulation (EC) No. 1946/2003 on *Transboundary Movements of Genetically Modified Organisms* the regulatory instrument that implements the provisions of the Cartagena Protocol on Biosafety within the European Union).
- *Declaration on Human Cloning* adopted on 24 February 2005 by the General Assembly of the United Nations.
- *Convention on the Rights of Persons with Disabilities* adopted on 13 December 2006 by the General Assembly of the United Nations (entered into force on 3 May 2008).
- The Conference of the Parties to the Convention on Biological Diversity, Decision XI/11, of 8–19 October 2012, Hyderabad, India, *New and Emerging Issues Relating to the Conservation and Sustainable Use of Biodiversity.*

UNESCO

- *Universal Declaration on the Human Genome and Human Rights*, adopted by the UNESCO General Conference in 11 November 1997 and subsequently endorsed by the United Nations General Assembly in 1998.
- *International Declaration on Human Genetic Data*, adopted by UNESCO General Conference on 16 October 2003.

- *Universal Declaration on Bioethics and Human Rights*, adopted on 19 October 2005.
- UNESCO – COMEST (World Commission on the Ethics of Scientific Knowledge and Technology) 2005: *The Precautionary Principle*, adopted in Paris on March 2005 by UNESCO.
- UNESCO (2006). *The Ethics and Politics of Nanotechnology*, Paris: United Nations Educational, Scientific and Cultural Organization.

FAO

- Food and Agricultural Organization (FAO), *International Treaty on Plant Genetic Resources for Food and Agriculture*, adopted on 2001, available at ftp://ext-ftp.fao.org/ag/cgrfa/it/ITPGRe.pdf [Accessed 10 August 2012].

FAO/WHO

- FAO/WHO [Food and Agriculture Organization of the United Nations/World Health Organization] (2009) *FAO/WHO Expert Meeting on the Application of Nanotechnologies in the Food and Agriculture Sectors: Potential Food Safety Implications: Meeting Report*. Rome: FAO and WHO, available at http://www.evira.fi/attachments/elintarvikkeet/elintarviketietoa/fao_who_nano_expert_meeting_report_final__2_.pdf [Accessed 29 August 2013].
- FAO/WHO [Food and Agriculture Organization of the United Nations/World Health Organization] (2010) FAO/WHO Expert Meeting on the Application of Nanotechnologies in the Food and Agriculture Sectors. Potential Food Safety Implications. Meeting Report. Rome: FAO and WHO, available at http://whqlibdoc.who.int/publications/2010/9789241563932_eng.pdf [Accessed 29 August 2013].
- FAO/WHO [Food and Agriculture Organization of the United Nations/World Health Organization] (2013) *Paper: State of the Art on the Initiatives and Activities Relevant to Risk Assessment and Risk Management of Nanotechnologies in the Food and Agriculture Sectors*. Genève: FAO and WHO, available at http://www.fao.org/docrep/018/i3281e/i3281e.pdf [Accessed 29 August 2013].

WTO

- World Trade Organization (WTO), SPS Agreement, reached in Marrakech, 1994.

WMA

- World Medical Association (WMA), Declaration of Helsinki, *Ethical Principles for Medical Research Involving Human Subjects*, Seoul, October 2008.

ICH

- International Conference on Harmonisation of Technical Requirements for Registration of Pharmaceuticals for Human Use (ICH), *ICH Harmonised Tripartite Guideline. Guideline for Good Clinical Practice*, ICH, E-6(R1) 1996 (as issued by CPMP in July 1996: see CPMP/ICH/135/95).

WADA

- World Anti-Doping Agency Foundation Broad (WADA) *World Anti-Doping Code*, first adopted in 2003, revisited on 17 November 2007 (effective from 1st January 2009), available at http://www.wada-ama.org/Documents/World_Anti-Doping_Program/WADP-The-Code/WADA_Anti-Doping_CODE_2009_EN.pdf [Accessed 19 May 2012].

UNODA

- United Nation Office for Disarmament Affairs (UNODA), *The Biological Weapons Convention* or *Convention on the Prohibition of the Development, Production and Stockpiling of Bacteriological (Biological) and Toxin Weapons and on their Destruction (BWC)*, adopted on 10 April 1972 and entered into force on 26 March 1975.

OPWC

- Organization for the Prohibition of Chemical Weapons (OPWC), *The Chemical Weapons Convention* or *Convention on the Prohibition of the Development, Production, Stockpiling and Use of*

Chemical Weapons and on their Destruction (CWC), adopted on 13 January 1993 and entered into force on 29 April 1997.

COUNCIL OF EUROPE

- *Convention for the protection on Human Rights and Fundamental Freedoms* (ECHR) (CETS n. 5), adopted in Rome on 11 November 1950 (entered into force on 3 September 1953).
- *European Agreement on the Exchange of Therapeutic Substances of Human Origin* (CETS n. 26) adopted in Paris on 15 December 1958 (entered into force on 1 January 1959).
- *Convention on Elaboration of a European Pharmacopeia* (CETS n. 50), adopted in Strasbourg on 22 July 1964 (entered into force on 22 May 1974).
- *European Social Charter* (ESC) (CETS n. 35), adopted in Turin on 3 October 1961 (entered into force on 26 February 1965), revised in Strasbourg on 3 May 1996 (CETS n. 163) (entered into force on 1 July 1999).
- *Convention on Unification of Certain Points of Substantive Laws on Patents for Invention* (CETS n. 84), adopted in Strasbourg on 27 November 1963 (entered into force on 1 August 1980).
- *Anti-doping Convention* (CETS n. 135), adopted in Strasbourg on 16 November 1989 (entered into force on 1 January 1990).
- *European Agreement on the Exchange of Tissue-Typing Reagents* (CETS n. 47), adopted in Strasbourg on 17 September 1974 (entered into force on 23 April 1977).
- *Convention for the Protection of Individuals with regard to Automatic Processing of Personal Data* (CETS n. 108), adopted in Strasbourg on 28 January 1981 (entered into force on 10 October 1985).
- *Additional Protocol amending European Social Charter* (CETS n. 142), adopted in Turin on 21 October 1991.
- *Additional Protocol to European Social Charter Providing for a System of Collectives Complaints* (CETS n. 158), adopted in Strasbourg on 11 November 1995 (entered into force on 1 July 1998).
- *Convention for the Protection of Human Rights and Dignity of the Human Being with regard to the Application of Biology and*

Medicine (Convention on Human Rights and Biomedicine or the Oviedo Convention) (CETS n. 164), adopted in Oviedo on 4 April 1997 (entered into force on 1 December 1999).

- *Additional Protocol to the Convention on Human Rights and Biomedicine on the Prohibition of Cloning Human Beings* (CETS n. 168), adopted in Paris on 12 January 1998 (entered into force on 1 March 2001).
- *Additional Protocol Regarding the Supervisory Authorities and Transborder Data Flows* (CETS n. 181), adopted in Strasbourg on 8 November 2001 (entered into force on 1 July 2004).
- *Additional Protocol to the Convention on Human Rights and Biomedicine concerning Transplantation of Organs and Tissues of Human Origins* (CETS n. 186), adopted in Strasbourg on 24 January 2002 (entered into force on 1 May 2006).
- *Additional Protocol to Anti-doping Convention* (CETS n. 188), adopted in Warsaw on 12 September 2001 (entered into force on 1 April 2004).
- *Additional Protocol to the Convention on Human Rights and Biomedicine concerning Biomedical Research* (CETS n. 195), adopted in Strasbourg on 25 January 2005 (entered into force on 1 September 2007.
- *Additional Protocol to the Convention on Human Rights and Biomedicine concerning genetic testing for health proposes* (CETS n. 203), adopted in Strasbourg on 11 November 2008 (not entered into force yet).

(i) Parliamentary Assembly of the Council of Europe

- Recommendation 934 of the Parliamentary Assembly of the Council of Europe *on the genetic engineering*, adopted on 26 January 1982.
- Recommendation *1160* of the Parliamentary Assembly of the Council of Europe *on the preparation of a convention on bioethics*, adopted on 28 June 1991.
- Recommendation 1425 of the Parliamentary Assembly of the Council of Europe *on the biotechnology and the intellectual propriety*, adopted on 20 September 1999.

- Recommendation 1468 of the Parliamentary Assembly of the Council of Europe *on the biotechnology*, adopted on 29 June 2000.
- Recommendation 1512 of the Parliamentary Assembly of the Council of Europe *on the protection for the human genome by the Council of Europe*, adopted on 20 April 2001.
- Recommendation 2017 (2013) of the Parliamentary Assembly of the Council of Europe *Nanotechnology: Balancing Benefits and Risks to Public Health and the Environment*, adopted on 26 April 2013 (final version).

(ii) Committee of Ministers of the Council of Europe

- Recommendation RE (81) 1 of the Committee of Ministers of the Council of Europe *on regulations for automated medical data banks*, adopted on 1981.
- Recommendation RE (90) 3 of the Committee of Ministers of the Council of Europe *on medical research on human being*, adopted on 1990.
- Recommendation RE (90) 13 of the Committee of Ministers of the Council of Europe *on prenatal genetic screening, prenatal genetic diagnosis and associated genetic counseling*, adopted on 1990.
- Recommendation RE (92) 1 of the Committee of Ministers of the Council of Europe *on the use of analysis of deoxyribonucleic acid (DNA) used within the framework of the criminal justice system*, adopted on 1992.
- Recommendation RE (97) 15 of the Committee of Ministers of the Council of Europe *on xenotransplantations*, adopted on 1997.
- Recommendation RE (2003) 10 of the Committee of Ministers of the Council of Europe *on xenotransplantations*, adopted on 19 June 2003.
- The Recommendation RE (2004) 10 of the Committee of Ministers of the Council of Europe *concerning the protection of the human rights and dignity of persons with mental disorder and its Explanatory Memorandum*, adopted on 22 September 2004.
- Recommendation RE (2006) 4 of the Committee of Ministers of the Council of Europe *on research on biological materials of human origin*, adopted on 15 March 2006.

(iii) Committee on Bioethics (DH-BIO, Previous CAHBI and CDBI)

- *Report on Human Artificial Procreation. Principles Set Out in the Report of the Ad Hoc Committee of Experts on Progress in the Biomedical Sciences (CAHBI)*, 1989.
- *Explanatory Report on Convention for the Protection of Human Rights and Dignity of the Human Being with regard to the Application of Biology and Medicine*, 17 December 1996.
- *Working Party on Human Genetics*, adopted in Strasbourg on 27 October 1997.
- *Medically Assisted Procreation and the Protection of the Human Embryo: Comparative Study on the Situation in 39 States*, adopted in Strasbourg on 4 June 1998.
- *Cloning Comparative Study on the Situation in 44 States*, adopted in Strasbourg on 4 June 1998.
- White Paper on *the Protection of Human Rights and Dignity of People Suffering from Mental Disorders Especially Those Placed as Involuntary Patients in Psychiatric Establishment*, 2000.
- *Explanatory Report to the State of Art in the Field of Xenotransplantation*, adopted in Strasbourg on 21 February 2003.
- *The Protection of the Human Embryo* in vitro. *Report by the Working Party on the Protection of the Human Embryo and Foetus*, adopted in 19 June 2003.
- *Explanatory Report to the Additional Protocol to the Convention on Human Rights and Biomedicine concerning the Biomedical Research*, adopted in Strasbourg on 25 January 2005.
- *The Court's Case-Law Concerning the Health in General*, 18 July 2008.
- *Report on Medical Decisions in End-Life Situations and Ethical Implications of the Available Options*, adopted in Strasbourg on 2–5 December 2008.
- *Bioethics and the Case-Law of the Court*, 14 October 2009.
- *Background Document on Preimplantation and Prenatal Genetic Testing: Clinical Situation, Legal Situation*, adopted in Strasbourg on 22 November 2010.
- *Statement on Genome Editing Technologies*, adopted in Strasburg 1–4 December 2015.

EUROPEAN UNION

- *Treaty on the European Union and Treaty on the Functioning of the European Union* (consolidated versions 2010/C 83/01), adopted in Lisbon on 13 December 2007 and entered into force on 1 December 2009.
- *Charter of Fundamental Rights of the European Union* (2010/C 83/02), proclaimed in Nice on 7 December 2000 and entered into force on 1 December 2009. It has been recognized by the Article I-6 of the *Lisbon Treaty*.
- *White Paper on Sport* COM(2007) 391 final of the European Commission of 11 July 2007, http://ec.europa.eu/sport/documents/wp_on_sport_en.pdf [Accessed 19 May 2012].

(α) Advanced Therapy/Cell Based Medicinal Products

- Directive 2001/20/EC of the European Parliament and of the Council of 4 April 2001 *on the Approximation of the Laws, Regulations and Administrative Provisions of the Member States Relating to the Implementation of Good Clinical Practice in the Conduct of Clinical Trials on Medicinal Products for Human Use, Official Journal of the European Union*, L121, 1.5.2001, p. 34, http://ec.europa.eu/health/files/eudralex/vol-1/dir_2001_20/dir_2001_20_en.pdf [Accessed 7 April 2015].
- Directive 2002/98/EC of the European Parliament and of the Council of 27 January 2003 *Setting Standards of Quality and Safety for the Collection, Testing, Processing, Storage and Distribution of Human Blood and Blood Components* and amending Directive 2001/83/EC, *Official Journal of the European Union*, L33, 8.2.2003, p. 30 http://eur-lex.europa.eu/LexUriServ/LexUriServ.do?uri=OJ:L:2003:033:0030:0040:EN:PDF. [Accessed 7 April 2015].
- Directive 2004/23/EC of the European Parliament and of the Council of 31 March 2004 *on Setting Standards of Quality and Safety for the Donation, Procurement, Testing, Processing, Preservation, Storage and Distribution of Human Tissues and Cells, Official Journal of the European Union*, L102, 31.4.2004, pp. 48–58

- Commission communication of 3/12/2009 *Detailed Guidelines on Good Clinical Practice Specific to Advanced Therapy Medicinal Products* (ENTR/F/2/SF/dn D(2009) 35810, Brussels: European Commission, http://ec.europa.eu/health/files/eudralex/vol-10/2009_11_03_guideline.pdf [Accessed 7 April 2015].
- Regulation (EU) No. 536/2014 of the European Parliament and of the Council of 16 April 2014 *on Clinical Trials on Medicinal Products for Human Use* and Replacing Directive 2001/20/EC, *Official Journal of the European Union*, L158/1, 27.5.2014 http://ec.europa.eu/health/files/eudralex/vol-1/reg_2014_536/reg_2014_536_en.pdf [Accessed 1 May 2015].

(*β*) Biotechnologies

- Commission Decision of 27 September 2000 *Concerning the Guidance Notes for risk Assessment Outlined in Annex III of Directive 90/219/EEC on the Contained Use of Genetically Modified Microorganisms* (notified under document number C(2000) 2736), *Official Journal of the European Union*, L258, 12.10.2000, pp. 43–48, http://eur-lex.europa.eu/legal-content/EN/TXT/PDF/?uri=CELEX:32000D0608&from=EN [Accessed 13 April 2015].
- Directive 2001/18/EC of the European Parliament and of the Council of 12 March 2001 *on the Deliberate Release into the environment of genetically modified organisms* and repealing Council Directive 90/220/EEC, *Official Journal of the European Union*, L106, 17/4/2001, http://eur-lex.europa.eu/resource.html?uri=cellar:303dd4fa-07a8-4d20-86a8-0baaf0518d22.0004.02/DOC_1&format=PDF. [Accessed 24 March 2015].
- Regulation (EC) 1829/2003 of the European Parliament and the Council of 22 September 2003 on *Genetically Modified Food and Feed*, *Official Journal of the European Union*, L268/1, 18.10.2003, http://ec.europa.eu/food/food/animalnutrition/labelling/Reg_1829_2003_en.pdf [Accessed 9 May 2015].
- Regulation (EC) No. 1830/2003 of the European Parliament and of the Council of 22 September 2003 *Concerning the Traceability and Labelling of Genetically Modified Organisms and the Traceability of Food and Feed Products Produced from Genetically Modified Organisms and Amending Directive 2001/18/EC, Official Journal*

of the European Union, L165, 30.4.2004, pp. 1–141, http://eur-lex.europa.eu/legal-content/EN/TXT/PDF/?uri=CELEX:32004R0 882&from=EN. [Accessed 18 June 2015].

- Commission Recommendation of 4 October 2004 on technical guidance for sampling and detection of genetically modified organisms and material produced from genetically modified organisms as or in products in the context of Regulation (EC) No. 1830/2003, *Official Journal of the European Union*, L348, 24.11.2004, pp. 18–26, http://eur-lex.europa.eu/legal-content/EN/TXT/PDF/?uri=CELEX:32004H0787&from=EN [Accessed 18 June 2015].

- Commission Regulation (EC) No. 1981/2006 of 22 December 2006 *on detailed rules for the implementation of Article 32 of Regulation (EC) No. 1829/2003 of the European Parliament and of the Council as Regards the Community Reference Laboratory for Genetically Modified Organisms*, *Official Journal of the European Union*, L368, 23.12.2006, pp. 99–109, http://eur-lex.europa.eu/legal-content/EN/TXT/PDF/?uri=CELEX:32006R1981&qid=1434625654447&from=EN. [Accessed 18 June 2015].

- Directive 2009/41/EC of the European Parliament and of the Council of 6 May 2009 *on Contained Use of Genetically Modified Micro-organisms*, *Official Journal of the European Union*, L125, 21/05/2009, pp. 75–97, http://eur-lex.europa.eu/legal-content/EN/TXT/PDF/?uri=CELEX:32009L0041&from=EN. [Accessed 24 March 2015].

- Commission Recommendation 2010/C200/01 of 13 July 2010 on *Guidelines for the Co-existence Measures to Avoid the Unintended presence of GMOs in Conventional and Organic Crops*, *Official Journal of the European Union* C200/1, 22/7/2010, pp. 0036–0047, http://ec.europa.eu/food/plant/docs/plant_gmo-agriculture_coexistence-new_recommendation_en.pdf [Accessed 2 July 2015].

- Commission Implementing Regulation (EU) No. 120/2014 of 7 February 2014 *Amending Regulation (EC) No. 1981/2006 on Detailed Rules for the Implementation of Article 32 of Regulation (EC) No. 1829/2003 of the European Parliament and the Council as Regards the Community Reference Laboratory for Genetically Modified Organisms*, *Official Journal of the European Union*, L39, 8.2.2014, pp. 46–52, http://eur-lex.europa.eu/legal-

at ftp://ftp.cordis.europa.eu/pub/fp7/docs/nanocode-recommendation.pdf [Accessed 19 May 2012].

- Commission communication of 17/06/2008 *on regulatory aspects of nanomaterials*, COM(2008) 366 final, Brussels: European Commission, http://eur-lex.europa.eu/LexUriServ/LexUriServ.do?uri=COM:2008:0366:FIN:en:PDF [Accessed 1 March 2015].
- Parliament resolution of 24/04/2009 *on regulatory aspects of nanomaterials* (2008/2208(INI)), Brussels: European Parliament, http://www.europarl.europa.eu/sides/getDoc.do?pubRef=-//EP//NONSGML+TA+P6-TA-2009-0328+0+DOC+PDF+V0//EN [Accessed 29 February 2015].
- Recommendation *on a code of conduct for responsible nanosciences and nanotechnologies research: 1st revision. Analysis of results from the public consultation*, http://ec.europa.eu/research/consultations/nano-code/results_en.pdf [Accessed 24 February 2015].
- Recommendation of the European Commission of the 18 October 2011 *on the definition of nanomaterial*, 2011/696/EU, 2011, *Official Journal of the European Union*, L275/38, http://eur-lex.europa.eu/legal-content/EN/TXT/PDF/?uri=CELEX:32011H0696&from=EN [Accessed 1 March 2015].
- Regulation (EC) No. 1169/2011 of the European Parliament and of the Council of 25 October 2011 *on the provision of food information to consumers*, amending Regulations (EC) No. 1924/2006 and (EC) No. 1925/2006 of the European Parliament and of the Council, and repealing Commission Directive 87/250/EEC, Council Directive 90/496/EEC, Commission Directive 1999/10/EC, Directive 2000/13/EC of the European Parliament and of the Council, Commission Directives 2002/67/EC and 2008/5/EC and Commission Regulation (EC) No. 608/2004, *Official Journal of the European Union*, L304/18.
- Directive 2011/65/EU of the European Parliament and of the Council of 08/06/2011 *on the restriction of the use of certain hazardous substances in electrical and electronic equipment*, *Official Journal of the European Union*, L174/88, http://eur-lex.europa.eu/legal-content/EN/TXT/PDF/?uri=CELEX:32011L0065&from=EN [Accessed 28 February 2015].

- Regulation (EU) No. 528/2012 of the European Parliament and the Council of 22/05/2012 *Concerning the Making Available on the Market and Use of Biocidal Products, Official Journal of the European Union*, L167/1, 27/6/2012, pp. 1–123, http://eur-lex. europa.eu/legal-content/EN/TXT/PDF/?uri=CELEX:32012 R0528&from=EN [Accessed 14 March 2015].
- Directive 2012/19/EU of the European Parliament and the Council of 04/07/2012 *on Waste Electrical and Electronic Equipment (WEEE), Official Journal of the European Union*, L197/38, 24/07/2012, http://eur-lex.europa.eu/legal-content/ EN/TXT/PDF/?uri=CELEX:32012L0019&from=EN [Accessed 14 March 2015].
- Resolution of the European Parliament of 19 January *2012 on the Council's position at first reading with a view to the adoption of a regulation of the European Parliament and of the Council, concerning the making available on the market and use of biocidal products* (05032/2/2011 – C7-0251/2011-2009/0076 (COD)), P7_TA-PROV(2012)0010.
- Commission communication of 3/10/2012 *Second regulatory review on nanomaterial*, COM (2012) 572 final. Brussels: European Commission, http://eur-lex.europa.eu/legal-content/ EN/TXT/PDF/?uri=CELEX:52012DC0572&from=EN [Accessed 20 February 2015].
- Regulation (EU) No. 528/2012 of the European Parliament and the Council of 22/05/2012 *Concerning the Making Available on the Market and Use of Biocidal Products, Official Journal of the European Union*, L167/1, 27/6/2012, pp. 1–123, http://eur-lex. europa.eu/legal-content/EN/TXT/PDF/?uri=CELEX:32012R052 8&from=EN [Accessed 20 February 2015].
- Commission communication of 3/10/2012 *second regulatory review on nanomaterial*, COM (2012) 572 final. Brussels: European Commission, http://eur-lex.europa.eu/legal-content/EN/ TXT/PDF/?uri=CELEX:52012DC0572&from=EN [Accessed 20 February 2015].
- Regulation (EU) No. 334/2014 Amending Regulation (EU) No. 528/2012 of the European Parliament and the Council of 22/05/2012 *concerning the making available on the market and*

Union, L287, 5/11/2003, http://eur-lex.europa.eu/legal-content/ EN/TXT/PDF/?uri=CELEX:32003R1946&from=EN [Accessed 24 March 2015].

- Directive 2009/41/EC of the European Parliament and of the Council of 6 May 2009 *on Contained Use of Genetically Modified Micro-organisms, Official Journal of the European Union*, L125, 21/05/2009, pp. 75–97, http://eur-lex.europa.eu/legal-content/EN/TXT/PDF/?uri=CELEX:32009L0041&from=EN [Accessed 24 March 2015].

(i) European Group on Ethics in Science and New Technologies (EGE, Former GAEIB)

- **Opinion n° 1** – 12/03/1993 – The Ethical Implications of the Use of Performance Enhancers in Agriculture and Fisheries.
- **Opinion n° 2** – 12/03/1993 – Products Derived from Human Blood or Human Plasma.
- **Opinion n° 3** – 30/09/1993 – Opinion on Ethical Questions Arising from the Commission Proposal for a Council Directive for Legal Protection of Biotechnological Inventions.
- **Opinion n° 4** – 13/12/1994 – The Ethical Implications of Gene Therapy.
- **Opinion n° 6** – 20/02/1996 – Ethical Aspects of Prenatal Diagnosis.
- **Opinion n° 8** – 25/09/1996 – Ethical Aspects of Patenting Inventions Involving Elements of Human Origin.
- **Opinion n° 9** – 28/05/1997 – Ethical Aspects of Cloning Techniques.
- **Opinion n° 11** – 21/07/1998 – Ethical Aspects of Human Tissue Banking.
- **Opinion n° 12** – 23/11/1998 – Ethical Aspects of Research Involving the Use of Human Embryo in the Context of the 5th Framework Programme.
- **Opinion n° 13** – 30/07/1999 – Ethical Issues of Healthcare in the Information Society.
- **Opinion n° 14** – 14/11/1999 – Ethical Aspects Arising from Doping in Sport.

- *Draft Charter on Fundamental Rights of European Union*, Bruxelles, 15 June 2000, http://www.europarl.europa.eu/charter/civil/pdf/con233_en.pdf [Accessed 17 May 2012].
- **Opinion n° 15** – 14/11/2000 – Ethical Aspects of Human Stem Cell Research and Use.
- **Opinion n° 16** – 07/05/2002 – Ethical Aspects of Patenting Inventions Involving Human Stem Cells.
- **Opinion n° 17** – 04/02/2003 – Ethical Aspects of Clinical Research in Developing Countries.
- **Opinion n° 18** – 28/07/2003 – Ethical Aspects of Genetic Testing in the Workplace.
- **Opinion n° 19** – 16/03/2004 – Ethical Aspects of Umbilical Cord Blood Banking.
- **Opinion n° 20** – 16/03/2005 – Ethical Aspects of ICT Implants in the Human Body, http://www.pedz.uni-mannheim.de/daten/edz-du/gew/EGE%20Opinion20.pdf. [Accessed 18 June 2015].
- **Opinion n° 21** – 17/01/2007 – Ethical Aspects of Nanomedicine.
- **Opinion n° 22** – 13/07/2007 – Recommendation on Ethical Review of hESC FP7 Research Projects.
- **Opinion n° 25** – 17/11/2009 – Ethics of Synthetic Biology, http://www.coe.int/t/dg3/healthbioethic/COMETH/EGE/20091118%20finalSB%20_2_%20MP.pdf. [Accessed 18 June 2015].
- **Opinion n° 26** – 22/02/2012 on Ethics of Information and Communication Technologies, http://ec.europa.eu/bepa/european-group-ethics/docs/publications/ict_final_22_february-adopted.pdf [Accessed 12 June 2015].
- **Opinion n° 27** – 16/01/2013 An Ethical Framework for Assessing Research, Production, and Use of Energy, http://konyvtar.eski.hu/tmpimg/815490358_0.pdf [Accessed 18 June 2015].
- **Opinion n° 28** – 20/05/2014 – Ethics of Security and Surveillance Technologies, http://www.statewatch.org/news/2014/jun/eu-com-opinion-ethics-security-surveillance-technologies.pdf [Accessed 18 June 2015].
- **Opinion n° 29** – 13/10/2015 – Ethical implications of New Health Technologies and Citizen Participation, http://ec.europa.eu/research/ege/pdf/opinion-29_ege.pdf [Accessed 19 October 2017].

(ii) European Food Safety Authority (EFSA)

- European Food Safety Authority (EFSA) (2010) Guidance on the Environmental Risk Assessment of Genetically Modified Plants, *EFSA Journal 2011*, **8**(11), 1879, pp. 1–111. http://www.efsa. europa.eu/it/scdocs/doc/1879.pdf [Accessed 29 March 2015].
- European Food Safety Authority (EFSA) (2011) Guidance for Risk Assessment of Food and Feed from Genetically Modified Plants, *EFSA Journal 2011*, **9**(5), 2150, pp. 1–37. http://www.efsa. europa.eu/en/search/doc/2150.pdf [Accessed 29 March 2015].
- European Food Safety Authority (EFSA) Panel on Genetically Modified Organisms (GMOs) (2012) Scientific Opinion Addressing the Safety Assessment of Plants Developed through Cisgenesis and Intragenesis, *EFSA Journal*, **10**(2), p. 2561 (1–33).
- European Food Safety Authority (EFSA) (2015) Annual report of the EFSA Scientific Network of Risk Assessment of Nanotechnologies in Food and Feed for 2014. EFSA supporting publication 2014: EN-762, pp. 1–11.

(iii) European Medicines Agency (EMA)

- European Medicines Agency (EMA) (2006) *Reflection Paper on Nanotechnology-Based Medicinal Products for Human Use*, EMA/CHMP/79769/2006, London: European Medicines Agency, http://www.ema.europa.eu/docs/en_GB/document_library/ Regulatory_and_procedural_guideline/2010/01/ WC500069728.pdf. [Accessed 9 April 2015].

Scientific Committee on Consumer Safety (SCCS)

- Scientific Committee on Consumer Safety (SCCS) (2012) *Guidance on the Safety Assessment on Nanomaterials in Cosmetics*, Brussels: European Commission.

(iv) Scientific Committee on Emerging and Newly Identified Health Risks (SCENIHR)

- Scientific Committee on Emerging and Newly Identified Health Risks (SCENIHR) (2009) *Risk Assessment of Products of Nano-*

technologies – Opinion of 19 January 2009 – Brussels: European Commission.

- Scientific Committee on Emerging and Newly Identified Health Risks (SCENIHR) (2014a) *Opinion on Synthetic Biology I: Definition – Opinion of 23, 24, 25 September 2014*, Luxembourg: European Commission.
- Scientific Committee on Emerging and Newly Identified Health Risks (SCENIHR) (2014b) *Preliminary Opinion on Synthetic Biology II: Risk Assessment – Opinion of 20, 28 November 14 December 2014*, Luxembourg: European Commission.

Table of Cases of the European Court of Human Rights and the Court of Justice of European Union

1. Human dignity

Status and protection of embryo

European Court of Human Rights (ECtHR):

- *Costa and Pavan v. Italy* (Appl. 54270/10), judgement of 28 August 2012.
- *Evans v. the United Kingdom [GC]* (App. 6339/05), judgement of 10 April 2007, *Reports Judgements and Decisions*, 2007-I.
- *Tysiąc v. Poland* (App. 5410/03), [4th section], judgement of 20 March 2007, *Reports Judgements and Decisions*, 2007-I.
- *Vo v. France* (App. 53924/00), judgement of 8 July 2004, *Reports Judgements and Decisions*, 2004-VIII.
- *X v. The United Kingdom* (Appl. 8416/78), decision of the Commission of 13 May 1980, *Decision and Reports*, 19, also cited as *Paton v. UK, D.R, 19.*

Court of Justice of European Union (CJEU):

- Judgement of the Court of Justice (Grand Chamber), *International Stem Cell Corporation v. Comptroller General of Patents and designs and Trade Marks* (Case C-364/13), 18 December 2014, not published yet.
- Judgement of the Court of Justice (Grand Chamber), *Oliver Brüstle v. Greenpeace eV* (Case C-34/10), 18 October 2011, not yet published.

- Judgement of the Court of Justice *The Netherlands v. Parliament and Council* (Case C-377/98), 9 October 2001 *ECR I-7079*.

2. Right to health

(i) Medically assisted procreation

European Court of Human Rights (ECtHR):

- *Costa and Pavan v. Italy* (Appl. 54270/10), judgement of 28 August 2012.
- *Evans v. the United Kingdom [GC]* (App. 6339/05), judgement of 10 April 2007, *Reports Judgements and Decisions*, 2007-I.
- *Dickson v. the United Kingdom [GC]* (App. 44362/04), judgement of 4 December 2007 selected for publication in *Reports of Judgements and Decisions*.
- *S.H. and others v. Austria* (App. 57813/00), judgement of 3 November 2011.

Court of Justice of European Union (CJEU):

- Judgement of the Court of Justice (Grand Chamber), Case C-506/06, *Sabine Mayr v. Bäckerei und Konditorei Gerhard Flöckner OHG*, 26 February 2008 *European Court Reports*, 2008, p. I-01017.

(ii) Prenatal diagnosis

European Court of Human Rights (ECtHR):

- *Draon v. France [GC]* (App. 1513/03), judgement of 6 October 2005, *Reports Judgements and Decisions*, 2006-IX.
- *Costa and Pavan v. Italy* (Appl. 54270/10), judgement of 28 August 2012.

(iii) Surrogacy

European Court of Human Rights (ECtHR):

- *Menneson v. France* (Appl. 65192/11), judgement of 26 June 2014, *Reports Judgements and Decisions*, 2014.

(iv) Free and informed consent

European Court of Human Rights (ECtHR):

- *Acmanne and others v. Belgium* (Appl. 10435/83) decision of the Commission of 10 December 1984.
- *Bogumil v. Portugal* (App. 35228/03), judgement of 7 October 2008, inadmissible.
- *Evans v. the United Kingdom [GC]* (App. 6339/05), judgement of 10 April 2007, *Reports Judgements and Decisions*, 2007-I.
- *Gennadi Naoumenko v. Ukraine* (App. 42023/98) [2nd section], judgement of 10 February 2004.
- *Glass v. the United Kingdom* (App. 61827/00) [4th section], judgement of 9 March 2004, ECHR 2004-II.
- *Jalloh v. Germany* [GC] (App. 54810/00), judgement of 11 July 2006, *Reports Judgements and Decisions*, 2006-IX.
- *Hoffmann v. Austria* (App. 12875/87), judgement of 23 June 1993, *Series A* no. 255-C, n° 12875/87.
- *Juhnke v. Turkey* (Appl. 1620/03), judgement of 23 September 2010, *Reports of Judgements and Decisions*, 2010.

Court of Justice of European Union (CJEU):

- Judgement of the Court of Justice *The Netherlands v. Parliament and Council* (Case C-377/98), 9 October 2001 *ECR I-7079*.

(v) Right to information

European Court of Human Rights (ECtHR):

- *Öneryildiz v. Turkey* (App. 48939/99), judgement of 30 November 2004, *Reports of Judgement and Decisions*, 2004-XII.
- *Open Door Counselling Ltd and Dublin Well Woman v. Ireland* (Appl. 14234/88, 14235/88), judgement of 29 October 1992, *Series A*, No. 246-A.
- *Roche v. The Unite*d Kingdom (App. 32555/96), judgement of 19 October 2005, *Reports of Judgements and Decisions*, 2005-IX.
- *Tysiąc v. Poland* (Appl. 5410/03), judgement of 3 March 2007, *Reports Judgements and Decisions*, 2007-I.

(vi) Personal integrity

European Court of Human Rights (ECtHR):

- *Acmanne and others v. Belgium* (Appl. 10435/83) decision of the Commission of 10 December 1984.
- *Grimailovs v. Latvia* (Appl. 6087/03), judgement of 25 June 2013.
- *Juhnke v. Turkey* (Appl. 1620/03), judgement of 23 September 2010, *Reports of Judgements and Decisions*, 2010.
- *Tysiąc v. Poland* (Appl. 5410/03), judgement of 24 September 2007, *Reports Judgements and Decisions*, 2007-I.
- *X v. The United Kingdom* (Appl. 8416/78), decision of the Commission of 13 May 1980, *Decision and Reports*, 19, also cited as *Paton v. UK, D.R, 19*, p. 244.
- *Zarzycki v. Polan* (15351/03), judgement of 3 March 2013.

Court of Justice of European Union (CJEU):

- Judgement of the Court of Justice *The Netherlands v. Parliament and Council* (Case C-377/98), 9 October 2001 *ECR I-7079*.

(vii) Retention and use of biological samples

European Court of Human Rights (ECtHR):

- *Van der Velden v. the Netherlands* (App. 29514/05), judgement of 7 December 2006, *Reports of Judgements and Decisions*, 2006-XV.
- *S. and Marper v. the United Kingdom* [GC] (App. 30562/04 and 30566/044), judgement of 4 December 2008, *Reports Judgements and Decisions*, 2008.
- *W. v. the Netherlands* (App. 20689/08), judgement of 20 January 2009.

Court of Justice of European Union (CJEU):

- Judgement of the Court of Justice Cause (Third Chamber), *Belgian State v. De Fruytier* (Case C-237/09), 3 June 2010 *OJ C 209, 31.7.2010*, p. 10–10.
- Judgement of the Court of Justice Cause (Third Chamber), *CopyGene A/S v Skatteministeriet* (Cause C-262/08), 10 June 2010 *OJ C 221, 14.8.2010*, pp. 3–4.

- Judgement of the Court of Justice Cause (Second Chamber), *Future Health Technologies Limited v. The Commissioners for Her Majesty's Revenue and Customs* (Cause C-86/09), 10 June 2010, *OJ C 221, 14.8.2010*, pp. 11–12.
- Judgement of the Court of Justice Cause (First Chamber), *Finanzamt Leverkusen v. Verigen Transplantation Service International* (Cause C-156/09), 18 November 2010, *OJ C 13, 15.1.2011*, pp. 8–9.

(viii) Confidentiality of personal information concerning health

European Court of Human Rights (ECtHR):

- *I v. Finland* (App. 20511/03) [4th section], judgement of 17 July 2008.
- *MacGinley and Egan v. The United Kingdom* (App. 21825/93 and 23414/94), judgement of 26 November 1996, *Reports*, 1998-III.
- *MacGinley and Egan v. The United Kingdom* (App. 21825/93 and 23414/94), judgement (revision) of 28 January 2000, *Reports of Judgements and Decisions*, 2000-I.
- *S. and Marper v. the United Kingdom* [GC] (App. 30562/04 and 30566/044), judgement of 4 December 2008, *Reports Judgements and Decisions*, 2008.
- *Roche v. The United Kingdom* (App. 32555/96), judgement of 19 October 2005, *Reports of Judgements and Decisions*, 2005-IX.
- *Z. v. Finland* (App. 22009/93) [Chamber], judgement of 25 February 1997, *Reports*, 1997-I.

(ix) Assisted suicide

European Court of Human Rights (ECtHR):

- *Haas v. Switzerland* (Appl. 31322/07) judgement of 20 January 2011, *Reports of Judgements and Decisions*, 2011.
- *Pretty v. United Kingdom* (Appl. 2346/02) judgement of 29 April 2002, *Reports of Judgements and Decisions*, 2002-III.

(x) Interruption of artificial nutrition and hydration

European Court of Human Rights (ECtHR):

- *Lambert and others v. France* (Appl. 46043/14) judgement of 5 June 2015 *Reports of Judgements and Decisions*, 2015.

(xi) Vaccines

European Court of Human Rights (ECtHR):

- *Association X v. The United Kingdom* (Appl. 7154/75), decision of the Commission of 12 July 1978, *Decision and Reports*, 14.

3. Self-determination

European Court of Human Rights (ECtHR):

- *Christine Goodwin v. The United Kingdom* (App. 28957/95), judgement of 11 July 2002, *Reports of Judgements and Decisions*, 2002-VI.
- *Juhnke v. Turkey* (Appl. 1620/03), judgement of 23 September 2010, *Reports of Judgements and Decisions*, 2010.
- *Lambert and others v. France* (Appl. 46043/14), judgement of 5 June 2015 *Reports of Judgements and Decisions*, 2015.
- *Pretty v. United Kingdom* (Appl. 2346/02), judgement of 29 April 2002, *Reports of Judgements and Decisions*, 2002-III.
- *Affaire Y.Y. c. Turquie* (Requête no 14793/08) arrêt 10 mars 2015.

Court of Justice of European Union (CJEU):

- Judgement of the Court of Justice *P.V. and Cornwall County Council* (Case C-13/94), 30 April 1996, *Report of cases of Court of Justice*, 1996, 3, I pp. 2143–2147.
- Judgement of the Court of Justice *K.B. and National Health Service* (Case C-117/01), 7 January 2004, *Report of cases of Court of Justice*, 2004, I-00541.

4. Right to a healthy environment

European Court of Human Rights (ECtHR):

- *Arrondelle v. The United Kingdom* (Appl. 7889/77), decision of the Commission of 15 July 1980, *Decision and Reports*, 19.
- *Athanassoglou and others v. Switzerland* (Appl. 27644/95), judgement of 6 April 2000, *Reports of Judgements and Decisions*, 2000-IV.

Court of Justice of European Union (CJEU):

5. Privacy

European Court of Human Rights (ECtHR):

- *Halford v. the United Kingdom* (Appl. 20605/92) judgement of 25 June 1997, *Reports of Judgements and Decisions*, 1997-III.
- *I v. Finland* (App. 20511/03) [4th section], judgement of 17 July 2008.
- *Affaire Knauth c. Allemagne* (Requête 41111/98), décision de 22 novembre 2001 *Recueil des arrêts et décisions*, 2001-XII.
- *S. and Marper v. the United Kingdom* [GC] (App. 30562/04 and 30566/044), judgement of 4 December 2008, *Reports Judgements and Decisions*, 2008.
- *Niemietz v. Germany* (App. 13710/88), judgement of 16 December 1992 *Serie-A*, 251-B.
- *Affaire Schüth c. Allemagne* (Requête n° 1620/03), Arrêt de 23 septembre 2010 *Recueil des arrêts et décisions*, 2010.
- *Roche v. The United Kingdom* (App. 32555/96), judgement of 19 October 2005, *Reports of Judgements and Decisions*, 2005-IX.
- *Smith and Grady v. The United Kingdom* (App. 33985/96 and 33986/96), judgement of 27 September 1999 *Reports of Judgements and Decisions*, 1999-VI.
- *Van deer Velden v. The Netherlands* (App. 29514/05), judgement of 7 December 2006, *Reports of Judgements and Decisions*, 2006-XV.
- *Z. v. Finland* (Appl. 22009/93), judgement of 25 February 1997.

Court of Justice of European Union (CJEU):

- Judgement of the Court of Justice (Grand Chamber) *Google Spain SL, Google Inc. / Agencia Española de Protección de Datos, Mario Costeja González* (C-131/12) 13 May 2014.
- Judgement of the Court of Justice (Grand Chamber) *Tele2 Sverige AB v. Post- och telestyrelsen, and Secretary of State for the Home Department* (Joined Cases C-203/15 and C-698/15) 21 December 2016.
- Judgement of the Court of Justice *Rechnungshof v. Österreichischer Rundfunk and Others and Christa Neukomm and Joseph Lauermann v. Österreichischer Rundfunk* (Case C-465/00; C-138/01; C-139/01), 20 Mai 2003, *European Court Reports*, 2003, p. I-04989.
- Judgement of the Court of Justice (Grand Chamber) *Maximillian Schrems c. Data Protection Commissioner* (Case C-362/14) 6 October 2015.

- Judgement of the Court of Justice (Grand Chamber) *Volker und Markus Schecke GbR, Hartmut Eifert v Land Hessen* (Joined Cases C-92/09; C-93/09), *OJ C 13, 15.1.2011*, pp. 6–7.

6. Principle of non-discrimination

European Court of Human Rights (ECtHR):

- *Christine Goodwin v. The United Kingdom* (App. 28957/95), judgement of 11 July 2002, *Reports of Judgements and Decisions*, 2002-VI.
- *Hoffmann v. Austria* (App. 12875/87), judgement of 23 June 1993, *Series A*, no. 255-C, n° 12875/87, *Serie A*, 255-C.
- *Van deer Velden v. The Netherlands* (App. 29514/05), judgement of 7 December 2006, *Reports of Judgements and Decisions*, 2006-XV.
- *Y.Y. v. Turkey* (App. 14793/08), pending case.

Court of Justice of European Union (CJEU):

- Judgement of the Court of Justice (Grand Chamber), Case C-506/06, *Sabine Mayr v Bäckerei und Konditorei Gerhard Flöckner OHG*, 26 February 2008 *European Court Reports*, 2008, p. I-01017.
- Judgement of the Court of Justice *K.B. and National Health Service* (Case C-117/01), 7 January 2004, *Report of cases of Court of Justice*, 2004, I-00541.
- *Mary Brown v. Rentokil Initial UK Ltd* (Case C-394/96), judgement of 30 June 1998, *European Court Reports*, 1998, I-4185.
- Judgement of the Court of Justice *P.V. and Cornwall County Council* (Case C-13/94), 30 April 1996, *Report of cases of Court of Justice*, 1996, 3, I pp. 2143–2147.

7. Freedom of scientific research

European Court of Human Rights (ECtHR):

- *Guerra and others v. Italy* (App. 14967/89), judgement of 19 February 1998, *Reports*, 1998-I.
- *Hertel v Switzerland* (Appl. 59/1997/843/1049), judgement of 25 August 1998, *Reports of Judgements and Decisions*, 1998.

- *MacGinley and Egan v. The United Kingdom* (App. 21825/93 and 23414/94), judgement of 26 November 1996, *Reports*, 1998-III.
- *S. and Marper v. the United Kingdom* [GC] (App. 30562/04 and 30566/044), judgement of 4 December 2008, *Reports Judgements and Decisions*, 2008.
- *Parrillo v. Italy* (Appl. 46470/11), judgement of 27 August 2015.
- *Roche v. The United Kingdom* (App. 32555/96), judgement of 19 October 2005, *Reports of Judgements and Decisions*, 2005-IX.

Court of Justice of European Union (CJEU):

- Judgement of the Court of Justice (Grand Chamber), of 18 October 2011 *Oliver Brüstle v. Greenpeace eV* (Case C-34/10), not yet published.
- Judgement of the Court of Justice of 9 October 2001 *The Netherlands v. Parliament and Council* (Case C-377/98), *ECR I-7079*.

8. Intellectual property

European Court of Human Rights (ECtHR):

- *Anheuser-Busch Inc. v. Portugal* (App. 73049/01), judgement of 11 January 2007, *Reports of Judgements and Decisions*, 2007-I.

Court of Justice of European Union (CJEU):

- Judgement of the Court of Justice (Grand Chamber), *International Stem Cell Corporation v. Comptroller General of Patents and designs and Trade Marks* (Case C-364/13), 18 December 2014, not published yet.
- Judgement of the Court of Justice (Grand Chamber), of 18 October 2011 *Oliver Brüstle v. Greenpeace eV* (Case C-34/10), not yet published.
- Judgement of the Court of Justice of 9 October 2001 *The Netherlands v. Parliament and Council* (Case C-377/98), *ECR I-7079*.

PART I

Introduction

Which criteria can drive the technoscientific advance? Can human rights represent these criteria, and how can we draw down from them a concrete guidance for research and innovation? This work attempts to answer these questions and to provide a robust theoretical framework, as well as a concrete tool, for governing a technoscientific progress boosted by emerging technologies.

Nowadays human rights are targets of increasing distrust, especially with regard to their capability to tackle most challenging issues raised by the run of technoscience. This seems to lead to a clear-cut separation between the realm governed by principles, on the basis of our democracies, and that of science and technologies. They seem to adhere to two different logics.

I am of the opinion, instead, that human rights not only are a powerful factor of inspiration for our modern consciences, but they can also be a concrete tool for steering our decision making in fields where they did not play any role yet or they have not gained full attention yet.

I am convinced that there is space to play an increasing part in governance frameworks for human rights and that they are far from being an obstacle for technoscientific advance. For this reason, there is the need to provide a tool able to show how we can derive from human rights an ensemble of practical criteria which can be taken into account by policymakers for taking decisions related to research innovation, by regulators for devising robust governance frameworks able to resist possible claims in defense of breached rights that may abruptly stop trajectories of governance, as well as by judges for deciding hard cases rendered even more complex by the technoscientific progress in societies.

Human rights are at the centre of our political communities (Gewirth, 1996) and they inform our moral argumentation also in technoscientific issues, representing a reference point for legal reasoning, especially in common law countries (Brownword, 2008). Human rights, though, have also a legal dimension which is particularly strong at least in Europe (Ruggiu, 2015). This dimension can be an opportunity for developing in Europe a governance framework which can be taken into consideration also out of it. It is true that the technoscientific advance needs excellence in science and technology, but as noted by some, if this is reached through the respect of individual rights, this progress is stronger and more durable (Krupp and Holliday, 2005).

This innovation is what we need.

There is therefore the need of tools that address possible and concrete pathways for the implementation of human rights in governance frameworks, as well as to make the European context a model able to face other governance arrangements that have the leadership at the global level. To reach the same outcomes with the protection of rights of persons is not only a moral or a legal duty: it is feasible. There are many signals that this route is concrete. This choice would transform European governance by making the inheritance of human rights an added value of our technoscientific advance.

After the entry of human rights in our legal orders, not only our political communities are non-neutral any longer (Gewirth, 1996). Also human rights are non-neutral, they express a precise axiological preference (Viola, 2000) and force therefore to modify the internal architecture of our political communities: policies, regulation, governance. However, since there are interests, such as rights, at the centre of our legal frameworks that need to be protected, irrespective of the actual course taken by a given technoscientific field, they can be also considered as 'technology neutral', that is, independent from the technoscientific field concerned (Leenes et al., 2017, p. 43). Rights can be applied in any technological field. This also means that in principle any field (nanotechnologies, biotechnology, synthetic biology, geoengineering, etc.) can be developed according to an approach attentive to the protection and the implementation of human rights. It is just a question of will and vision.

Emerging technologies[1] are impacting largely our communities. Some applications increasingly rapidly enter into the market with a great baggage of uncertainty: scientific, ethical, social and regulatory. Are the available data robust enough? What do we ignore? What social transformation are we creating? How can we overcome raised ethical concerns? Is the existing regulation adequate or do we need a new one? If yes, when acting?

It needs therefore to contribute to the development of these technologies by promptly adopting strategies which can show being robust in the future, even far. In this enterprise we need some guiding ideas. We need not only descriptive works, therefore, in order to understand where our present is located. But also works able to transform the present by using what we already have: rights.

If we take a picture of the current state of the technoscientific development, we would have an image accurate of what is going but partial. In the shadow there will be many of those who contribute to its success, as well as those who are affected by it. These can represent an unexpected in governance. Both because they can show that the taken way was wrong and because once their voice might be heard. Probably we should begin to change our belief that implementing human rights is a cost for innovation (Holmes and Sunstein, 1999). The lack of implementation is instead a cost which is irrationally distributed among anonymous individuals and borne by the society as whole (through taxes) affecting the architecture of our advances. Human rights not only shed some light on the position of these people but contribute strengthen the course undertaken by our progress, giving it a more rational structure.

The book is articulated in two parts. The first is of theoretical nature. The second of practical stance and aims at giving law schools a further tool for the study of the European human rights law via-à-vis challenges of the technoscientific progress.

---

[1] According to the European Commission's definition (2009) emerging technologies have two key features. They are *enabling technologies* (they enable process, goods and service innovation throughout the economy) and are *converging technologies* (thanks to their integration they give rise to new disciplines and sectors). See http://eur-lex.europa.eu/legal-content/EN/TXT/PDF/?uri=CELEX:52009DC0512&from=EN. Accessed 1 December 2017.

Part I deals with governance issues and human rights. Chapter 1 means to report the growth of human rights as subject matter in the debate over the technoscience. Chapter 2 tackles theoretical issues concerning governance and the crisis of the command-and-control model for the rise of the 'new governance' paradigm. In this context, soft regulation and forms of Corporate Social Responsibility began to take place in order to extend the possibilities of governance involving also private actors for a common goal. Chapter 3 deals with concrete governance arrangements with a particular attention to the European context. In this framework several specific cases related to the European continent are tackled in the light of the protection of human rights: governance of biotechnology, governance of nanotechnologies and the rising governance of synthetic biology. The final part of the chapter focuses on the *Responsible Research and Innovation* model which is affirming at the European level. The issue of the transformation of this model according to an approach sensitive to human rights is deepened. This part can be deemed the core of the book and represents the possible response to quest for responsible governance.

Part II aims at mapping the content of some human rights in the context of the technoscientific progress. The basic idea, *inter alia* indirectly suggested by Phil Macnaghten, is that the content of each right must illuminated with regard to the specific technological context. Therefore, Chapter 4 attempts to analyse, form the legal standpoint, the idea of human dignity in front of the challenges of human genetic modifications. This chapter basically follows the insight of Deryck Beyleveld and Roger Brownsword with regard to human dignity (in particular what they call the 'empowerment conception' of dignity). Chapter 5 attempts to develop the meaning of the right to health related to the context created by nanotechnologies. Chapter 6 represents a detailed study of one relevant aspect of the right to health: self-determination. This subject matter is dealt with by referring to the case of performance-enhancing technologies. Also Chapter 7 deepens the regulatory framework related to the right to bodily integrity in front to challenges of human enhancement. Chapter 8 tackles a right closely linked to the right to health with regard to geoengineering: the right to a healthy environment. Chapter 9 analyses in depth privacy with regard to the spread of Artificial Intelligence systems, in particular, in the workplace. Chapter 10, lastly, deals with the right to scientific research in particular with regard to the rise of participatory experiences in medicine.

References

Beyleveld, D., and Brownsword, R. (2001). *Human Dignity in Bioethics and Biolaw*, Oxford: Oxford University Press.

Brownsword, R. (2008). *Rights, Regulation and the Technological Revolution*, Oxford: Oxford University Press.

European Commission (2009). Commission communication 'Preparing for our future: Developing a common strategy for key enabling technologies in the EU' {SEC(2009) 1257}, Brussels, 30.09.2009 COM(2009) 512 final.

Gewirth, A. (1996). *Community of Rights*, Chicago: University of Chicago Press.

Holmes, S., and Sunstein, C.R. (1999). *The Cost of Rights. Why Liberty Depends on Taxes*, New York: W.W. Norton.

Krupp, F., and Holliday, C. (2005). Let's Get Nanotech Right, *Wall Street Journal*, Management Supplement, p. B2.

Leenes, R., Palmerini, E., Koops, B.-J., Bertolini, A., Salvini, P., and Lucivero, F. (2017). Regulatory Challenges of Robotics: Some Guidelines for Addressing Legal and Ethical Issues, *Law, Innovation & Technology*, **9**(1), pp. 1–44.

Ruggiu, D. (2015). Anchoring European Governance: Two versions of Responsible Research and Innovation and EU Fundamental Rights as 'Normative Anchor Points', *Nanoethics*, **9**(3), pp. 217–235.

Viola, F. (2000). *Etica e metaetica dei diritti umani*, Torino: Giappichelli.

Chapter 1

Human Rights and Technoscientific Development

1.1 Introduction

1.1.1 Developments of science and technology have always gone
hand in hand with the history of mankind, but the awareness of
their potential impact on human beings is quite recent and is closely
linked to the rise of human rights[1] as autonomous deontic figure of
the legal world (Trujillo and Viola, 2014). With the emergence of the
need to protect some common interests of all human beings individually concerned, the international community has progressively

---

[1]I will refer to human rights as the ones, and exclusively those recognized by the
international law rules listed in documents like the Universal Declaration, the
Covenants of '66, the European Convention on Human Rights (ECHR) to individuals,
groups and associations to the State and enforceable in the face of judicial or quasi-
judicial organs like the European Court of Human Rights (ECtHR), UN Committees on
human rights (Ruggiu, 2012b, p. 7 nt. 1; 2013b, p. 207 nt. 12). In this regard human
rights should be deemed as legal rights with a right-holder (the individual), a duty-
holder (the State which is thus the recipient of a legal obligation of protecting human
rights stemming from international law), a determinate content (the due behaviour
of the State with active or omissive nature set forth by both international documents
on human rights, in particular the ECHR, and the jurisprudence of supranational
courts, in particular the ECtHR).

Human Rights and Emerging Technologies: Analysis and Perspectives in Europe
Daniele Ruggiu
Copyright © 2018 Pan Stanford Publishing Pte. Ltd.
ISBN 978-981-4774-93-2 (Hardcover), 978-0-429-49059-0 (eBook)
www.panstanford.com

acknowledged that, together with its enormous benefits, scientific and technological progress could entail unintended consequences that might affect those interests themselves. Despite the fact that scientific and technological development is due to a multitude of actors, only some of them strictly linkable to State activity and thus public (Pariotti, 2007), the international community has considered the State the best subject for granting the protection of those interests acknowledged as human rights. This State responsibility embraces both the field of action of public authority and, in certain circumstances, that of private actors. In this regard, nowadays, beyond the responsibility of the State for those activities of research and industry directly linked to the sphere of its sovereignty, in the market dimension '[t]he [S]tate is responsible for defining the risks of technologies under product authorization procedures and product liability law and ensuring market operator compliance' (von Schomberg, 2013, p. 54). According to the international community the State is the main subject which has primary responsibility for the protection of human rights in the face of threats from technoscience (Francioni, 2007), including the responsibility to ensure respect for human rights by business enterprises (Weschka, 2006, p. 654). Beside States, the individual has become a subject of international law as a holder of rights (e.g. human rights) and duties (e.g. the duty to respect human rights with regard to transnational corporations), which means that the individual can both claim violations to his/her own rights by the State through supranational courts (e.g. the European Court of Human Rights) and, in certain cases, be called on to answer for their own violations (e.g. with regard to international crimes before the International Penal Court either in the case of transnational corporations having assumed special self-obligations to a code of conduct, or through third-party certification arrangements) (Pariotti, 2007, 2013). It is in this legal framework that we must understand the phenomenon of technoscientific development, its limits and its latent threats to the human being. Thus the emergence of clear State responsibility in the question of research and development (R&D) can be deemed an acquired result of the international framework, whose path can be seen as a succession of steps towards the complete affirmation of human rights law in this matter and should be considered a phenomenon still in progress.

1.2 The International Framework on Human Rights and Scientific and Technological Developments[2]

1.2.1 Contrary to what one might think, the international community turned its attention to the potential impact of R&D on human rights quite early, at the end of the Second World War to be precise. In this regard, the debate on the impact of technological and scientific development on human rights began with the birth of human rights as legal rights. The discovery of the destruction caused by the atomic bombs on Hiroshima and Nagasaki and the atrocity of the genetic experiments in the Nazi concentration camps showed the disturbing side of scientific research. At that time there arose an awareness that human rights needed to be enforced by the legal obligation of the State in the face of the international community. Thus, the establishment of a core of inviolable rights seemed to be the only means of avoiding the devastating consequences of the war and limiting the negative effect of scientific and technological development. The progressive extension of the scope of R&D and the phenomenon of technological globalization presented the organization of the United Nations (UN) with the need to affirm a set of common principles among States for maintaining peace, preserving international security and respecting human rights and fundamental freedoms (Beghé Loreti, 2006). Thus, human rights have been recognized internationally as the benchmark against which States can measure the legitimacy of their policy choices and applications concerning modern science (Francioni, 2007, p. 4).

Within the Preamble of the Charter of the United Nations,[3] member States reaffirmed their faith in fundamental human rights, in the dignity and the worth of all human beings, connecting in

---

[2]This paragraph takes up and examines in more depth Ruggiu (2011, 2012a, 2013a).
[3]ONU (1945) *Charter* of the United Nations, adopted in San Francisco on 26 June 1945 (entered into force on 24 October 1945).

this way the enjoyment of human rights with the basic ideas of equality and non-discrimination. The Second World War, indeed, clearly showed that scientific research may give rise to new forms of discrimination among individuals that can go towards the complete negation of individuals' dignity. In this sense the link between human rights and R&D was posed quite early. The concepts of human dignity and the idea of equality among persons are the two guiding principles in this matter and were both expressed in the Preamble of the Universal Declaration of Human rights (UDHR)[4] of 1948. This instrument which is, as known, not binding, became the starting point of many binding international documents and affirmed the natural, indivisible and inviolable values of human beings (Viola, 2000).

Here the principle of autonomy has been strictly linked to the principle of personal integrity and individual corporeity. As is known, autonomy is the core principle of biomedical research as it emerged a few years later with the two International Covenants of 1966. Thus, Article 7 of the International Covenant on Civil and Political Rights,[5] which has binding force, in the context of the general prohibition of inhuman and degrading treatments, which is the core of personal integrity, decreed the requirement of free, informed consent to medical and scientific experimentation. In the context of biomedical research, States have recognized the individual's right to enjoy the benefits of scientific progress and applications (Art. 15 of International Covenant on Economic, Social, and Cultural Rights[6]), which has been understood by some to be the beginning of the freedom of scientific research (Francioni, 2007, p. 4). In this ambit the World Medical Association Declaration of Helsinki of 1964, subsequently amended in Tokyo in 1975, in Venice in 1983, in Hong Kong in 1989, in Somerset West (South Africa)

---

[4] ONU (1948) *Universal Declaration on Human rights* (UDHR), adopted in New York on 10 December 1948 by the General Assembly of the United Nations.

[5] ONU (1966) *International Covenant on Civil and Political Rights* (ICCPR), adopted in New York on 16 December 1966 (entered into force on 23 March 1976) and the *International Covenant on Economic, Social, and Cultural Rights* (ICESCR), adopted in New York on 16 December 1966 (entered into force on 3 January 1976).

[6] ONU (1966) *International Covenant on Economic, Social, and Cultural Rights* (ICESCR), adopted in New York on 16 December 1966 (entered into force on 3 January 1976).

in 1996, in Edinburgh in 2000 and finally in Seoul in 2008,[7] set forth some basic principles for biomedical research (i.e. free informed consent; complete and adequate information on risks, goals, methods, benefits; safeguard of patient's integrity etc.).

The close link between human rights and scientific and technological development has been reaffirmed by the General Assembly of the United Nations. In this regard a fundamental date should be pinpointed in 1968. During the spring of 1968 at the International Conference on Human rights held in Teheran, member States posed the question of the impact of technoscientific development on the human rights system. The Proclamation of Teheran,[8] which concluded the meeting, recognized that although progress in science and technology opens up a wide range of possibilities for society, the economy, culture and human knowledge, this can, nevertheless, endanger the rights and freedom of individuals. For this reason the Proclamation suggested the development of interdisciplinary studies at both a national and supranational level and underlined the following concerns: the respect for privacy, the protection of individuals' integrity both physical and intellectual, the need to set limits on the use of electronics within democratic countries and the need to keep a more balanced view among scientific and technological developments on the one hand and the intellectual, spiritual and cultural advancement of humanity on the other (Ogata, 1990).

The subsequent Declaration on the Use of Scientific and Technological Progress in the Interests of Peace and Mankind adopted with Resolution 3384 (XXX) of 1975[9] deviated significantly from the previously trodden path because attention was focused on the role of States rather than on the position of individuals who need to be protected. Here nations and peoples took the place of individuals. The concern of the Soviet bloc and China was that

---

[7] The World Medical Association (WMA) Declaration of Helsinki (2008) *Ethical Principles for Medical Research Involving Human Subjects*, adopted in Seoul in October 2008.

[8] ONU (1968) *Proclamation of Tehran, Final Act of the International Conference on Human rights*, 22 April to 13 May 1968.

[9] The Declaration was adopted without the support of the Western countries, all of which abstained from voting.

human rights could be used as a pretext to intervene in the domestic policy of a country. The Declaration, indeed, obliged States to ensure that scientific and technological development was in the interest of international peace and security, freedom and independence of States, and for the purposes of the development of peoples (point 1). Member States must refrain from using scientific and technological progress with a view to violating the sovereignty and integrity of other member States, interfering in their affairs, waging aggressive acts of belligerency, suppressing national freedom movements, or pursuing policies of racial discrimination, although not directly in contrast with the Charter of UN and with the other principles of international law (point 4) (Ogata, 1990). It is worth noting that in the subsequent period the issue of human rights and technoscientific development was marked by a strong polarization in East-West relations.

After 1983 the United Nations reached a turning point. First, with regard to the biomedical ambit, it adopted two resolutions. Resolution 38/112 and 113 on Human Rights and Scientific and Technological Developments[10] and Resolution 1999/63 on Human Rights and Bioethics[11] prescribed that respect for basic human rights increased in scientific and technological development thanks to the adoption of rules, ethical codes and forms of cooperation among 'life sciences' both at a national and supranational level. In this regard, governments, specialized bodies of the UN and other non-governmental international organizations should inform the Secretary General of the UN about the measures for the respect of human rights they intended to adopt.

Later in 1986 awareness was generally reached on the negative and incontrollable effects that science could have. In that period a resolution was adopted by the Commission on Human rights[12] which invited member States of the United Nations to take any measure to use development in science and technology for

---

[10]ONU (1983) Resolution 38/112 e 113 of General Assembly *on Human rights and scientific and technological developments*, 16 December 1983.

[11]ONU (1999) Resolution 1999/63 of General Assembly *on Human rights and Bioethics*, 28 April 1999.

[12]ONU (1983) Resolution 1986/9 of the Commission on Human rights, *Use of the scientific and technological developments for the promotion and the realisation of human rights and fundamental freedoms*, 27 March 1983.

the promotion and realization of human rights and fundamental freedoms. Starting from the conviction that the positive and negative effects are two sides of the same coin, maximum vigilance was recommended of the possible implications of technoscientific development on human rights.

The relationship between human rights and scientific and technological progress is highly complex. On the one hand, to achieve the well-being of human beings the goals of science and technology need to be expanded to the greatest extent. On the other, science and technology can produce risks that may endanger human rights. This happens when the development of biotechnologies and electronics may intrude into individuals' privacy, when an increase in medicine can lead not only to new and unexpected cures for human diseases, but also to modifications of the human body inside its DNA. Here human identity and integrity are in question. Moreover, in the face of the links between industry and scientific research, controlling new sophisticated forms of mass production becomes increasingly difficult, given the possibilities of creating new dangers and risks for the environment.

In this regard the action of the UN in protecting the environment can acquire a strategic role in anticipating the protection of human rights. Indeed, threats to the environment can turn into risks for human life and health. Thus, the UN Convention on Biological Diversity of Rio de Janeiro[13] of 1992 again posed concerns over sustainable development and a precautionary approach, which were embedded in Principle 15 of the Rio Declaration.[14] In this context the right of States to exploit the environment and its natural resources, including both animal and human genetic resources, is limited by the UN Charter and the principles of international law, among which we can count human rights. Finally, the precautionary principle was established in the field of genetic engineering and biotechnology by the Cartagena Protocol.[15]

---

[13] ONU (1992) *Convention on Biological Diversity* of Rio de Janeiro, adopted on 5 June 1992.

[14] The United Nations Conference on Environment and Development (UNCED) (1992) *The Rio Declaration on Environment and Development*, adopted on 3 to 14 June 1992.

[15] ONU (2000) *Cartagena Protocol on Biosafety* to the *Convention on Biological Diversity* on 29 January 2000. The EU and all EU member states have ratified the protocol. Regulation (EC) No 1946/2003 of the European Parliament and of

The possibilities opened up by genetic engineering soon came to the centre of international attention during the '90s due to their great ethical implications. While the biogenetic resources relevant to the food and agricultural sectors are the object of the sovereign rights of States, human genetic resources have been increasingly perceived as a common good of the whole of mankind (Francioni, 2007, p. 10ff.). The advances in genetic engineering with the complete mapping of the human genome opened up the possibility of applying genetics to life sciences with the consequent hope of improving health and increasing human longevity and welfare. The availability of this new material led to concerns over its exploitation and the rights around it.

Here we face two different legal conceptions: one which conceives State sovereign rights over all natural resources, and one which removes those rights from States with a view to transferring them to the international community as a whole. In this sense international practice has evolved towards the affirmation of the standards of human rights by extending the principle of the common heritage of mankind from the domain of natural resources to the new concept of the human genome. Thus in 1997 the UNESCO adopted a non-binding document, the Universal Declaration on Human Genome and Human Rights, subsequently endorsed by the General Assembly of the UN with the resolution of 9 December 1998.[16] Here the human genome was solemnly declared as a common heritage of humankind (Art. 1). The term 'heritage' is neither legal nor technical, but it underlines the fact the genome cannot be treated as a natural resource subject to individual or collective appropriation. Its value for the whole of mankind lies in the ethical obligation to preserve and safeguard the human species. In the Preamble, any manipulation of the human genome for social and political purposes in ways incompatible with the inherent dignity of all the members of human family is rejected. Article 4

---

the Council of 15 July 2003 *on transboundary movements of genetically modified organisms* is the regulatory instrument that implements the provisions of the *Cartagena Protocol on Biosafety* within the European Union.

[16]UNESCO (1997) *Universal Declaration on the Human Genome and Human rights*, adopted by the UNESCO General Conference in 11 November 1997 and subsequently endorsed by the United Nations General Assembly in 1998.

states that the human genome in its natural state shall not give rise to financial gain. Article 24 considers germline intervention as contrary to human dignity and Article 6 prohibits the discrimination of individuals on the basis of one's genetic features. Thus the Declaration requires that genomic research in biology and medicine respects human dignity and fundamental individuals' and peoples' rights. Finally, the document prescribes respect for international norms and commitment to international cooperation with regard to the evaluation of risks and benefits deriving from genomic research. In 1999 UNESCO adopted a resolution for the implementation of measures designed to facilitate the interpretation of the Declaration in domestic law (Francioni, 2007, p. 12).

The deepening of concerns regarding genetic engineering led to the UNESCO Declaration of Human Genetic Data[17] of 2003, which recognized that each individual has a proper genetic make-up, but, also that a person's identity cannot be reduced to the genetic characteristics of the individual as it involves educational, environmental, emotional and personal factors. With the possibilities created by cloning techniques, fear spread to the possible extension of those to human beings. These fears led the UN towards the adoption of a specific Declaration on Human Cloning[18] in 2005, which prohibited human cloning as incompatible with human dignity and human life. This document also invited all member States to adopt any measure necessary to avoid the development of such applications of genetic engineering.

2005 is the year of the most relevant document on bioethics: the UNESCO Universal Declaration on Bioethics and Human Rights.[19] This document acknowledged the need to lay down some international guidelines common to all States in the field of medicine and the life sciences. What is notable is that it connects the freedom of scientific research and the benefits of science and technology with

---

[17]UNESCO (2003) *International Declaration on Human Genetic Data*, adopted by UNESCO General Conference on 16 October 2003.

[18]ONU (2005) *Declaration on Human Cloning*, adopted on 24 February 2005 by the General Assembly of the United Nations.

[19]UNESCO (2005) *Universal Declaration on Bioethics and Human rights*, adopted by acclamation by the 33rd session of the General Conference of the UNESCO in 19 October 2005.

the protection of human rights, stating that scientific research can never clash with human dignity and respect for human rights. The pillars of the UNESCO declaration are: consent, personal integrity, equality, privacy and confidentiality, health, protection of the future generation and protection of the environment, the biosphere and biodiversity. The great diversity of ethical perspectives among States and individuals led to the recognition of the close relationship between cultural pluralism and the respect for human rights, while also recognizing the limits of cultural diversity in the existence of a set of universal non-negotiable standards such as human rights. The Declaration recalled States' responsibility for spreading and sharing scientific knowledge and adopting transparency, professionalism, honesty and integrity in the processes of decision making, fostering public participation and the engagement of ethical committees. According to some, the precautionary principle should be *indirectly* recalled by the Declaration at Article 17 (Brownsword, 2008, pp. 34, 105).

The European landscape of human rights law is polarized by the two systems of the European Union (EU) and the Council of Europe. With regard to human rights and scientific and technological development, these two systems count, respectively, upon the Charter of Fundamental Rights of the EU and the Oviedo Convention and its Protocols. Both are integrated by an efficient judicial mechanism of enforcement: on the one hand, the Strasbourg Court (or ECtHR) with regard to the Council of Europe (and the European Convention on Human Rights or ECHR), and, on the other, the Luxembourg Court (or CJEU) with regard to the EU (and the EU Charter). These judicial mechanisms coexist with a plurality of mechanisms protecting human rights in Europe (Bultrini, 2004). We may mention by way of an example: the European Ombudsman of the European Union,[20] the OCSE Representative of Freedom of Media[21] and the UN Committees for those member countries of the two International Covenants of 1966 (the ICCPR and the ICESCR), etc. The EU and the Council of Europe systems are mutually

---

[20] Article 228 of the Treaty of European Union (TEU-L) (Ex Art. 195 TEC).

[21] The OSCE Representative on Freedom of Media provides early warning on freedom of expression and promotes full compliance with OSCE press freedom commitments.

intertwined and represent a unique model of enforcement of human rights which, thanks the efficiency of the two respective courts, has, until now, no equal in the world.

1.3 The EU Framework of the Charter of Fundamental Rights[22]

1.3.1 Within the EU framework, the Charter of Fundamental Rights[23] is the only text specifically dedicated to the interests corresponding to human rights. First of all, it is to say that human rights and fundamental rights need to be conceptually distinguished (Pariotti, 2013, p. 3ff.; Ruggiu, 2012b, p. 920. Indeed, although they have many points of coincidence, from the legal standpoint, the rights set forth by the EU Charter are distinguished from human rights by content (there are some rights such as freedom of research that are not affirmed by human rights law in the ECHR), structure (the EU fundamental rights are set at the same level as other EU interests such as the free circulation of goods, public health, counterterrorism[24]) and trait (human dignity is an autonomous right and not the pillar, the hidden concept of the human rights system in Europe which is present in several other rights such as the prohibition of inhuman and degrading treatments, the equality principle, etc., and in this form it is used in all the international documents on human rights).

The main characteristics of the text are that it has paid greater attention to the issues of technological and scientific innovation and it has incorporated some social rights together with civil and political ones. The purposes of the ethical advisory body of the Union (EGE), which collaborated in writing the draft of the Charter, was to distinguish the text from the previous experience of the European Convention on Human Rights (ECHR) of the Council of

---

[22]This paragraph was written with the support of the European Commission FP7 Science in Society funded project, Ethics in Public Policy Making: The Case of Human Enhancement (EPOCH), grant number SIS-CT-2010-266660.

[23]European Union (2000) *Charter of Fundamental Rights of the European Union*, adopted in Nice on 7 December 2000.

[24]Robles Morchón 2001, 263.

Europe by underling the possible impact of technology on rights, and by considering a range of rights broader than the mere group of civil and political rights.[25] For the first aspect we need to consider, for instance, the autonomous relevance of the concept of dignity (Art. 1), the right to integrity with regards to the field of medicine and biology (Art. 3), the wording of the equality principle under Article 21 (which takes into account the possible threats of genetics in this regard), the protection of personal data (Art. 8) and the express reference to the freedom of scientific research (Art. 13). Instead, for the second aspect, we must mention the recognition of the right to access to preventive health and the right to benefit from medical treatment under national law (Art. 35), the provision of the high level for consumers' protection by the EU authorities (Art. 38) and for environmental protection (Art. 37), which are a clear novelty in a document with this extent.

It must be remembered that the CJEU was already adjudicating human rights as general principles of EU law and the constitutional traditions common to the member States before the Charter became legally binding at the beginning of 2010.[26] After the Charter of Fundamental Rights of the EU[27] came into force, those rights became more visible with the formal appearance of fundamental rights. Indeed, with the entry into force of the Lisbon Treaty (TEU-L)[28] on 1 December 2009, fundamental rights can now be *directly* applied

---

[25] On this see the draft of the EU Charter drawn up by the EGE (2000).

[26] The CJEU acknowledged that: '... fundamental rights form an integral part of the general principles of law the observance of which the Court ensures, and that, for that purpose, the Court draws inspiration from the constitutional traditions common to the Member States and from the guidelines supplied by international treaties for the protection of human rights on which the Member States have collaborated or to which they are signatories. The European Convention on Human rights and Fundamental Freedoms has special significance in that respect.' See judgement of the Court of Justice *Omega Spielhallen- und Automatenaufstellungs-GmbH v. Oberburgermeisterin der Bundesstadt Bonn* (Case C-36/02), 14 October 2004 *European Court Reports*, 2004, p. I-09609, §33.

[27] European Union (2000) *Charter of Fundamental Rights of the European Union*, proclaimed in Nice on 7 December 2000.

[28] European Union (2007) *Lisbon Treaty* (Treaty of Lisbon amending the Treaty of the European Union and the Treaties establishing the European Community), adopted in Lisbon on 13 December 2007 and entered into force on 1 December 2009.

by referring to the Charter and are expressly endowed with a legal obligation of the member States and EU public authorities.

Although the Charter has not been incorporated, the Charter is formally recognized by the TEU-L. This fact gives the Charter the significance of a constitutionally separate document, almost a *per se* existence. Underlining the constitutional level of these new Treaties, the language of the EU addresses individuals directly as the concern for their freedoms and rights, holding the political powers legally responsible for their protection. In this sense, thanks to the Charter, the fundamental rights of the EU are directly linked to individuals and their power of petition in the face of the CJEU.

Nevertheless, the Charter could have different degrees of efficacy among member States. We need to bear in mind that Protocol No. 30 on the Application of the Charter of Fundamental Rights of the European Union to Poland and Britain prevents the CJEU from assessing the incompatibility of the Polish and British national laws or practices with the fundamental rights of the Charter (Pernice, 2008, p. 244ff.). This fact led some to think that from the legal standpoint the Charter should not have created any new rights for EU citizens (Pernice, 2008). This conclusion is based on the opinion which sees fundamental rights as a mere declination of human rights previously applied by the CJEU. As said above, this opinion cannot be shared. Behind this particular view there is the fear that through the entry into force of the Charter, social rights, which were one of the main features of the text, would also have to be introduced into the EU framework. The analysis of the rights set forth by the Charter leads to the conclusion that there is no coincidence between fundamental rights and human rights, but a mere convergence with regard to the list and content.

Therefore, the most relevant consequence of Article 6 of TEU-L, which produced the entry into force of the EU Charter are a reshaping of the legal sources of the EU by giving the EU Charter the same legal force as the treaties; the fact that the CJEU can *directly* apply the fundamental rights of the Charter[29]; the possible implementation of those rights (through the above-

---

[29]It is not necessary to adjudicate human rights as general principles of the EU law and the constitutional traditions common to the member States any longer.

mentioned process of EU accession to ECHR[30]); the reshaping of the relationship between the CJEU and the national courts; and the transformation of the application of national judges, who can evaluate the compatibility of national law with the Charter, cease to apply the contrasting national law, and address a preliminary ruling to the Court (Feliziani, 2011, p. 25). In this sense we must mention the fundamental *Kücükdeveci v. Swedex case*[31] with which the CJEU acknowledged the binding nature of the Charter with regard to the principle of equality by directly applying the rights set forth there (Feliziani, 2011).

1.3.2 In its Preamble the Charter places individuals at the centre of the activity of the EU authorities. It recognizes the indivisible, universal values of human dignity, freedom, equality and solidarity. This should have stressed the importance of the person in the overall architecture of both the Charter and EU law. In this regard, Article 1 decrees human dignity as inviolable. The centrality of the person is also preserved against the threats that come from biomedicine and biology. The special relevance of the concept of dignity leads one to think that the regulator intended to create an autonomous right claimable by individuals or group. In this sense the right to dignity, which is a novelty in the framework of international documents on human rights (for example the ECHR has no analogous provision), was applied for the first time in 2011 with the *Brüstel case*.[32]

Dignity is also protected in other places in the EU text. With regard to the value of dignity, the Charter prohibits making the human body and its parts a source of financial gain (Art. 3,2c). In this sense human dignity is strictly connected to personal integrity

---

[30] Recently the accession process of the EU to the ECHR has been stopped due to the contrary opinion of the Court of Justice of the European Union of the 18 December 2014 since in its current form, the draft agreement on the accession of the EU to the ECHR would fail to have regard to the specific characteristics of EU law with regard to the judicial review of acts, actions or omissions on the part of the EU by empowering the Strasbourg Court of the exclusive jurisdiction. See http://curia.europa.eu/juris/document/document.jsf?docid= 160882&doclang= EN. Accessed 1 December 2017.

[31] Judgement of the Court of Justice *Kücükdeveci v. Swedex GmbH & Co. KG* (Case C-555/07) 19 January 2010 *OJ C 63, 13.3.2010*, p. 4–4.

[32] Judgement of the Court of Justice (Grand Chamber), *Oliver Brüstle v. Greenpeace eV* (Case C-34/10) 18 October 2011, not yet published.

(Art. 3) and the use of the human body, which is increasingly at stake with new technologies such as nanomedicine. The same Article 3, 2 (a, b, d) grants both free and informed consent in the field of medicine and biology and personal integrity by prohibiting eugenic practices (especially those aiming at the selection of persons) and the reproductive cloning of human beings. The modernity of its touch is also expressed by the non-discrimination principle. The development of genetic engineering, and, perhaps, in the remote future, synthetic biology, could lead to new and subtle forms of discrimination within our societies. In this context, with the equality principle, the Charter prohibits any discrimination on the basis of genetic features (Art. 21). Biotechnologies produce a huge mass of bio-information whose availability constitutes a source of power over individuals (Rodotà, 1995). In the future, the weight of this information will increase in importance. With regard to the activity of bio-banks, both the confidentiality of and the access to personal data are protected in the Charter (Art. 8). Finally, the Charter affirms the freedom of scientific research as an autonomous fundamental right which represents an interesting novelty in the framework of human rights law, especially vis-à-vis the ECHR.[33] Since the beginning of the second millennium the main goal of the European Union has been to create 'the most competitive and dynamic knowledge-based economy in the world'.[34] The plan was to reach the principle world economies of the US and Japan. In this framework a special role was granted to scientific research. In this sense the recognition made by the Charter should have boosted the development of the economies of the EU countries. Nevertheless, the interests of scientific research may clash with human rights, and in this sense we need to discover a new sensitivity to the international documents on human rights, where human rights are deemed as

---

[33] It follows Article 15 of the ICESCR which affirms 'the right of everyone to enjoy benefits of scientific progress and its application'. In addition, State parties to the ICESCR 'undertake to respect of the freedom indispensable for scientific research and creative activity' (Francioni, 2008, p. 4). Within the ECHR such freedom can be posed under the freedom of thought (Art. 9) and does not represent a specific and autonomous right.

[34] Lisbon European Council 23–24 March 2000: Presidency conclusion, available at http://www.europarl.europa.eu/summits/lis1_en.htm. Accessed 1 December 2017.

interests which cannot be balanced with other public interests of the State such as scientific research (Ruggiu, 2012b).

1.4 The Council of Europe Framework and the Oviedo Convention and Its Protocols[35]

1.4.1   The heart of the whole architecture created by the Council of Europe is the European Convention on Human Rights (ECHR) which only affirms civil and political rights. This means that some rights that are relevant to the applications of science and technology, such as the right to health and the right to a healthy environment, should not find an express legitimation here. Fortunately, thanks to the work of the European Court of Human Rights (the ECtHR) it is not true. Besides, although the Convention reserves no specific attention for scientific and technological development, thanks to the activity of the Strasbourg Court (the ECtHR), it maintains a central importance in this matter.

Within the Council of Europe regulatory framework, on the other hand, the Oviedo Convention[36] is a treaty expressly designed for issues concerning scientific and technological developments, although many other documents produced by the Council of Europe may be relevant in specific sectors. The Oviedo Convention is a legally binding document which addresses the protection of human dignity and the integrity of the person regarding the applications of biomedicine. It only sets forth general principles, while some specific aspects are dealt with further in other texts that should regulate them in detail. Thus, it has been followed by some more specific protocols which have already entered into force (those

---

[35]This paragraph takes up and analyses in greater detail (Ruggiu, 2012a).

[36]Council of Europe (1997) *Convention for the Protection of Human rights and Dignity of the Human Being with regard to the Application of Biology and Medicine* (Convention on Human rights and Biomedicine or the Oviedo Convention) (CETS n. 164), adopted in Oviedo on 4 April 1997 (entered into force on 1 December 1999). Austria, Belgium, Germany, Ireland, Liechtenstein, Russia, United Kingdom abstained, whereas France, Italy, The Netherlands, Poland, Ukraine, Sweden did not ratify the Convention.

on the prohibition of human cloning,[37] human transplantation[38] and biomedical research[39]), and others, which although signed, await a sufficient number of ratifications (i.e. on genetic testing for health purposes[40]). Meanwhile, other protocols are anticipated in another form (e.g. xenotransplantation,[41] on persons with mental disorder[42]) and one on nanotechnologies has recently been formally requested by the Parliamentary Assembly of the Council of Europe.[43] In this regard, the aim of the Council of Europe project was to elaborate a general framework where the protocols have the task of dealing with some specific matters in more detail.

The protection of the human being established by the Convention is structured in multiple layers like an onion. The first level protects individuals from the distorted use of science. The second level protects individuals as a part of mankind and society. The final level protects individuals in decision-making processes by creating a system of implementation of public participation (de Salvia C., 2000).

---

[37] Council of Europe (1998) *Additional Protocol to the Convention on Human rights and Biomedicine on the Prohibition of Cloning Human Beings* (CETS n. 168), adopted in Paris on 12 January 1998 (entered into force on 1 March 2001).

[38] Council of Europe (2002) *Additional Protocol to the Convention on Human rights and Biomedicine Concerning Transplantation of Organs and Tissues of Human Origins* (CETS n. 186), adopted in Strasbourg on 24 January 2002 (entered into force on 1 May 2006).

[39] Council of Europe (2005) *Additional Protocol to the Convention on Human rights and Biomedicine Concerning the Biomedical Research* (CETS n. 195), adopted in Strasbourg on 25 January 2005 (entered into force on 1 September 2007).

[40] Council of Europe (2008) *Additional Protocol to the Convention on Human rights and Biomedicine Concerning the Genetic Testing for health purposes* (CETS n. 203), adopted in Strasbourg on 11 November 2008 (it has not entered into force yet).

[41] Council of Europe (1997) Recommendation RE (97) 15 of the Committee of Ministers of the Council of Europe *on Xenotransplantations*, adopted on 1997; Council of Europe (2003) The Recommendation RE (2003) 10 of the Committee of Ministers of the Council of Europe *on Xenotransplantations*, adopted on 19 June 2003.

[42] Council of Europe (2004) Recommendation RE (2004) 10 of the Committee of Ministers of the Council of Europe *Concerning the Protection of the Human Rights and Dignity of Persons with Mental Disorder and its Explanatory Memorandum*, adopted on 22 September 2004.

[43] Council of Europe (2013) Recommendation 2017 (2013) of the Parliamentary Assembly of the Council of Europe *Nanotechnology: Balancing Benefits and Risks to Public Health and the Environment*, adopted on 26 April 2013 (final version). In this document either a Committee of Ministries recommendation or an additional protocol to the Oviedo Convention is requested (Point 5.6).

Compared with the EU Charter experience, the concept of human dignity has a different relevance here. Indeed, it represents the hidden pillar of the whole architecture of the human rights law of the Council of Europe. Thus, the express aim of the Convention is the protection of human dignity from the 'misuse of biology and medicine' and in this context all rights set forth serve to achieve this aim.[44] It is notable that this is the first text of international human rights law where human dignity is expressly mentioned with this relevance. In this regard, Article 2 decrees the primacy of 'the interests and welfare of the human being' 'over the *sole* interest of society or science' (italics mine). This means that the interests of the human being can only be postponed in certain conditions. In this sense, Article 15 mentions the freedom of research in the field of medicine and biology which is only subject to the provisions of the Oviedo Convention and the other norms protecting the human being. In this sense the welfare and the interest of the human being may be overshadowed by a cluster of public interests including scientific research, public health etc. on the condition that they are legitimate and provided by a law. Thus, Article 26 prescribes that the rights set forth by the treaty can only be superseded by restrictive measures provided by the law and by measures which 'are necessary in a democratic society in the interest of public safety, for the prevention of crime, for the protection of public health or for the protection of rights and freedom of others'. In addition to the provision of dignity, Article 21 prohibits the human body and its parts as such from giving rise to financial gain. But it must be considered that it does not prohibit the sale of pharmaceutical products of human origin (e.g. human blood and its derivatives) and private enterprise in this sector.

The notion of human dignity and the person is lacking from the Convention. The general opinion which emerges within the ECtHR case law is that the term 'human being' also includes the fetus, while the word 'person' only refers to a person who is already born (de Salvia C., 2000, p. 92; Byk 1999, pp. 101–121).[45]

---

[44] See the *Preamble* of the Oviedo Convention.

[45] *X v. The United Kingdom* (Appl. 8416/78) decision of the Commission of 13 May 1980 *Decision and Reports*, 19, also cited as *Paton v. UK*; *Evans v. the United*

Within the issues related to human dignity some other provisions are also applicable, especially in the field of genetic engineering. The Additional Protocol[46] of 1998 prohibits the creation of human beings that are genetically identical to one another whether living or dead (Art. 1). Article 11 of the Oviedo Convention prohibits any form of discrimination of the person on the grounds of his or her genetic heritage, integrating in this regard Article 14 of the ECHR (equality principle) which does not mention genetic features within the list of the prohibited reasons for discrimination. It must be said that the list is open and that this kind of discrimination would have been prohibited anyway. The Protocol of 1998[47] integrates Article 11 with the prohibition of any stigmatization of persons or groups on the basis of their genetic characteristics (Art. 4). In this regard, unlike the UNESCO Declaration on the Human Genome, the Convention does not declare the human genome to be the heritage of mankind, showing a clear tendency towards a more practical approach and more awareness of the role of the private actor in the current market of genetic resources. With regard to the rights of the future generations, Article 13 states that an intervention seeking to modify the human genome may only be undertaken for preventive, diagnostic or therapeutic purposes and only if its aims are not to modify the genome of any descendants. This represents the insurmountable limit for genetic enhancement. Thus, any enhancing genetic therapy or any therapy directed to human enhancement that modifies the germline should be viewed as prohibited because

---

Kingdom [GC] (App. 6339/05) judgement of 10 April 2007 *Reports Judgements and Decisions*, 2007-I; *Vo v. France* (App. 53924/00) judgement of 8 July 2004, par. 18: "The Convention does not define the term 'everyone' (in French *'toute personne'*). These two terms are equivalent and found in the English and French versions of the European Convention on Human Rights, which however does not define them. In the absence of a unanimous agreement on the definition of these terms among member States of the Council of Europe, it was decided to allow domestic law to define them for the purposes of the application of the present Convention".

[46] Council of Europe (1998) *Additional Protocol to the Convention on Human rights and Biomedicine on the Prohibition of Cloning Human Beings* (CETS n. 168), adopted in Paris on 12 January 1998 (entered into force on 1 March 2001).

[47] Council of Europe (2008) *Additional Protocol to the Convention on Human rights and Biomedicine concerning the genetic testing for health purposes* (CETS n. 203), adopted in Strasbourg on 11 November 2008 (it has not entered into force yet).

such modifications would violate the rights of the persons who inherit this genetic modification. In this regard it should be noted that the techniques of medically assisted procreation cannot be allowed for the purpose of selecting the sex of a future child. Finally, an Additional Protocol[48] prohibits human reproductive cloning.[49] Article 12 of the Oviedo Convention prohibits any predictive genetic test unless it is for the purposes of health or for scientific research linked to the purposes of health, and subjected to appropriate genetic counselling. In this sense the preimplantation genetic diagnosis (PGD) in medically assisted procreation techniques is to be deemed as allowed in this framework.[50] In this regard, no insurance or employment contract can include a genetic test.

The principle of self-determination finds special relevance within the framework of the Convention. Article 5 of the Oviedo Convention decrees the free and informed consent to any medical treatment and states that it may be withdrawn at any time.[51] Thus, the same principle of autonomy is confirmed by the Protocol on Biomedical Research[52] (Art. 13), which protects the consent of diverse persons due to their vulnerable condition (e.g. persons without the capacity to consent,[53] pregnant women, persons in emergency clinical

---

[48] Council of Europe (1998) *Additional Protocol to the Convention on Human rights and Biomedicine on the Prohibition of Cloning Human Beings* (CETS n. 168), adopted in Paris on 12 January 1998 (entered into force on 1 March 2001).

[49] The Oviedo Convention prohibits the 'reproductive cloning', but not the 'therapeutic cloning'. In the first case the cloned embryo is transferred to a woman's uterus in the view of having a baby genetically identical to the cell donor. In the second case the embryo's inner mass is harvested and grown in culture *in vitro* for subsequent derivation of embryonic stem cells in view of therapeutic applications as the cure of Alzheimer's disease. In this case the human cloned stem cell derives from a human embryo which, if transferred to a uterus, would be able to become a foetus, and then a baby. Here the outcome is not an identical human being, but the destruction of the embryo for therapeutic purpose. See Andorno, 2002, 961; Colombo, 2003.

[50] *Costa and Pavan v. Italy* (Appl. 54270/10) judgement of 28 August 2012.

[51] *Yhunke v. Turkey* (Appl. 1620/03) judgement of 23 September 2010 *Reports of Judgements and Decisions*, 2010.

[52] Council of Europe (2005) *Additional Protocol to the Convention on Human rights and Biomedicine concerning the Biomedical Research* (CETS n. 195), adopted in Strasbourg on 25 January 2005 (entered into force on 1 September 2007).

[53] *Lambert et autres contre la France* (Requête no 46043/14) introduite le 23 juin 2014.

situations,[54] persons deprived of liberty,[55] etc.). It is worthwhile noting that this protocol protects the nationals of States not party, in cases where scientific research attempts to avoid complying with the provisions of the instrument by involving individuals not covered by its norms (e.g. citizens of Developing Countries). The Protocol also sets forth the right to the confidentiality of personal data[56] and the accessibility of the information by the persons involved in biomedical research[57] (Arts. 25 and 26). In addition, the Protocol affirms the principle of proportionality between risks and benefits (risks should not to be disproportionate to the prospective benefits). Finally, it decrees the principle of multidisciplinary examination by scientific and ethical committees on the merit, the aim and the ethical implication of the research (Arts. 7, 8 and 9).

Access to healthcare is also regulated,[58] giving rise to the question of the centrality of the right to health *vis-à-vis* the new applications of science and technology in the field of medicine. Article 3 states that access to healthcare should be equitable. Article 4 rules that any intervention in the health ambit, including research, must be carried out in accordance with professional obligations and standards (including, thus, the norms of ethical codes and guidelines).

With regard to the further implementation of the provision of the Treaty and its protocols, Article 28 of the Convention obliges States to ensure that the fundamental questions raised by the developments of biology and medicine are subjected to public debate.

1.4.2 The Oviedo Convention fails to include a compulsory mechanism. Indeed, within the Oviedo Convention, the Secretary General of

---

[54] *Pretty v. United Kingdom* (Appl. 2346/02) judgement of 29 April 2002 *Reports of Judgements and Decisions*, 2002-III.

[55] *Yhunke v. Turkey* (Appl. 1620/03) judgement of 23 September 2010 *Reports Judgements and Decisions*, 2010.

[56] *Z. v. Finland* (Appl. 22009/93) judgement of 25 February 1997.

[57] *Roche v. The United* Kingdom (App. 32555/96) judgement of 19 October 2005 *Reports of Judgements and Decisions*, 2005-IX.

[58] *Affaire Vincent c. France* (Requête 6253/03) arrêt de 24 octobre 2006.

the Council of Europe determines the application of the Convention and the Strasbourg Court can only give 'advisory opinions on legal questions concerning the interpretation of the text', but it cannot take any decision by directly applying its norms (Art. 29). In this sense, the Court has only an interpretative role which cannot be confused with the advisory task provided within the ECHR by Article 47 (ECHR). This soft form of link between the ECtHR and the Oviedo Convection is quite rare and it is the sole example of this cooperation within the conventions of the Council of Europe (Gitti, 1998, p. 721). In fact, the ECtHR is not a body of the whole Council of Europe, but only of the ECHR, thus it is infrequent that an organ provided for another treaty (i.e. the ECHR) acts in support of the application of the norms of another convention (such as in the case of the Oviedo Convention).

As said above, within the regulatory framework of the Council of Europe on technical and-scientific development, we also need to consider the following set of regulatory documents of soft law that might guide the ECtHR jurisprudence: the European Social Charter[59] (ESC) with regard to the right to health; the European Agreement on the Exchange of Therapeutic Substances of Human Origin[60]; the European Agreement on the Exchange of Tissue-Typing Reagents[61]; the Convention on Unification of Certain Points of Substantive Laws on Patents for Invention[62] with regard to the IP issues; the Convention on Elaboration of a European Pharmacopeia[63];

---

[59]Council of Europe (1961/1996) *European Social Charter* (ESC) (CETS n. 35), adopted in Turin on 3 October 1961 (entered into force on 26 February 1965) revised in Strasbourg on 3 May 1996 (CETS n. 163) (entered into force on 1 July 1999).

[60]Council of Europe (1958) *European Agreement on the Exchange of Therapeutic Substances of Human Origin* (CETS n. 26), adopted in Paris on 15 December 1958 (entered into force on 1 January 1959).

[61]Council of Europe (1974) *European Agreement on the Exchange of Tissue-Typing Reagents* (CETS n. 47), adopted in Strasbourg on 17 September 1974 (entered into force on 23 April 1977).

[62]Council of Europe (1963) *Convention on Unification of Certain Points of Substantive Laws on Patents for Invention* (CETS n. 47), adopted in Strasbourg on 27 November 1963 (entered into force on 1 August 1980).

[63]Council of Europe (1964) *Convention on Elaboration of a European Pharmacopeia* (CETS n. 50), adopted in Strasbourg on 22 July 1964 (entered into force on 22 May 1974).

the Convention for the protection of individuals with regard to automatic processing of personal data,[64] and its protocols[65] with regard to confidentiality and privacy; the Convention on the Protection of the Environment through Criminal Law[66]; and the Anti-doping Convention[67] and its protocol.[68] There is no compulsory mechanism included in these documents either.

We should also mention the activity of the Parliamentary Assembly and the Committee of Ministers[69] which represents mere guideline within the Council of Europe regulatory framework, but can be taken in account especially by the ECtHR.

---

[64] Council of Europe (1981) *Convention for the Protection of Individuals with Regard to Automatic Processing of Personal Data* (CETS n. 108), adopted in Strasbourg on 1 January 1981 (entered into force on 1 October 1985).

[65] Council of Europe (2001) *Additional Protocol to the Convention for the Protection of Individuals with Regard to Automatic Processing of Personal Data* (CETS n. 50), adopted in Strasbourg on 8 November 2001 (entered into force on 1 July 2005).

[66] Council of Europe (1998) *Convention on the Protection of the Environment through Criminal Law* (ETS no. 172), adopted in Strasbourg on 4 November 1998 (it has not entered into force yet).

[67] Council of Europe (1989) *Anti-doping Convention* (CETS n. 153), adopted in Strasbourg on 16 November 1989 (entered into force on 1 March 1990).

[68] Council of Europe (2002) *Additional Protocol to Anti-doping Convention* (CETS n. 188), adopted in Warsaw on 12 September 2002 (entered into force on 1 April 2004).

[69] See e.g. Recommendation 934 of the Parliamentary Assembly of the Council of Europe *on the genetic engineering*, adopted on 26 January 1982; Recommendation 1425 of the Parliamentary Assembly of the Council of Europe *on the biotechnology and the intellectual propriety*, adopted on 20 September 1999; Recommendation 1468 of the Parliamentary Assembly of the Council of Europe *on the biotechnology*, adopted on 29 June 2000; Recommendation 1512 of the Parliamentary Assembly of the Council of Europe *on the protection for the human genome* by the Council of Europe, adopted on 20 April 2001; Recommendation 2017 (2013) of the Parliamentary Assembly of the Council of Europe *Nanotechnology: Balancing Benefits and Risks to Public Health and the Environment*, adopted on 26 April 2013 (provisional version); Recommendation RE (2003) 10 of the Committee of Ministers of the Council of Europe *on xenotransplantations*, adopted on 19 June 2003; Recommendation RE (2006) 4 of the Committee of Ministers of the Council of Europe *on research on biological materials of human origin*, adopted on 15 March 2006.

1.5 The Issue of Compulsory Jurisdiction of the Strasbourg Court for the Economic and Social Aspects of Human Rights[70]

1.5.1 The Oviedo Convention is a soft law instrument and it has, as is known, no mechanism of enforcement for rights included in it. This means that there is no way of making infringing States comply with the norms of the treaty. The ECHR could provide a compulsory mechanism, but it only affirms civil and political rights, and the rights involved in scientific and technological development are above all economic and social rights, such as the right to health and the right to the environment. In this context the European Social Charter[71] (ESC) exists. This instrument serves to protect the social and economic aspects of human rights in the framework of the conventions of the Council of Europe. This text contains a series of economic and social rights that States *can* accept as goals of their social policies. Within the first part, the State parties undertake to create the political conditions for putting rights into effect. In the second part, the economic and social rights are listed. With regard to technoscientific progress, it is worthwhile mentioning specifically the right to health (Art. 11), the right to safe and healthy work conditions (Art. 3) and the right to medical assistance (Art. 13).

The importance of this document rests on the fact that there are two aspects that are often affected by technical and-scientific development: health and the environment. In this regard it would be advisable for those aspects to be implemented and enforced by compulsory mechanisms.

Nevertheless, the limit of the Charter is its low degree of efficacy (Foà, 1998, p. 71; Olivieri, 2008, p. 505). According to point A Part three of the revised Charter, the State party must take appropriate measures to ensure that at least six articles are chosen among a

---

[70]This paragraph takes up and analyses in more detail (Ruggiu, 2012a).
[71]Council of Europe (1961/1996) *European Social Charter* (ESC) (CETS n. 35), adopted in Turin on 3 October 1961 (entered into force on 26 February 1965) revised in Strasbourg on 3 May 1996 (CETS n. 163) (entered into force on 1 July 1999).

series of rights.[72] Besides, States must consider themselves bound by a supplementary number of articles of Part II of the ESC.[73] There is no mechanism of enforcement provided with regard to the ESC either (Foà, 1998, p. 71; Olivieri, 2008, p. 532). Indeed the State party should communicate a periodical report to the Secretary General on the application of the Charter.[74] In 1995 an additional protocol to the ESC[75] provides a set of collective complaints by the international organizations of employers and trade unions (Art. D of Part three of the revised Charter). National and supranational organizations of employers and trade unions, and the international NGOs can submit complaints alleging the unsatisfactory application of the ESC to a Committee of Independent Experts who shall draw up a report which is submitted to the Committee of Ministers. Then, on the basis of the report, the Committee of Ministers will adopt a resolution by the majority of its members. The resolution is not binding, but it carries out only some measures of political nature (Olivieri, 2008, p. 531).

The lack of any compulsory mechanism in the Oviedo Convention and the other instruments of the Council of Europe such as the ESC risks leaving individuals without any protection in the face of technoscientific progress. The fact that the Strasbourg Court can only give advisory opinions on the application of the norms of the Oviedo Convention means that it cannot apply any norm of the treaty directly. In this sense, the ECtHR can recall those norms, as well

---

[72] This is the list of rights from which a State should choose: the right to work (Art. 1), the right to organize (Art. 5), the right to collective bargaining (Art. 6), the right to children and young persons' protection (Art. 7), the right to social security (Art. 12), the right to social and medical assistance (Art. 13), the right of the family to social, legal and economic protection (Art. 16), the right of migrant worker and their families to protection and assistance (Art. 19), the right to equal opportunities and equal treatment in the question of employment and occupation (Art. 20).

[73] The series of Articles by which a State is bound shall be not less than sixteen (in the case of articles) or sixty-three (in the case of paragraphs).

[74] See Council of Europe (1991) *Additional Protocol amending European Social Charter* (CETS n. 142), adopted in Turin on 21 October 1991, Article 1.

[75] Council of Europe (1995) *Additional Protocol to European Social Charter Providing for a System of Collectives Complaints* (CETS n. 158), adopted in Strasbourg on 11 November 1995 (entered into force on 1 July 1998).

as any other norm from each convention of the Council of Europe, to support its decisions *interpretatively*.[76] In this way, soft law documents can influence by reflex (*par ricochet*) the jurisprudence of the Strasbourg Court. For example, the ECtHR could extend the content of the principle of equality of Article 14 (ECHR) to include discrimination on the grounds of an individual's genetic features by recalling Article 11 of the Oviedo Convention. But it cannot apply them in a direct manner. It is not possible to give Article 29 of the Oviedo Convention (on the interpretative competence of the ECtHR) a different meaning by adding compulsory jurisdiction to the task of giving advisory opinions which the treaty recognizes the ECtHR. This would be an erroneous way to use the articles of the Oviedo Convention. The reason for this is that if within the Oviedo Convention a compulsory jurisdiction was given to the Court, the non-member States of the Council of Europe who were party of the Oviedo Convention, would find a judicial body without having signed any explicit protocol on this point. Indeed, the Oviedo Convention was meant as a universal instrument and in this sense, it was open to the signatures of both member States of the Council of Europe and non-member States (Art. 33 Oviedo Convention). This is the reason why signing Article 29 (Oviedo Convention) could never be interpreted as the acceptance of any compulsory jurisdiction (Gitti, 1998). This does not mean that any violation of human rights in the field of biology and medicine can never be punished. Notwithstanding the above-mentioned limits *ratione materiae* and *ratione personae*, the Court is recognized as having the possibility to intervene in this matter.

In this regard, the Strasbourg Court can intervene not only by recalling norms of soft law from the interpretative standpoint, as said, by reflex. As human rights can have both economic and social implications, the ECtHR can *indirectly* protect (par ricochet) the economic and social aspects of human rights such as those raised by

---

[76] *Demir and Baykara v. Turkey* (Appl. 34503/97) judgement of 12 November 2008 *Reports Judgements and Decisions*, 2008, par. 81: 'the Court took account, in interpreting Article 8 of the Convention, of the standards enshrined in the Oviedo Convention on Human Rights and Biomedicine of 4 April 1997, even though that instrument had not been ratified by all the States Parties to the Convention'.

social rights.[77] This means that economic and social rights cannot be recalled by the Court alone (i.e. directly), but in conjunction with articles of the ECHR that have allegedly been violated, they can (indirectly) be applied. In this way some social rights, such as the right to health and the right to a healthy environment, can find a kind of protection within the ECHR framework (Foà, 1998, p. 60; Olivieri, 2008, p. 537; Gitti, 1998, p. 730; Ruggiu, 2012a, p. 13). Only in this sense can one affirm those rights existing in the ECHR. This conclusion is nowadays a consolidated acquisition of the jurisprudence of the Strasbourg Court. In this sense social rights can be protected either with the application by reflex (*par ricochet*) of all the norms of the Oviedo Convention and its protocols, as well as any other convention of the Council of Europe (such as ESC), or (most of all) as economic and social consequences of the violation of the rights and freedoms enshrined by the ECHR. This makes that provided by the ECHR an efficient mechanism of enforcement in this matter (Ruggiu, 2012b; 2013b).

1.6 The Rise of a Renewed Interest for Emerging Technologies

1.6.1 Recently the international community has shown a renewed interest in emerging technologies and human rights, expressing the need to widen the scope of the old tools and build new ones. The focus of this new course of the international community's attention to emerging technologies is in particular on nanotechnologies[78] and the phenomenon of the convergence of new technologies.

A first sign of this renewed attention is the report on 'The ethics and politics of nanotechnology' published by the UNESCO in 2006.[79] In this text the experts of the organization of the UN address several

---

[77]See *Airey v. Ireland* (Appl. 6289/73) judgement of 9 October 1979, *Series A*, No. 32, par. 26 'Whilst the Convention sets forth what are essentially civil and political rights, many of them have implications of a social or economic nature'.

[78]As regards the interest of the international community on synthetic biology see the Chapter 3.

[79]UNESCO (2006) *The Ethics and Politics of Nanotechnology*, Paris: United Nations Educational, Scientific and Cultural Organization.

aspects concerning the development of nanotechnologies[80] with particular attention to the international level and questions of global justice. Two main concerns are addressed. The first is the knowledge shared among the most industrialized countries and developing nations, which might call into question the right to development, but it relates to more general questions of justice (UNESCO, 2006, p. 13).

The risk at stake is that nanotechnologies could be the next opportunity to enlarge the economic gap between States and to consolidate the current framework in which some countries produce high-tech products and others provide them with the raw materials. Developing nations risk being distanced by a knowledge divide if they cannot find ways to participate on an equal footing with others. It is clear that the question which arises here is one of justice and equal access to the resources of knowledge. Here human rights are only in the background. This knowledge divide is deemed being based on both funding and the legal implications of intellectual propriety. In the international community the need of scientists and experts to find ways of closing this knowledge gap both within their own countries and between nations thus appears urgent (ibid., p. 15). Here comes the second concern which involves the role of science. In the twentieth century, science increasingly came under new pressures that guided the creation, publication and sharing of public information and questioned both its transparency and its independency. One of these elements of pressure on science is the expanding system of intellectual property. Another is the increasing demand that science become accountable to the public. The third is the abuse of scientific information by governments given the increasing secrecy and the demand for antiterrorism measures. All three of these sources of pressure on science have a negative effect on the kind and quality of scientific research by introducing incentives that are contrary to the values of objectivity and disinterestedness (ibid., p. 17). This second concern is closely linked to the first since systems of intellectual

---

[80]Given the fact that there is not a *single* nanotechnology, but a *plurality* of nanomaterials and nanoparticles with different properties, functions and legal and ethical implications, it would be better to speak of nanotechnologies in the plural.

property rights can introduce prohibitive costs for those who are interested in researching nanomaterials and create barriers of great complexity in terms of competing and overlapping patent claims. 'Rather than introducing incentives for more rewards, it introduces anxiety concerning the legality and liability of using what might be perceived as products of nature or natural processes' (ibid., p. 20). In this context the experts of UNESCO encouraged member States to develop a common definition of nanomaterial in order to avoid solutions which could widen the economic divide among nations, to set voluntary standards for commercial production and to promulgate new ethical standards for commercial as well as university research practices.

Other UN bodies have also been involved with regard to nanotechnologies in food and agriculture.

Recently the Food and Agriculture Organization of the United Nation (FAO) and the World Health Organization (WHO) delivered two meeting reports on the use of nanotechnologies in food and agriculture: one in 2009,[81] whose findings and recommendations were summarized in a further report in 2010,[82] and the second in 2012.[83] Finally, in 2013 a technical paper was issued on the state of the art of ongoing initiatives concerning risk assessment and risk management with regard to nanotechnological applications in food and agriculture.[84] It is interesting to note that the term 'human

[81] FAO/WHO [Food and Agriculture Organization of the United Nations/World Health Organization] (2009) *FAO/WHO Expert Meeting on the Application of Nanotechnologies in the Food and Agriculture Sectors: Potential Food Safety Implications: Meeting Report.* Rome: FAO and WHO. 104 pp.

[82] FAO/WHO [Food and Agriculture Organization of the United Nations/World Health Organization] (2010) *FAO/WHO Expert Meeting on the Application of Nanotechnologies in the Food and Agriculture Sectors. Potential Food Safety Implications.* Meeting Report. Rome: FAO and WHO. 130 pp. http://apps.who.int/iris/bitstream/10665/44245/1/9789241563932_eng.pdf. Accessed 17 May 2016.

[83] FAO/WHO [Food and Agriculture Organization of the United Nations/World Health Organization] (29 March 2012) *Joint FAO/WHO Seminar on Nanotechnologies in Food and Agriculture.* Rome: FAO and WHO. http://www.fao.org/3/a-au206e.pdf. Accessed 17 May 2016.

[84] FAO/WHO [Food and Agriculture Organization of the United Nations/World Health Organization] (2013) *Paper: State of the Art on the Initiatives and Activities Relevant to Risk Assessment and Risk Management of Nanotechnologies in the Food and Agriculture Sectors.* Geneve: FAO and WHO http://www.fao.org/docrep/018/i3281e/i3281e.pdf. Accessed 17 May 2016.

rights' does not occur in all these documents. Instead the term 'ethical principles' (which should probably encompass only human rights) occurs only once in the 2009 text and once in the 2010 text of 2010.

On 1–5 June 2009 the FAO and the WHO convened a Meeting Report on applications of nanotechnologies in food and agriculture, focusing on three main areas: the use of nanotechnologies in food production and processing; the potential human health risks associated with this use; and the elements of transparent and constructive dialogues among stakeholders. Seventeen experts from relevant disciplines such as food technology, toxicology and communication met at the FAO headquarters. The relevance of com-munication issues is noteworthy, while there is an absence of legal issues, especially those concerning health and the environment. As anticipated above, this experts' meeting gave rise to a report in 2009 and a second one in 2010 which summarized the findings and recommendations of the first meeting.

The 2010 document underlines, on the one hand, the adequacy of the current risk assessment approaches used by FAO/WHO concerning nanotechnologies used in food and agriculture, though novel risk assessment strategies, such as a tiered approach, are also encouraged (FAO/WHO, 2010, pp. 66–67). On the other, it stressed the need to pay adequate attention to transparency 'so that the public can be made aware of the products in which nanotechnology is being used (labelling), workplace disclosure and protections, and the public release of all data used to make decisions on safety' (ibid., p. 49). In this regard it notes that '[e]ven if the public chooses not to take advantage of its oversight option, the existence of the right to do so proves reassuring' and certified transparency as an instance of best practice (ibid.). In the meanwhile, greater engagement by stakeholders is suggested.

The 2013 report is a study of the state of the art of arrangements of risk management and risk assessment developed all over in the world with regard to nanotechnological applications in food and agriculture. The overview provides a brief but quite complete picture of the comparative nature of the initiatives and the current debate on the question.[85] From this overview it emerges that the regulations specifically designed for nanotechnologies are very few and nearly null,[86] whilst *vis-à-vis* an increasing market and the spread of nanotechnologies in commercial products the existing regulations are generally deemed suitable also with regard to nanotechnologies. The conclusion of the report is that, given a situation where inventories that register nanotechnologies in consumer products are scarce (p. 34), notwithstanding 'information on this topic is limited' (ibid., p. 30), so that, considering the increasing process of commercialization, we are facing to a sort of 'collective experimentation' (Wynne, Felt, 2007, p. 68), 'current risk assessment approaches were suitable' in 2009 and so they are now (FAO/WHO, 2013, p. 33). Nevertheless, mandatory labelling such as that provided by the EU 'would lead to greater transparency for the consumer and enable consumer freedom of choice' (ibid., p. 34).

The focus on new applications of science and technology has been recently increased at the Council of Europe level too.

A further signal of this renewed attention to emerging technologies is a recommendation of the Parliamentary Assembly of the Council of Europe.[87] The importance of this document lies in the fact that it is a tool of soft law (the first one on this ambit within the Council of Europe) and in the fact that it expresses the clear need for new tools concerning nanotechnologies and human rights, albeit not binding. The document was the conclusion of an intense process of discussion within the Council of Europe on the impact of new technologies in the field of human rights, which was anticipated by a Council of Europe Declaration of 2005 (Flament, 2013).

---

[85] An interesting feature of the work is that it also provides an overview of the main activities of governmental and nongovernmental organizations that will likely influence forthcoming regulation. Its limit is that the European continent is reduced to the EU and no profile of any European country is provided, while it would have been interesting to compare policy and regulatory initiatives of each country too.

[86] An exception is the European Union.

[87] Council of Europe (2013) Recommendation 2017 (2013) of the Parliamentary Assembly of the Council of Europe *Nanotechnology: Balancing Benefits and Risks to Public Health and the Environment*, adopted on 26 April 2013 (provisional version).

With the Warsaw Declaration, the Council of Europe stated that the member States were 'determined to ensure security to [their] citizens in the full respect of human rights and fundamental freedoms and [their] relevant international obligations. [... They] will further develop [their] activities in combating [...] the challenges attendant on scientific and technological progress. [They] shall promote measures consistent with [their] values to counter those threats'.[88] The interest for nanotechnologies was anticipated by a motion for a resolution on this matter, dated 2010 by the Monegasque parliamentarian Bernard Marquet. After this act, in November 2012 the Russian parliamentarian Valeriy Sudarenkov, member of the Committee of Social Affairs, Health and Sustainable Development of the PACE (Parliamentary Assembly of the Council of Europe), presented a report entitled 'Nanotechnology: Balancing Benefits and Risks to Public Health and the Environment' (Flament, 2013) with a draft recommendation which was entirely accepted by the Assembly. The report was adopted and subsequently followed by an expert report by the lawyer Ms. Ilise L. Feitshans on 13 January 2013. Accordingly, the Parliamentary Assembly recommendation followed on April 2013.

The recommendation requested (a) the adoption of a precautionary approach based on the precautionary principle, while taking into account freedom of scientific research and encouraging innovation; (b) an attempt to harmonize regulatory frameworks with regard to risk assessment and risk management, protection of researchers and workers in the nanotech industry, protection of patients and consumers including labelling requirements taking into account informed consent imperatives; (c) the adoption of guidelines according to a multiple stakeholders' approach (able to integrate national governments, international organizations, the Parliamentary Assembly, civil society, experts and scientists); (d) the adoption of either a Council of Ministers recommendation or a binding legal document which should take the form of an additional protocol to the Oviedo Convention (Art. 5). Moreover, it requested that the DH-BIO to carry out a feasible study on the elaboration

---

[88]Council of Europe (2005) *Warsaw Declaration*, adopted on 16–17 May 2005, point 8.

of standards in this area. In this regard, in the framework of its prerogatives under the Oviedo Convention, in December 2012 the DH-BIO decided to launch a work 'On the Major Challenges for Human Rights of Emerging Technologies' in the biomedical field, with special attention to nanotechnologies with the aim of publishing a white paper in 2015 (Flament, 2013). This work gave rise to an international conference organized by the DH-BIO entitled 'Emerging technologies and human rights' in 2015.[89] At the same time, in December 2012, the Committee on the Convention for the Protection of Individuals with Regard to Automatic Processing of Personal Data decided to write a report entitled 'Nanotechnology, Ubiquitous Computing and the Internet of the Things' which was compiled by Georgia Miller and Matthew Kearnes[90] and is still in draft form (Flament, 2013).

In 2014 the Rathenau Instituut delivered a report to the DH-BIO with a preface of Anne Forus of the Committee of the Council of Europe (van Est et al., 2014). This work aims at exploring, beyond the state of the art, what kind of ethical and legal challenges converging technologies posed to human rights and human dignity. The growing integration of nanotechnologies, biotechnology, information technology and cognitive science, as a factor of innovation, especially in the biomedical field, (Roco and Bainbridge, 2003) gives also rise to several concerns as regards the implications on individuals where various new types of interventions are being introduced in the human brain and body. In this framework synthetic biology could not take place since 'it is mainly involved with research on the microbial level and as such does not raise bioethical issues that are relevant from a human rights' perspective' (ibid., p. 9). Instead, here, the spread of devices that are 'intimately integrated into our lives and which can be used to monitor our mental and physical health conditions or actively stimulate and coach us to adopt a healthy lifestyle' needs

---

[89] http://www.coe.int/t/dg3/healthbioethic/Activities/12_Emerging% 20technolo-
gies/PREMS185414%20GBR%202038%20ProgrammeConférence%20TEXTE%
2016X24.pdf.

[90] https://www.coe.int/t/dghl/standardsetting/dataprotection/Reports/Miller%20
Kearnes%20-%20Nano%20privacy%20Draft%20report%20%2017%2005%
202013.pdf.

the rapid redefinition of our approach (ibid., p. 8). Through the technological convergence of mainly nanotechnologies, biology and information technology in biomedical field 'the number of ways to intervene in the human body will greatly increase' (ibid., p. 13). The report analyses in particular what ethical concerns are raised by applications of cognitive sciences, nanotechnologies, information technology in daily life, by offering an ethical framework over which the DH-BIO will develop its action in future. Furthermore, it addresses applications of those converging technologies within and outside the medical domain that provides an almost unknown terrain for the DH-BIOP, such as human enhancement technologies, lab-on-a-chip, integrated epidemiological systems, e-health and wearable health monitoring systems (e-coaching), gaming, etc. (ibid., p. 36). This work has thus the merit to stress the role of the progressive commercialization of products and applications in the increase of the impact of those technologies on society. Yet, the analysis of these implications from the individual rights perspective is lacking. Thus further developments of the Council of Europe of interest for emerging technologies are expected.

References

Andorno, R. (2002). Biomedicine and International Human Rights Law: In Search of Global Consensus, *Bulletin of the World Health Organization*, **80**, pp. 959–963.

Beghé-Loreti, A. (1998). Brevi considerazioni sulla protezione giuridica delle informazioni genetiche nel diritto internazionale comunitario, *Rivista internazionale dei diritti dell'uomo*, **11**(1), pp. 12–19.

Beghé-Loreti, A. (2001). Osservatorio sulla bioetica. Bioetica e diritto comunitario: le prospettive dischiuse dalla Carta dei diritti fondamentali dell'Unione Europea, *Rivista internazionale dei diritti dell'uomo. Atti del Seminario internazionale Assistenza umanitaria e diritto internazionale umanitario (Milano, 24, maggio 2000)*, **14**(1), pp. 157–160.

Beghé-Loreti, A. (2006). Normativa internazionale e ricerca biomedica. Conquiste attuali e prospettive future, in: De Dios Vial Carrea, J., Sgreccia E. (eds.), *Etica della ricerca biomedica. Per una visione cristiana. Atti della IX Assemblea Generale della Pontificia Academia*

Pro Vita (Città del Vaticano, 24–26 Febbraio 2003), Città del Vaticano: Libreria Editrice Vaticana, available at http://www.academiavita.org/index.php?option= com_content&view= article&id= 238%3Aa-loreti-beghe-normativa-internazionale-e-ricerca-biomedica&catid=54%3 Aatti-della-ix-assemblea-della-pav-2003&Itemid=66&lang=it. Accessed 12 May 2012.

Beghé-Loreti, A., and Marin, L. (1999). La tutela della persona umana nella sperimentazione clinica dei farmaci e il ruolo dei comitati etici tra regole internazionali di disciplina e normativa italiana, *Rivista internazionale dei diritti dell'uomo. Atti del Seminario internazionale Diritti dell'uomo e profili etnico-religiosi: una prospettiva globale (Milano, 3–4, dicembre 1998)*, **12**, pp. 641–675.

Blasi, A. (1999). La protezione dei dati personali nella giurisprudenza della Corte europea dei diritti dell'uomo, *Rivista internazionale dei diritti dell'uomo. Atti del Seminario internazionale Diritti dell'uomo e profili etnico-religiosi: una prospettiva globale (Milano, 3–4, dicembre 1998)*, **12**, pp. 543–559.

Brownsword, R. (2008). *Rights, Regulation and the Technological Revolution*, Oxford: Oxford University Press.

Bultrini, A. (2004). *La pluralità dei meccanismi di tutela dei diritti dell'uomo in Europa*, Torino: Giappichelli.

Byk, C. (1999). Bioéthique et Convention européenne des droits de l'homme, in: Pettiti, L.E., Decaux, E., and Imbert, P.H. (eds.), *La Convention européenne des droits de l'homme. Commetraire article par article*, Paris: Economica, pp. 101–121.

Cassese, A. (1988). *I diritti nel mondo contemporaneo*, Roma-Bari: Laterza.

Colombo, R. (2003). La clonazione umana, in: VV.AA. *Il divieto di clonazione umana nel dibattito internazionale. Aspetti scientifici etici e giuridici*, Città del Vaticano: Libreria Editrice Vaticana available at http://www.academiavita.org/index.php?option=com_content&view=article&id= 120%3Ala-clonazione-umana&catid=43%3Ail-divieto-della-clonazione-nel-dibattito-internaz&Itemid= 69&lang= it. Accessed 12 May 2012.

Daloiso, V., and Spagnolo, A.G. (2009). Nanotecnologie: la riflessione etica in alcuni paesi, *Medicina e morale*, **1**, pp. 11–29.

de Salvia, C. (2000). La Convenzione del Consiglio d'Europa sui diritti dell'uomo e la biomedicina, *I diritti dell'Uomo. Cronache e battaglie*, **XI**(1–2) gennaio-agosto, pp. 99–109.

de Salvia, M. (1997). Ambiente e Convenzione europea dei diritti dell'uomo, *Rivista internazionale dei diritti dell'uomo*, **10**, pp. 246–257.

Del Vecchio, A.M. (2001). Considerazioni sulla tutela dell'ambiente in dimensione internazionale ed in considerazione con la salute umana, *Rivista internazionale dei diritti dell'uomo Atti del Seminario internazionale Assistenza umanitaria e diritto internazionale umanitario (Milano, 24, maggio 2000)*, **14**(2), pp. 339–364.

European Group on Ethics in Science and New Technologies (EGE) (2000). *Citizens' Rights and New Technologies: A European Challenge. Report of the European Group on Ethics in Sciences and New Technologies on the Charter on Fundamental Rights related to technological innovation as requested by President Prodi on 3 February 2000*, Bruxelles, 3, Mai, available at http://www.europarl.europa.eu/charter/civil/pdf/con233_en.pdf. Accessed 17 May 2012.

Errico, M. (2007). Diritto alla salute e sperimentazioni sull'uomo: la 'reasonable availability' dei farmaci nei paesi in via di sviluppo, *Sociologia del diritto*, **34**(1), pp. 27–49.

FAO/WHO [Food and Agriculture Organization of the United Nations/World Health Organization] (2010). *FAO/WHO Expert Meeting on the Application of Nanotechnologies in the Food and Agriculture Sectors. Potential Food Safety Implications. Meeting Report.* Rome: FAO and WHO http://whqlibdoc.who.int/publications/2010/9789241563932_eng.pdf. Accessed 29 August 2013.

FAO/WHO [Food and Agriculture Organization of the United Nations/World Health Organization] (2013). *Paper: State of the Art on the Initiatives and Activities Relevant to Risk Assessment and Risk Management of Nanotechnologies in the Food and Agriculture Sectors.* Geneva: FAO and WHO http://www.fao.org/docrep/018/i3281e/i3281e.pdf. Accessed 29 August 2013.

Flament, G. (2013). *NIA Briefing Note: Nanotechnology and the Council of Europe*, Brussels: Nanotechnology Industrial Association, pp. 1–4.

Foà, S. (1998). Il fondamento europeo del diritto alla salute. Competenze istituzionali e profili di tutela, in: C.E. Gallo, B. Pezzini (eds.), *Profili attuali del diritto alla salute*, Milano: Giuffrè, pp. 57–93.

Francioni, F. (2007). Genetic Resources, Biotechnology and Human Rights, in: Francioni, F. (ed.) *Biotechnologies and International Human Rights*, Oxford and Portland, Oregon: Hart Publishing, pp. 3–32.

Gargiulo, P. (1998). Meccanismi di tutela dei diritti umani nel sistema delle Nazioni Unite, *Annali della Pubblica Istruzione. Rivista bimestrale del Ministero dell'Istruzione, dell'Università e della Ricerca*, **3–4**, available at http://www.annaliistruzione.it/riviste/annali/pdf/030498/030498 ar05.pdf. Accessed 12 May 2012.

Giannini, S. (2001). La brevettabilità del vivente: i diritti coinvolti dalle biotecnologie, *Diritto & Diritti*, available at http://www.diritto.it/articoli/civile/giannini.html. Accessed 12 May 2012.

Gitti, A. (1998). La Corte europea dei diritti dell'uomo e La Convenzione sulla biomedicina, *Rivista internazionale dei diritti dell'uomo*, **11**(3), pp. 719–735.

Jacobs, F., White, R., and Ovey, C. (2002). *The European Convention on Human Rights*, 3rd ed., Oxford: Oxford University Press.

Handl, G. (2001). *Human Rights and the Protection of the Environment*, in Eide, A., Krause, C., and Rosas, A. (eds.), *Economic, Social and Cultural Rights. A Textbook*, Dordrecht, Boston, London: Martinus Nijhoff Publishers, pp. 303–328.

Manfrellotti, R. (2005). Biotecnologie e regolazione tecnica, in A. D'Aloia (eds.), *Bio-teconologie e valori costituzionali. Atti del seminario di Parma svoltosi il 19 marzo 2004*, Torino: Giappichelli, pp. 371–385.

Mizzon, M. (2001). Principi generali in materia probatoria nei procedimenti dinanzi alla Corte di Strasburgo, *I diritti dell'Uomo. Cronache e battaglie*, XII, pp. 40–49.

Murphy, T. (ed.) (2009). *New technologies and Human Rights*, Oxford: Oxford University Press.

Oddedino, A. (2001). La tutela della salute è e resta ancillare rispetto alle politiche della Comunità, *Diritto pubblico comparato ed europeo*, I, pp. 379–383.

Ogata, S. (1990). Introduction: United Nations approach to Human Rights and Scientific and Technological Developments, in: Weeramantry, C.G. (ed.), *Human Rights and scientific and technological development*, Tokyo: United Nation University Press, available at http://archive.unu.edu/unupress/unupbooks/uu06he/uu06he00.htm. Accessed 12 May 2012.

Olivieri, F. (2008). La Carta sociale europea tra enunciazioni dei diritti, meccanismi di controllo e applicazioni delle corti nazionali. La lunga marcia verso l'effettività, *RDSS. Rivista del diritto della sicurezza sociale*, **8**(3), pp. 509–539.

Pariotti, E. (2007). 'Effetto orizzontale' dei diritti umani e imprese transnazionali nello spazio europeo, in Trujillo, I., and Viola, F. (eds.), *Identità, diritti, ragione pubblica in Europa*, Bologna: Il Mulino, pp. 171–201.

Pariotti, E. (2013). *I diritti umani: concetto, teoria, evoluzione*, Padova: CEDAM.

Pariotti E., and Ruggiu D. (2012). Governing Nanotechnologies in Europe: Human Rights, Soft Law, and Corporate Social Responsibility, in: Van Lente, H., Coenen, C., Konrad, K., Krabbenborg, L., Milburn, C., Seifert, F., Thoreau, F., and Zülsdorf T. (eds.), *Little by Little. Expansions of Nanoscience and Emerging Technologies*, Heidelberg: IOS Press/AKA-Verlag, pp. 157–168.

Pernice, I. (2008). The Treaty of Lisbon and Fundamental Rights, in: Griller, Ziller, S.J. (eds.), *The Lisbon Treaty: EU Constitutionalism without a Constitutional Treaty?*, Wien, New York: Springer, available at http://www.judicialstudies.unr.edu/JS_Summer09/JSP_Week_1/Pernice%20Fundamental%20Rights.pdf. Accessed 12 May 2012.

Piciocchi, C. (2001). La Convenzione di Oviedo sui diritti dell'uomo e la biomedicina: verso una bioetica europea?, *Diritto pubblico comparato ed europeo*, III, pp. 1301–1311.

Robles Morchón, G. (2001). La protezione dei diritti fondamentali nell'Unione Europea, *Ars interpretandi. Annuario di ermeneutica giuridica*, **6**, pp. 249–269.

Roco, M., and Bainbridge, S. (eds.) (2003). *Converging Technologies for Improving Human Performance. Nanotechnology, Biotechnology, Information Technology and Cognitive Science*, Kluwer Academic Publishers (currently Springer): Dordrecht, The Netherlands.

Rodotà, S. (1995). *Tecnologie e diritti*, Bologna: Il Mulino.

Ruggiu, D. (2011). Diritti umani e nanotecnologie in Europa: sul ruolo della Corte di Strasburgo, in: Guerra, G., Muratorio, A., Piccinni, M., Pariotti, E., and Ruggiu, D. (eds.), *Forme di responsabilità, regolazione e nanotecnologie*, Bologna: Il Mulino, pp. 647–676.

Ruggiu, D. (2012a). Synthetic Biology and Human Rights in the European Context: Health and Environment in Comparison of the EU and the Council of Europe Regulatory Frameworks on Health and Environment, *Biotechnology Law Report*, **31**(4), pp. 337–355.

Ruggiu, D. (2012b). *Diritti e temporalità. I diritti umani nell'era delle tecnologie emergenti*, Bologna: Il Mulino.

Ruggiu, D. (2013a) Nanotecnologie e diritti umani: sul ruolo della Corte europea dei diritti dell'uomo, in: Bauner, C., and Durante, V. (eds.), *Ética ambiental e bioética: la proteção jurìdica da biodiversidade*, Caxias do Sul (Brasil): EDUCS.

Ruggiu, D. (2013b). Temporal Perspectives of the Nanotechnological Challenge to Regulation. How Human Rights Can Contribute to the

Present and Future of Nanotechnologies, *Nanoethics*, **7**(3), pp. 201–215.

Somsen, H. (ed.) (2007). *The Regulatory Challenge of Biotechnology. Human Genetics, Food and Patent Law*, Cheltenham (UK), Northampton, Ma (USA): Edward Elgar Publishing.

Spielmann, D. (1999). Human Rights Case Law in the Strasbourg and Luxembourg Courts: Conflicts, Inconsistencies, and Complementarities, in: Alston, P. (ed.), *The EU and Human Rights*, Oxford: Oxford University Press.

Szyszczack, E. (2001). Protecting Social Rights in the European Union, in: Eide, A., Krause, C., and Rosas, A. (eds.), *Economic, Social and Cultural Rights. A Textbook*, Dordrecht, Boston, London: Martinus Nijhoff Publishers, pp. 493–513.

Tallacchini, M. (2003). Retorica dell'anonimia e proprietà dei materiali biologici umani, in: D'Agostino, F. (ed.), *Corpo esibito, corpo violato, corpo venduto, corpo donato. Nuove forme di rilevanza giuridica del corpo*, Milano: Giuffrè, pp. 171–192.

Toebes, B. (2001). The Right to Health, in: Eide, A., Krause, C., and Rosas, A. (eds.), *Economic, Social and Cultural Rights. A Textbook*, Dordrecht, Boston, London: Martinus Nijhoff Publishers, pp. 169–190.

Trujillo, I., and Viola, F. (2014). *What Human Rights Are Not (or Not Only). A Negative Path to Human Rights Practice*, New York: Nova Science Publ. Inc.

UNESCO (2006). *The Ethics and Politics of Nanotechnology*, Paris: United Nations Educational, Scientific and Cultural Organization.

van Est, R., Stemerding, D., Rerimassie, V., Schuijff, M., Timmer, J., and Brom F. (eds.) (2014). *From Bio to NBIC convergence – From Medical Practice to Daily Life. Report written for the Council of Europe, Committee on Bioethics*, Rathenau Instituut: The Hague.

Van Overwall, G. (2010). *Human Rights' Limitations in Patent Law*, in: Grosheide, W. (ed.), *Intellectual Property and Human Rights: A Paradox*, Cheltenham (UK), Northampton, Ma (USA): Edward Elgar Publishing, pp. 236–271.

Warbrick, C. (2007). Economic and Social Interests and the European Convention on Human Rights, in: Baderin, M.A., and McCorquodale, R. (eds.), *Economic, Social and Cultural Rights in Action*, Oxford, New York: Oxford University Press.

Weschka (2006). Human Rights and Multinational Enterprises: How Can Multinational Enterprises Be Held Responsible of Human Rights

Violations Committed Abroad? *Zaitschrift für ausländisches öffentlisches Recht Völkerrecht,* **66**, pp. 652–661.

Wynne, B., and Felt, U. (2007). *Taking European Knowledge Society Seriously,* Office Publications of the European Communities; trad. it. di M. Tallacchini, *Scienza e governance. La società europea della conoscenza presa sul serio,* Catanzaro: Rubettino.

Xenos, D. (2003). Asserting the Right to Life (Art. 2 ECHR) in the Context of Industry, *German Law Journal,* **8**, pp. 231–254.

Zanghì, C. (2002). *La protezione internazionale dei diritti dell'uomo,* Torino: Giappichelli.

Chapter 2

Governance: A Theoretical Framework

2.1 Introduction

2.1 With the increasing importance of science and technology for the growth of the economies of the most industrialized countries, we can observe a progressive shift from modes of government that are still state-centric, based on a 'top-down legislative approach' that attempts to regulate the behaviour of people and institutions in detailed and compartmentalized ways, to modes of governance characterized by the diffusion of actors (public and private) who contribute to steer processes of economic, financial and technoscientific development through the creation of self-regulating and self-coordinated ecosystems (Roco, 2006, p. 3). Since the desired and unintended effects of technoscientific progress cross the boundaries of each single nation (Beck, 1986) and the globalized dimension of the market also involves latest results of science and technology,[1] it is a fact that the rise of emerging technologies cannot

---

[1] For example, nanotechnologies are used in a wide number of commercial products that reach store shelves worldwide. We can think of OLED technology (organic light-emitting diode) that is used in television screens, computer monitors, mobile phones, and gaming consoles. Silver nanospheres are used in sportswear, toys, toothpastes, air conditioning systems, and washing machines. Titanium dioxide

Human Rights and Emerging Technologies: Analysis and Perspectives in Europe
Daniele Ruggiu
Copyright © 2018 Pan Stanford Publishing Pte. Ltd.
ISBN 978-981-4774-93-2 (Hardcover), 978-0-429-49059-0 (eBook)
www.panstanford.com

be restricted to a given territory or region. The rapid development of technological innovation and the increasing size and power of transnational corporations are leading to the globalization of production and market systems boosted by pressures for further trade liberalization (Lyall and Tait, 2005, p. 8). In this context, the spread of emerging technologies has become a transnational phenomenon which needs forms of governing able to encompass: the action of several nations, the coordinating function of non-state organizations at regional, national, supranational level, the increasing importance of the role of firms and corporations in the evolution of a given scientific and technological sector. In this framework traditional arrangements of governing appear obsolete and clearly insufficient to coordinate a multitude of actors that can individually influence the trajectories of a single technological field. In this context, where no institution appears able to control the others, power is pulverized among a multitude of subjects (the State, non-state actors, private actors, both profit and non-profit entities) giving rise to phenomena of the progressive de-personalization of forms of governing, a blurring between public and private sectors, and the building of self-organizing networks (Stoke, 1998). Here governing and government are conceptually separated and tend to have a per se existence in the theoretical framework. In particular, the concept of governance seems even more capable of interpreting the novelties of this paradigm change in political theory.

This shift witnesses the progressive separation and independence of rules governing science and technology from the subjects regulating them, and the increasing presence of supranational and non-state actors (the EU, EGE, US OTA, WTO, WMA, the Council of Europe, transnational corporations, etc.) in the international landscape. What counts is the act of governing as the result of the interaction of a plurality of forces and institutions, not the law-

---

nanoparticles are widely used in cosmetics. Carbon nanotubes are used in some sport articles (e.g. bicycles), nanoparticles are also used in food and ingredients. In addition to the given scientific uncertainty concerning the risks of new technologies, this wide diffusion gives rise to the phenomenon of the spread of risks among people, connected to a sort of pulverization of responsibilities that are distributed among a multitude of unaware persons (workers, users, consumers, and simply unaware people in the case of mere human exposure) (Beck, 1986).

making subject itself (the government, the EU, etc.), that is considered as only one of the multiple actors of governance. Therefore, the essence of governance is that it focuses on mechanisms of governing which do not resort to traditional sanctions or the classic concept of authority (Stoke, 1998). The concept of governance refers to the creation of an order that cannot be externally imposed but is the result of the interaction of a multiplicity of governing forms and actors influencing one another (Kooiman and Van Vliet, 1993), p. 64). Thus governance (i.e. the act of governing) no longer stems from governments as such, but is the outcome of the intersection of multileveled and multi-subjective dimensions and structures (local, regional and supranational, public and private) (Stoke, 1998; Salter and Jones, 2002, p. 810; Pariotti and Ruggiu, 2012). Thus, this chapter intends to present the theoretical framework concerning the concept of governance necessary to understand current trajectories of governance, at European level in particular.

2.2 Governance and Meta-governance

2.2.1 There are several definitions of governance,[2] depending often on the ambit in which they were originated. It has been defined as a 'change in the meaning of government, referring to a new process of governing; or a changed condition of ordered rule; or the new method by which society is governed' (Rhodes, 1996, pp. 652–653); or as 'the development of governing styles in which boundaries between and within public and private sectors have become blurred' (Lyall and Tait, 2005, p. 4). These definitions underline some relevant aspects, but they lack the regulatory dimension[3] which pervades the concept of governance at its base

---

[2] For example, the Commission's White Paper on Governance defines it as 'rules, processes and behaviour that affect the way in which powers are exercised at European level, particularly as regards openness, participation, accountability, effectiveness and coherence'. European Commission (2001) *European Governance: a White Paper*, 25.7.2001, COM(2001) 428 final, p. 8, available at http://eur-lex.europa.eu/LexUriServ/site/en/com/2001/com2001_0428en01.pdf. Accessed 19.5.2012.

[3] In this work, by 'regulation' I intend a regulatory landscape of norms of hard and soft law which concur to regulate any aspect of social life with different degrees of legal

and is at the centre of this book insofar as it interacts with human rights law. The fact that the phenomenon of governance gives rise to forms of governing distributed among several subjects and deprived of a centre able to control and regulate everything, does not remove the fact that governance belongs to the normative dimension since it is characterized by the intersection of several flows of norms at different levels (international law, EU law, statutory law, judge made law, hard and soft law, ethical advice, technical norms), with different sources of production (the UN, UNESCO, the Council of Europe, the EU, national governments, national and supranational courts, ethical advisory boards, transnational corporations, etc.) and with different degrees of normativity (legal, ethical or technical norms). In this sense by governance I mean the network of processes with reticular character, diffuse among public and private actors both at national and supranational level, made up of norms of soft[4] and hard law, as well as ethical and technical norms, somehow coordinated and aimed at solving conflicts and making decisions in a particular technological, economic, or financial field (Ruggiu, 2012b, p. 156; Ruggiu, 2013a, p. 104).

With the crisis of the State-centric model of societal governing where just one actor, the State, has centralized the monopoly of all legal sources, the rise of governance has led not only to a multiplication of the sources of law and law-making subjects (Ferrarese, 2012, p. 124ff.), but also to the integration of further normative sources (ethical, technical) with those of law. In this sense, governance is broader than regulation since it encompasses the full range of research and innovation policies that together constitute broad structures for governing science (Roco, 2006, p. 3; Kearnes and Rip, 2009, p. 3). Thus regulation should be

---

normativity. Law has diverse degrees of normativity. There is a continuum between hard and soft law and possibly between other qualities of the law. (Peters, Pagotto, 2006, p. 8; Peters, 2001, p. 23). In this regard, regulation loses any state-centric connotation typical of hard law instruments and includes the activity of several actors both public and private.

[4]I intend the term 'soft law' as referring to guidelines, declarations or recommendations containing principles and standards, or voluntary measures (self-regulation tools, voluntary codes of conduct, third-party certification systems) not supported by formal legal sanctions, that can nevertheless have legal effects (Pariott and Ruggiu, 2012).

deemed as a one of a number of 'tools of governance' set up in the context of the overall governance of emerging technologies, which serves to interconnect technoscientific development and civil society (Eberlein and Kerwer, 2004, p. 131).

The rise of governance in scientific literature has made several profiles belonging to the complexity of the concept apparent.

First of all, this perspective is partially built on the challenge of the legal/constitutional tradition that dominated political theory up to the 1950s (Stoke, 1998, p. 19). The traditional model of power is built around a dominant agent, of a public nature, providing all services for a given community. In this framework power is legitimated through formal criteria of power delegation. In a context where there is no agent dominating others, in which, therefore, the mere interaction of a multitude of agents is relevant, the criteria of legitimation are deeply changed and, instead of (power conferring) rules, they rest on the effective provision of services, mobilizing resources and promoting cooperation, and on the capacity of agents to produce norms able to be followed by the other forces. In other words, accountability generates the legitimation of the action of different agents. In the governance framework, legitimation is able to produce accountability that determines the relationships of power among different actors and who is the recipient of regulation (Pariotti, 2011).

Another aspect of governance is the new modulation of responsibility relations in which the State takes a step back, while responsibilities are increasingly distributed into the private sectors on a voluntary base (Stoke, 1998, p. 21). There is a shift in the balance between the State and civil society, meaning that private forces increasingly accompany public ones. Within the framework of international relations, States are only one of the number of actors at play. Now several different agents act in the horizon of international relations. International and supranational organizations such as the WTO or the EU are even more important. Non-state actors, such as Greenpeace, ETC group, the International Life Science Institute (ILSI), which represents several industry members, or the Organisation for Economic Co-operation and Development (OECD), can influence policies of States and even policies of more complex political organizations such as the EU. Transnational corporations acting across several countries are able to determine the success or failure of policy choices. Third-party certification systems such as the ISO underline the role of corporations in the success of a given regulatory direction with regard to a particular technological ambit. In this instance, the effectiveness of regulation succeeds only if there is a spontaneous assumption of responsibility by the enterprises that play a leading role in the framework of international governance. In this sense, making stakeholders responsible can be crucial (Dorbeck-Jung and Shelley-Egan, 2013). In this context even the individual with the power of his/her internationally recognized rights (i.e. human rights) can affect directions of the governance of emerging technologies (Pariotti, 2007, 2013; Ruggiu, 2015). For example, in 2011, the patentability of isolated and purified neural progenitor cells for the treatment of neural defects (stem cells) was greatly limited by a decision of the CJEU (Ruggiu, 2013a; 2015).[5] Thus, contextually, national and supranational courts can also substantially affect choices of governance (Ruggiu, 2015). The rising landscape is one in which responsibilities are increasingly assumed by private actors and the role of the State is re-sized, even though it is far from disappearing or becoming secondary in this theoretical framework. In this context the concept of responsibility as such needs to be thoroughly redefined (von Schomberg, 2010; Owen et al., 2013; Ruggiu, 2015).

If in the current landscape of global governance there is no longer a single organization which dominates the others, one that can dominate a particular process of exchange (Stoke, 1998, p. 22). In this framework, governing is an interactive process since no single actor (public or private) has the monopoly of knowledge, the exclusive possession of all resources and the capability to tackle problems unilaterally (Lyall, Tait, 2005, p. 4). As has been rightly noted, it is a kind of 'invisible hand' of Adam Smith which guides 'the co-ordination of independent initiatives to a maximum advancement of science' (Polanyi, 1962). In this regard there is the need to create forms of systemic coordination among different agents through forms of information and knowledge exchange,

---

[5] Judgement of the Court of Justice (Grand Chamber), *Oliver Brüstle v. Greenpeace eV* (Case C-34/10) 18 October 2011, not yet published.

styles of partnership, and modes of joint-working able to generate self-regulating networks. A shift from the subject of governing to the quality of the relationship among the different subjects has been determined. It is the interactivity of relations that counts in a framework in which forms of cooperation are crucial in order to realize common purposes and goals. Systemic coordination leads to 'games *about* rules' instead of 'games *under* rules', meaning that regulation is the result of the interaction of several actors and not only the means by which collective action is steered and guided. In this context many consequences are produced, some desired and some unintended, but not all unintended consequences are undesired per se. In this sense, governance is a form of governing characterized by a degree of uncertainty and therefore it is suitable for steering fields, such as that of emerging technologies, dominated by a state of uncertainty with regard to risks, potentials, definitions, classifications, metrology, the applicability of existing norms and the production of new ones (Kearnes and Rip, 2009; von Schomberg, 2011).

Under this perspective governance implies the creation of autonomous and self-regulating networks of actors (Stoke, 1998, p. 23). This leads to the formation of a sort of 'regime' in which actors and institutions gain the capacity to blend and integrate resources, skills and purposes in a spontaneous and (lightly) coordinated long-term coalition. In a world where uncertainty in problem-solving processes increases due to the augmented complexity and the unavoidable diversity of life situations, a uniform solution is simply not possible (Scott and Trubek, 2002, p. 5ff.). Instead of adopting classical forms of government, various institutional arrangements are created to enable actors to cooperate over resources which are finite and to which they have open access. In certain contexts, self-organized systems of control among different actors can be more effective than regulations imposed by the government. The processes of integration among different levels of governance can be vertical or horizontal (Lyall, Tait, 2005, p. 10). The vertical ones act through the delegation of responsibilities from higher to lower levels and set up objectives within a clearly hierarchical mechanism. These integration systems often imply top-down control with some form of sanctions imposed. The horizontal ones take place across department boundaries in order to amalgamate policies of different institutions in a coordinated ecosystem. These institutional structures (e.g. interdisciplinary research in academia) are important but can be non- decisive with regard to the effectiveness of integration. The 'impact of effective horizontal integration is a loosening of control and the introduction of greater complexity into policy implementation processes' (ibid., p. 11). This gives rise to the problem of the accountability of these self-organized systems since, to a certain degree, they are driven by the self-interests of the members of the group. Thus if this systemic cooperation has to work, it needs to show it can realize common objectives and purposes in order to gain the trust of the other forces and participants. In this context, accountability is the main form of legitimation among the multitude of actors acting in the landscape of supranational governance. In a dimension where effectiveness is the main feature of the system, the ability to be followed by the other agents and to be recognized as accountable is the principal quality of all participants (Pariotti, 2011).

Finally, instead of traditional powers of governing aimed at commanding and using their own authority, new tools and techniques able to steer and orient the behaviour of agents have been created (Stoke, 1998, p. 24). Governance first identifies the main stakeholders and then develops the effective links among the most relevant parts. Thus it influences and guides relationships in order to reach the desired outcomes. In this sense, often public authorities are only led to redefine their role in this framework. Finally, it involves different sub-systems with a view to making them reflect and act through mechanisms of effective coordination aimed at reducing unintended effects. Old fashion command-control tools have lost their centrality in the international governance landscape, as well as their attractiveness. New softer forms of norms able to guide human behaviour have been developed. Forms of soft law such as guidelines, declarations, recommendations containing principles and standards or voluntary measures such as codes of conduct or third-party certification systems not supported by any legal sanction that can nevertheless have legal effect are spreading (Pariotti, 2011; Pariotti, Ruggiu, 2012). Furthermore, new forms of norms without any legal force such as ethical advice or technical standards have become even more important in this landscape. In this regard, governance appears as the result of the interaction of a number of norms only some of which are endowed with a legal nature, and it shows the capability of different modes of normativity to reach objectives increasingly relevant for the community. In this context human rights appear as a form in equilibrium among several expressions of normativity since they are referred to both in hard law documents of international law (such as the Council of Europe ECHR), and in soft law instruments (such as the Oviedo Convention), and even in ethical advice as principles of a merely ethical nature (e.g. the opinions of EGE, guidelines, codes of conduct, etc.). This fact makes them a suitable tool of governance able to act as the cornerstone of different dimensions of governance by giving them the necessary coherence (Ruggiu, 2013a, 2013b, 2015).

2.2.2 In literature, scholars increasingly distinguish the concept of governance from that of meta-governance. By meta-governance we mean that level constituted by organizations, structures and processes that produce (the conditions of) governance (Jessop, 2003) through the arrangement of normative tools (of a legal, ethical or technical nature) (Ruggiu, 2013a, p. 104). In this sense governance can be dissociated into its basic elements (such as codes of conduct, consultation processes, guidelines, ethical committees, research laboratories, research programmes, etc.) which can be studied at a more analytical level with a view to recomposing them within a clearer perspective. At the meta-governance level, relations among organizations and governance structures are basically characterized by heterarchy instead of a kind of framing of a hierarchical type. This fact deprives processes of governance of a centre which can determine all the relations of the system by giving a merely eccentric and diffuse character. In this regard governance appears *de facto* to be distributed among several subjects and, thus, to be effective, any model of governance should be able to involve a wide range of stakeholders (Pariotti and Ruggiu, 2012; Ruggiu, 2013a). The analytical study of these structures, organizations and institutions within the wide conceptual framework of governance aims to clarify their role, functions and individual contribution to governance in a given field. Consider the role of transnational corporations, or the action of NGOs within worldwide governance, or, again, the increasing importance of organs without legislative power but absolutely central in Europe, such as ethical committees, especially at the Community level.

These governance structures and processes can be further analysed with regard to their external elements. In this context we can distinguish formalized structures and processes producing governance from informal ones (Ruggiu, 2013a, p. 105). Those of the first type encompass soft and hard law instruments such as EU directives, recommendations, self-regulatory tools including codes of conduct, third-party certification systems that can have nevertheless legal effects and, in this regard, we can also include in this class human rights. Those of the second type can be, for example, reports and opinions of ethical advisory boards such as the EGE, internal documents of corporations without any legal effect, or acts of NGOs, or, again, scientific results of academic networks inasmuch as they can influence policy-making processes in some way. In other words, through the respect for formal procedures and norms of competence the structures and processes of governance produce in the first instance legal norms with binding force (hard law norms) or legal norms without binding force but able to have some legal effect (soft law norms); in the second case they produce norms (moral, technical, etc.) without any legal effect but able to influence decision-making and law-making processes (e.g. official reports, opinions at the EU level). It is clear that it is easier to analyse the behaviour of formalized structures and processes since the reference to the parameters of their existence, such as legal norms, are more easily identifiable than informal ones are. But since the separation of legal norms from other types of norms is not so definite, it is often (though not always) possible to follow the numerous intersections of norms of different type that governance produces. For example, corporations often resort to forms of self-regulation such as codes of conduct that are public and, once adopted, can have legal effects in any case. In this way they can be studied from a legal perspective. For example, ethical advisory boards such as the EGE act under a constellation of norms which constitute their legal basis, they are then subject to changes of remit, which influence the scope of their opinions and finally they produce a shift in regulation that can be detected. In this regard, the life of both formalized and informal structures of governance can be analysed and compared.

In sum, governance is the result of the intersection of several different plans and dimensions, normatively organized, with different degrees of formalization (often blurred), giving rise to a normative landscape in which norms of a given kind coexist with norms of a different type without a clear-cut separation, but within a process of stratification.

2.3 The 'New Turn of Governance'

2.3.1 The rise and consolidation of the concept of governance has occurred with the 'New Governance turn', when new forms of soft arrangements were developed within the EU. Modes of governance can be called new when their function of steering is characterized by (i) informality meant as non-typicality (they mainly use old tools in a non-typical fashion) (ii) weak hierarchic relations (public institutions maintains a coordinating role in this model) (iii) the presence of private actors that are systematically involved in policy formulation and (iv) the anticipatory nature (they adopt tools aimed at anticipating risks and triggering voluntary behaviour of stakeholders) (Peters and Pagotto, 2006; Scott and Trubek, 2002; Lyall, and Tait, 2005; Eberlein and Kerwer, 2004). It can be seen as a recipe to use old tools in a novel way (Smismans, 2008). New governance has been defined as 'a construct which has been developed to explain a range of processes and practices that have a normative dimension but do not operate primarily or at all through the formal mechanism of traditional command-and-control legal institutions [. . .] [it] signals a shift away from monopoly of traditional politico-legal institutions, and implies either the involvement of actors other than classically governmental actors, or indeed the absence of any traditional framework of government, as is the case in the EU and in any trans-national context' (de Búrca and Scott, 2002, p. 2). The increasing state of uncertainty in solving problems that appear even more complex and irreducibly diverse may simply not allow a uniform solution according to traditional modes of governing (Scott and Trubek, 2002, p. 5ff.). In these instances new forms of governance can better tackle issues generated by the increasing complexity, diversity and uncertainty of policy-making contexts by bringing together actors from various levels of governments (localities, subnational regions, national or European) in modes that facilitate dialogue and coordination among several levels of government by privileging, when it is feasible, the lowest possible level in order to extend deliberation among stakeholders and provide some degree of democratic legitimation according to a flexible, revisable, experimental approach which is also able to produce knowledge. New governance is thus characterized by being multilevel and decentralized since it gives rise to mechanisms which leave final policy-making not to the highest level but to the lowest possible one. These mechanisms often have a participatory nature by privileging distributed forms of problem solving aimed at implementing their legitimation (Eberlein and Kerwer, 2004, p. 122). They rely on less formal rules and open-ended standards, flexible and revisable guidelines and other forms of soft law able to adapt to diversity, tolerate alternative approaches, experiment problem-solving processes and create new knowledge by exchanging results, benchmarking performance and sharing best practices. In other words, even in the new governance framework, law does still 'play a role, but more as procedural framework than as a "policy instrument"' (ibid., 131).

To understand the concept of 'new governance' we need to refer to another (and counterpoised) governance model: the classic 'Community Method' (CCM). This model mainly follows command-and-control style and traditional forms of regulation. Here, at the EU level, the exclusive right to legislative initiative belongs to the European Commission, while the legislative and budgetary powers are the competence of the Council of Ministers and the European Parliament. In this context, qualified majority voting is identified by the European Commission as an essential requirement for ensuring the effectiveness of this method, and the Luxembourg Court (CJEU) as the central organ in guaranteeing respect for of the rule of law.[6]

---

[6] European Commission (2001) *European Governance: a White Paper*, 25.7.2001, COM (2001) 428 final, p. 8, available at http://eur-lex.europa.eu/LexUriServ/site/en/com/2001/com2001_0428en01.pdf. Accessed 19 May 2012.

In this regard, the CCM tends to give rise to binding legislative and executive acts at the EU level by imposing uniform rules for all member States (Scott and Trubek, 2002, p. 6). The main problem of this model is the lack of flexibility and the incapacity to quickly adapt to the fast course of the technoscientific development. Its regulatory structures are exposed to fast processes of obsolescence.

New governance can be identified in several categories of tool that can be used to implement the effectiveness of classic forms of governing. Some of them are also present in traditional forms of governance. For example, an initial flexibility was also established in the CCM with comitology. In this sense, the CCM itself contains the seeds of the rising method of new governance. Comitology is not distinguished by the character of norms which it produces but by the institutional structure which produce them. The novelty of the new governance method consists in the implementation of committees in the decision-making processes in order to facilitate the executive functions of both the Council and the Commission (ibid., p. 3).

Another instance of new governance consists in the great range of actors involved in the decision-making processes. In this regard the White Paper on European Governance emphasizes the Community aim of enhancing the participation of civil society throughout the 'policy-chain'.[7] In scientifically controverted issues there is an increasing demand 'for greater public involvement in assessing the costs and benefits' (Jasanoff, 2003, p. 236). This need has been slowly acquired at the institutional levels of the EU. The adoption of the new governance method (NGM) within the classic Community structures of the CCM occurred with regard to environmental protection in which the flexibility in setting norms accompanied the 'market proceduralization' of Community law.[8] This happened in the case Consultative Forum on Sustainable Development, for example.[9]

---

[7] Ibidem, 10.

[8] On the limits of this proceduralization of the Community market in the environmental field with regard to the right to a healthy environment see Ruggiu (2012a).

[9] This body is no longer extant. See Commission decision of 26 September 2001 repealing decision 97/150/EC *on the setting-up of a European Consultative Forum on Sustainable Development.*

A third category of new governance emerges from the partnership existing in the context of Community structural funding allocation through Community multi-annual programmes (e.g. FP6, FP7, Horizon, 2020). These partnership committees, in which the European Commission, member States, social and economic bodies and other expressions of civil society at local and regional level are represented, have great power in selecting research projects. In this instance we have the establishment of a case of multilevel and multi-actor governance which leads to inedited forms of governance.

Another instance of NGM is the figure of social dialogue established with the Maastricht Treaty which allows, in particular, the officially recognized representatives of employees and employers to enter into voluntary agreements that will subsequently be enacted as directives by the Council (Scott and Trubek, 2002, p. 4; Peters and Pagotto, 2006, p. 19).

A fifth category of new governance which deeply characterizes the model is the Open Method of Coordination (OMC) that has been developed with regard to the European Employment Strategy (Scott and Trubek, 2002; Peters and Pagotto, 2006; Pariotti, 2011). The OMC has been used since the European Council of Lisbon. Within the OMC member States agree on a set of policy goals but remain free to pursue them in the national context as regards the appropriate means and measures to be adopted. Mere cooperation is thus a less intrusive strategy than harmonization of national law and legislation. Its legal basis can be found in the provisions of the Article of the Treaty on the Functioning of the European Union (TFEU) ex-Article 99 TEC (member States' economic policies as a matter of common concern) and in Article TFEU ex-Article 128 TEC (member States' employment policies taking into account Council guidelines) (Peters and Pagotto, 2006). Related to the OMC is the concept of 'Environmental Policy Integration' (Art. 11 TFEU, ex-Article 6 TEC) which aims to horizontally integrate policy objectives of the different member States within the implementation of other areas of Community policy.

2.3.2 In the ambit of emerging technologies, with regard to nanotechnologies for example, the NGM is shaping the governance landscape (Kearnes and Rip, 2009, 16; Rip, 2002). In this framework, emerging approaches to the governance and regulation of science and technology arise at both European and national levels, leading to forms of negotiation of regulation and standard setting on a supranational level in the ISO and OECD through the use of voluntary, principle-based forms of soft law (Kearnes and Rip, 2009; Mandel, 2009; Marchant et al., 2008). In this context we can observe the proliferation of non-state initiatives and the development of forms of voluntary regulation through codes of conduct, recommendations, guidelines, certification systems, as well as comitology and agency networking accompanying the emergence of a discourse on responsible technological development (Kearnes and Rip, 2009, p. 4). At both national and EU level new approaches to science and technology issues able to engage private actors are rising (Kurath et al., 2014). Through codes of conduct, for example, the regulator pursues the goal of setting rules that govern the adoption and the development of forms of self-regulation among the principal stakeholders (e.g. Ruggiu, 2014). These forms of regulation of processes of self-regulation are also known as a case of 'meta-regulation'[10] and are aimed at implementing the participatory nature of regulatory processes (Parker, 2009; Coglianese and Mendeson, 2010). In the Community ambit a good example of an inclusive process developed by the European Commission is the Code of conduct for responsible nanosciences and nanotechnologies research (EC CoC)[11] (Dorbeck-Jung and Shelley-Egan, 2013; Ruggiu, 2014). It could be also deemed as a case of 'meta-regulation in action'. The EC CoC was anticipated and followed by two consultation processes, one

---

[10]By the term 'meta-regulation' we mean the mere descriptive fact of the State regulating its own regulation as a consequence of policies applying transparency, efficiency and market competition to itself. Meta-regulation can also entail any other form of regulation (whether by tools of State law or other mechanisms) that regulates any other forms of regulation. In this regard it may include legal regulation of self-regulation (Peters and Pagotto, 2006, pp. 6–7).

[11]European Commission (2008) Commission recommendation of 7/2/2008 *on a code of conduct for responsible nanoscience and nanotechnology research*, C (2008) 424 final, available at http://ec.europa.eu/research/consultations/pdf/nano-consultation_en.pdf. Accessed 18 February 2015. On this see Chapter 3 in Part I.

held in 2007[12] and the other in 2009/2010[13] (Ruggiu, 2014). In the consultation paper for a European Code of conduct for responsible nanosciences and nanotechnologies research (2007) the Commission emphasized that it was part of an 'ambition to promote a balanced diffusion of information on nanosciences and nanotechnologies' and that the code 'would offer those following it recognition of a responsible approach towards nanosciences and nanotechnologies research, making their action more visible at the European level' (European Commission, 2007, p. 2). In this context the aim of the Community authorities was to involve member States, industry, universities, research organizations, research funding organizations and other parties to take concrete actions for the safe use of nanotechnologies. As expressly declared by the Commission: 'Good governance of nanosciences and nanotechnologies implies an open and transparent dialogue addressing possible risks and realistic expectations'. A second consultation process involving 304 European and international experts held in 2009/2010 aimed to implement the adoption of the EC CoC (Meili et al., 2011).

As shown above, new governance is a label which encompasses both mere departures from the CCM and new forms of governing. From the analysis provided above some characteristics of the concept clearly emerge (Scott and Trubek, 2002, p. 5ff.). These are: inclusion, heterarchy, flexibility and adaptability.

(i) Different approaches use new modes for fostering public participation by parts of civil society in policy-making processes. In this way policy-making becomes a process of mutual problem-solving among stakeholders.[14] New governance processes tend to encourage the participation of affected actors, instead of

---

[12] http://ec.europa.eu/research/science-society/document_library/pdf_06/ consultation-nano-sinapse-feedback_en.pdf. Accessed 28 September 2017.

[13] http://ec.europa.eu/research/consultations/nano-code/results_en.pdf. Accessed 28 September 2017.

[14] Limits and shortcomings of the participatory approach are expressed by Simismans (2008). In particular, in the face of some questions, such as the case of emerging technologies, which implies the use of expertise and much technical knowledge, participatory arrangements are severely limited so that, even after the establishment of participatory paths, only experts can concur in defining the set of rules in those fields.

involving merely representative actors (de Búrca and Scott, p. 3). In this context public authorities do not disappear but maintain coordination competences by fostering inclusion and ensuring the preconditions of public engagement. In this regard transparency is a means of information sharing and learning. The social dialogue seems to respond to problems of democratic deficit. This often implies the delegation of problem-solving competences to the recipients themselves of Community norms (Scott and Trubek, 2002, p. 8).

(ii) The NGM tends to develop forms of coordination and integration of the diverse actors at play with multilevel arrangements able to integrate several dimensions such as localities, subnational, national and European regions according to modes that facilitate dialogue among parties. Thus, rather than hierarchically operating through the structure of the authority of central government, NGM fosters the emergence of infrastructures of governance that ensure coordination or exchange among constituent parts (de Búrca and Scott, p. 3). While traditional approaches to law look for hierarchy and put courts at the centre of systems of accountability, the NGM searches for heterarchy and often looks outside courts to find the real accountability of governance processes and structures (Scott and Trubek, 2002, p. 8). Here heterarchy means the abandon of hierarchical relations based on command-and-control logic in order to foster voluntary behaviour of stakeholder. Thus in this framework the coordination by public institutions is still strategic. For this reason, the regulation (hard and soft law) is still a means to reach to goal of obtaining the voluntary cooperation of stakeholders.

This model relies on the assumption that diversity and decentralization are values worthy of being pursued inasmuch as it ensures effectiveness of the system. In this regard it privileges forms of power delegation in favour of the lowest level possible when it is feasible. This fact relies on the subsidiarity principle according to which decisions should preferably remain at national level by amplifying the trend to accept diversity, allow flexibility and encourage decentralized experimentation. While a traditional conception of law searches a unitary and ultimate source of authority, the NGM looks for the fragmentation and dispersion of authority and is based upon fluid systems of power sharing (Scott and Trubek, 2002, p. 8).

In this context deliberation processes (e.g. consultations) are conceived for extending participation to stakeholders. In this way the capability for problem solving can be ameliorated and processes are endowed with a certain degree of democratic legitimation (Pariotti, 2011). In this regard, several mechanisms are designed for implementing cooperation among member States instead of increasing uniformity at Community law level.

(iii) This approach implies the flexibility and revisability of decision-making processes. A certain degree of elasticity in decisions, strategies and standards is essential to promote cooperation between the different parties involved in the legislative process. This implies abandoning forms of governance which rely on formal standards and hard legislations in favour of forms of soft law and interaction between different dimensions of normativity (ethical, technical, etc.). While a traditional conception of law relies on a clear distinction between law making on the one hand, and rule application and implementation on the other, the NGM tends to privilege indeterminate and flexible rules as more suited to meeting the challenges of modernity (Scott and Trubek, 2002, p. 8).

The NGM is characterized by strong forms of experimentalism (tentative governance) which tend to create knowledge by fostering informal modes of information exchange, benchmarking performance, and sharing best practices, which rely on multilateral surveillance instead of rigid forms of control among the parties involved. While the traditional conception of law underlines the linkage with existing knowledge, the NGM is a continuous process of generating new knowledge and data with a view to preparing new solutions and needed (albeit abrupt) changes of direction (ibid., p. 9). In emerging technologies field, some talk of 'tentative governance' meaning that these forms of governance develop according to a case-by-case logic (Stoke and Bowman, 2012) and encompass 'provisional, flexible, revisable, dynamic and open approaches that include experimentation, learning, reflexivity, and reversibility' (Kuhlmann et al., 2012).

We must not think that the NGM implies surpassing classic modes of regulation completely. In this context the traditional forms of regulation remain in the background, though they do not disappear (Scott and Trubek, 2002; Eberlein and Kerwer, 2004; Smismans, 2008). It is true that not only old instruments, such as agency and comitology, concur with new forms of governance (consultations processes, social dialogue), but traditional forms of regulation can pursue the goals of fostering public participation and flexibility within the new governance paradigm. In this regard, a hybrid dimension of governance, mixing new and classic patterns, has been reached at the EU level in the field of nanotechnologies.[15]

(iv) Finally, the NGM is characterized by adaptability. The complexity, which needs to be governed, causes great uncertainty regulatory, ethical, scientific. In this context the regulator does not have any prefixed recipe. If the aim of adapting previous regulatory structures. In a given field which is rapidly developing, there are most chances to control at early stage, with far fewer costs but too little information to decide how to act or not to act (Kearnes and Rip, 2009, p. 99). In this framework a too anticipated regulation risks of stopping opportunities of development especially in the field of emerging technologies where some unknown and unforeseen risks are often integral part of innovation (Owen et al., 2013). In this context scholars and policymakers prefer to rely on existing regulation and eventually adopt new rules when risks become apparent (Stoke and Bowman, 2012). This is the reason why in the meanwhile some flexible tools can better cope with challenges of rising fields that are just at their infancy. Yet, to apply previous legislations on novel fields of innovation is quite problematic since they could not foresee risks and harms that can be known only today.

---

[15] See Chapter 3 of Part I on European governance of emerging technologies.

A sub-specie of the new governance model is that of the self-governance. It can be deemed as a variation of the NGM since it stresses only some features and partially diverges from new governance paradigm. It is characterized by (i) autarkic nature, (ii) deep spontaneity, (iii) surrogatory character, (iv) flexibility and (v) provisory nature. Governance in this context tends to be a complex process of co-governance that involves a number of private subjects (sometimes also public authorities) through a redistribution of tasks and the establishment of distributed poles of self-government (Sørensen and Triantafillou, 2009). In self-governance stakeholders spontaneously take the initiative in order to arrange by themselves a flexible governance framework. Stakeholders take the initiative without any inputs stemming from public authorities. It is a spontaneous initiative. Thus, like NGM it tends to enlarge the participation. In this sense it is a form of self-organization in a given field where the initiative is taken by a given group of stakeholders where public institutions are only a part and play a peripheral role. It uses flexible and not legally binding tools in order to self-arrange a minimal governance framework. This process of self-coordination tends to develop through the adoption of self-regulatory instruments, forms of voluntary cooperation, shared programming and participated negotiation among different parts (ibid.). Differently from new governance patterns self-governance arrangements do not try to adapt any existing regulatory framework. It mainly develops when a technological field is at its infancy and public institutions do not provide any structure in this field yet. In this sense, in front of the inertia of public institutions self-governance aims at substituting the public sector in providing a regulatory framework. From this perspective it testifies the increasing importance of the private sector. The risk here is of being only a partial representation of the interest at stake. It is destined to be replaced by an institutional governance arrangement. 'Community self-governance provides a realistic and potentially powerful complement or alternative to regulation, legislation, treaties and other interventions by outside entities. [...] While self-governance tends to be less stringent than legislation and cannot change existing laws or institutions, it also offers significant advantages. First, self-governance is the right thing to do [. . . since] biologists need to "take responsibility" for "preventing potential misuses of their work." Second, almost always faster than other methods. Third, it derives from consent and is therefore frequently more elegant than externally imposed solutions. Finally, it is inherently international. This can be a crucial advantage in a world where science and commerce routinely span national boundaries' (Maurer et al., 2006, p. 4). The advantages of this should be greater efficacy, a greater efficiency and a greater democratization of the process (Sørensen and Triantafillou, 2009). Like the NGM, self-governance aims at providing efficacy to governance structures and at filling the lack of legitimation of the action of governance.

In face of the novelty of the NGM several attitudes stemming from the new governance model have been adopted by the CJEU. In one instance NGM has been clearly recognized by the Community jurisprudence. In *Stanley and Metson*[16] the Luxembourg Court recognized that the Nitrates Directive could be applied by member States in different ways since it does not intend to seek uniformity in national laws, but rather to get them to create the measures needed for ensuring protection against pollution by granting them wide discretion in identifying the waters covered by the Directive. Thus the Court acknowledged that the Directive pursued the objective through the instrument of social dialogue, but according to modes identified by the NGM. In this regard it assessed the capability to represent the several parties involved and the concept of 'collective representativity'. In this framework the participation of any party becomes essential in configuring this representativity (Scott and Trubek, 2002, p. 13).

The NGM has been endorsed also by EU institutions. According to the White Paper on Governance, five principles underpin good governance. In fact, in the Commission's view, the EU should implement its ability to pursue openness, participation, accountability,

---

[16]Judgement of the Court of Justice (Fifth Chamber), *The Queen v Secretary of State for the Environment and Ministry of Agriculture, Fisheries and Food, ex parte H.A. Standley and Others and D.G.D. Metson and Others*, of 29 April 1999 (Case C-293/97) [1999] *ECR I-2603*.

effectiveness and coherence.[17] First of all, Community institutions should work in a more open fashion. Thus, they, together with member States, must communicate what they do and the decisions that they take in an accessible language. Then, the relevance and quality of EU policy effectiveness depends on the implementation of a participatory path able to engage a broad part of civil society in decision-making processes. In this regard central governments must follow an inclusive approach when they implement EU policies. Thus, each Community institution must explain its role in legislative and executive processes and take responsibility for what it does. This process of the development of accountability must be taken at every level by member States too, therefore. Policies must be effective and timely, delivering what is needed, a clear array of objectives, an evaluation of their future impact and past experiences. Finally, all EU policies must be coherent and easily understood. In this framework we can clearly see the influence of new forms of governing on trajectories of EU policies.

2.4 The Theoretical Framework of the Relationship between Law and 'New Governance'

2.4.1 The rise of the NGM can be an opportunity for rethinking the relationship between law, as well as those recent expressions of law such as human rights, and the emerging processes of governance at the European level. The 'new governance turn' with its bag of flexibility, revisability, decentralization and new participatory approaches tends to strongly distinguish itself from law.

There is scepticism in the research community on the capability of the regulation of interpreting the new paths of modern times. In this sense some criticized the capability of regulation in general, and human rights in particular, to cope with the fast course of innovation with its bag of unforeseen risks and unpredictable harms, especially

---

[17]European Commission (2001) *European Governance: a White Paper*, 25.7.2001, COM (2001) 428 final, p. 10, available at http://eur-lex.europa.eu/legal-content/EN/TXT/PDF/?uri= CELEX:52001DC0428&rid= 2. Accessed 29 September 2017.

in the face of emerging technologies (Groves, 2015). For example, to justify resort to governance arrangements, Mihail Roco (2006, p. 13) stated that '[b]eyond very simple principles, no single set of rules of ethical behaviour is universally accepted'. In this context human rights could maintain only a residual role. For instance, René von Schomberg (2013) who defends a version of governance which entails EU fundamental rights, argues that within the EU framework human rights can be reflected in Community international policies by demonstrating European Union solidarity with the poorest on earth. Governance structures, especially in their novel form, tend to put in question the adequacy of traditional expression of law to cope with rapid changes triggered by innovation that need to found policy decisions on large consent. For example, some addressed a tendential inconsistency between tools of public engagement and those prefixed goals such as individual rights (Heydelbrand, 2003, p. 234). If processes of participation aim at identifying purposes and values of innovation, the value of the democratic participation, according to which the society decides what are its priorities, need to be preserved. In this context individual rights could alter the prioritization established by the majority in the face of technoscientific progress. According to some no decision even by a Constitutional court could subvert what people have democratically decided through processes of participation (Waldron, 1999, p. 94). In this framework the relationship between governance and regulation appears problematic. Thus, it is worthwhile asking whether these two phenomena (i.e. law and the NGM) represent two forms of a compatible or antagonist existence, and in this contest how rights can be structured in the face of new governance.

With a view to answering this question, it is useful to articulate the conceptual framework of this controversial relationship.

The relationship between law and new governance has been variously interpreted in literature (de Búrca and Scott, 2006). These different interpretations must not be deemed as necessarily alternative, but, to some degree, complementary since they variously contribute to an understanding of the controversial relationship between law, especially constitutional law with its bag of individual rights protected at national and supranational levels, and new rising forms of governance.

2.4.2   In this theoretical landscape we can identify three theses: the 'gap thesis' (with its sub-thesis), the 'hybridity thesis' (with a further three articulations), and the 'transformation thesis'.

According to the 'gap thesis' there is a hiatus between formal law and the practice of new governance (ibid., p. 4ff.). In this view, formal law is broadly blind to new governance trends and, in turn, governance tools are completely extraneous to law. In fact, legal texts, including constitutional ones, cancel the relevance and presence of new forms of governing. In this regard, law either cannot not keep pace with the developments of new governance or ignores them as they do not conform to its requirements, presuppositions and structures. We can address two distinct tendencies within the 'gap thesis'. One argues that law is resistant to the new governance phenomenon and in this regard it should be deemed as impermeable to these new trends. The other argues that law is facing a reduction of its capacity. The 'resistance argument' holds that law is an obstacle or an impediment to new governance. It would not only be blind to this trend, but it would also tend to inhibit this experimentation (Trubeck, 2006; Lobel, 2006; Sturm, 2006). The argument of 'law reduced capacity' is instead preoccupied not with what law does, but with what it can no longer do. The mismatch between its fundamental premises and those of new governance would put in peril the law's capacity to steer and inform normative directions of policy and to ensure the accountability of diverse actors in governance. Evidence of this blindness could be deemed the non-ratification of the EU Constitution due to the prevalence of new forms of governance that privilege the involvement of civil society (Kilipatrick, 2006).

The 'hybridity thesis' deals with the issue of the relationship between law and governance in a more optimistic and constructive fashion, viewing them as mutually interrelated and mutually sustaining (de Búrca and Scott, 2006, p. 6ff.). Each sphere potentially shares its own strong points by mitigating its own shortcomings. This thesis has both a descriptive and a normative version. According to some, the hybridity of law and new governance is a transitory phenomenon towards the complete embrace of new governance styles. According to others, it is a long-term phenomenon and not simply a passing stage.

In this framework the role of soft forms of regulation comes into consideration. The simultaneous and mutually interdependent resort to hard and soft law is the main characteristic of EU governance. Within the context of the development of new trends of governance soft law is variously interpreted. According to some, soft law is the second-best choice, less effective and alternative to hard law (Kilpatrick, 2006). According to others, it is less a tool for directly constraining and more a transformative tool able to change behaviour (Trubeck, Cottrell and Nance, 2006).

De Búrca and Scott (2006) draw three versions of the 'hybridity thesis': the 'baseline or fundamental normative hybridity', 'functional/developmental hybridity', and 'default hybridity or governance in the shadow of law'.

According to the 'baseline or fundamental normative hybridity' the framework of traditional legal tools still plays a relevant, if not prevalent, role (ibid., p. 7ff.). In this framework, constitutional norms and established rights that remain binding and justiciable appear essential. While new governance arrangements can serve to enhance the efficaciousness of hard law instruments, this normative dimension of constitutional rights represents the bottom-line below which the experimentation of new governance cannot take place. In this context we can also consider human rights, meaning that they are a sort of 'forbidden field' over which governance can never go.

With 'instrumental/developmental hybridity' the use of new governance tools represents a means for developing or applying the existing legal norms (ibid., p. 8ff.). These instruments serve for the elaboration and continuous transformation of the old legal ones. In this regard, the EU context could be deemed as a clear example of this version of the hybrid relationship between law and the new processes of governance (Scott and Holder, 2006). Accordingly, the Union would combine new governance tools within binding framework directives which are binding with regard to their aim, while they leave discretion with regard to the choice of the best measures needed for their implementation. Thus new governance tools would serve for the implementation of traditional Community tools. In the experience of the 'instrumental or developmental hybridity' new governance arrangements would be considered as means for the applying, elaborating and ensuring respect for legal and constitutional rights, for both new rights, such as social rights, and old ones, such as classic civil and political rights (de Búrca, 2006; Strum, 2006; Harvey, 2006). In this context some propose making the fundamental rights of the EU the beginning of an internal transformation of Union policies on emerging technologies (von Schomberg, 2011, 2013). Indeed, EU fundamental rights, together with other Union goals, are established in the treaties or in acts having the same force as treaties (i.e. the EU Charter) and can work within the EU governance as 'normative anchor points' by steering and guiding the choices of Community authorities in a proactive and anticipatory style, even influencing in advance research funding decisions and deciding which research projects are to be funded (von Schomberg, 2013).[18]

The thesis of 'default hybridity' or 'governance in the shadow of law' argues that legal norms represent a form of 'default penalty' applicable only when new governance tools fail to conform to stakeholders' behaviour (de Búrca and Scott, 2006, p. 9). Indeed, an earlier use of penalty defaults produces compliance in underregulated areas in which law encounters difficulties of enforcement. This leads to forms of regulation that are voluntary instead of mandatory. The increasing importance and influence of transnational or multinational corporations is an instance where mechanisms of command-and-control can enter into crisis. In this context, the use of mandatory rules is no longer useful especially if these corporations operate abroad. Furthermore, even inside national borders, instead of complying with statutory rules, great corporations are able to avoid the application of rules and to influence law-making processes. For these reasons resort to voluntary forms of regulation can produce interesting outcomes. This case is well exemplified in the environmental governance of the US, where the threat of federal intervention can induce States to elaborate clean air implementation plans (Karkkainen, 2006; Lobel, 2006).

The 'transformation thesis' argues that new governance implies a reshaping of the concept of law by abandoning a more formalistic

---

[18]On this see Chapter 3 of Part I on RRI.

and positivistic conception of law (de Búrca and Scott, 2006, p. 9 ff.). We need to rethink structures and forms of law according to new trajectories of social, economic and, above all, technological governance. The basic premises, normative presuppositions and functions of law thus need to be rethought in the light of emerging new practices and of public law in particular (Simon, 2006; Pariotti, 2011). 'Law, as a social phenomenon, is necessarily shaped and informed by the practices and characteristics of new governance, and new governance both generates and operates within the context of a normative order of law' (de Búrca and Scott, 2006, p. 9).

The rise of human rights in the national dimension through judicial practices which necessarily transcend the boundaries of each single State re-propose the question of the relationship between rights and new developments of governance with special regard to the European context. Notwithstanding the efforts of the scientific community, this relation still remains controversial. By paraphrasing de Búrca and Scott (2006), these new forms could either attempt an instrumental use of legal norms and human rights law practices or they could remain untouched by higher practices of European human rights law, or they could stand in the face of human rights law in an antagonistic manner by eroding one another's ambit of existence.

2.5 Soft Law: Nature, Justification, Functions, and the Issue of Its Legitimation

2.5.1 With the openness of law to soft forms of regulation we are in the territory of new governance. As acknowledged by Peters and Pagotto (2006, p. 4) 'both in international law and on the EU-level, governance by means of soft law is not new at all'. By 'soft law' we mean any set of norms and mechanisms that, even if not binding or enforced by virtue of a formal legal sanction or mechanism, have legally relevant effects (Pariotti and Ruggiu, 2012, p. 157 nt. 2; Ferrarese, 2010, p. 36). The resort to novel types of acts by political authorities represents a means to overcome the lack of formal law-making capacity (Peters and Pagotto, 2006, p. 5). The increasing climate of uncertainty and the complexity of global contexts suggest adopting less demanding and more flexible regulatory instruments. Furthermore, when there are concerns about the possibility of non-compliance the use of non-binding tools can be a more prudent strategy of regulation. Finally, the absence of legal sanctions makes resort to soft law more rapid and simpler than the conclusion of a binding treaty (ibid., p. 24). With regard to States, it is easier to adopt obligations stemming from soft law norms since they do not threaten the States' sovereign identity. With regard to enterprises, which are more sensitive to egoistic interests such as profit, makes it is easier to adopt these obligations since they need to augment their credibility in the business sphere, by adopting self-regulatory codes and thus assuming the image of an 'ethical enterprise' (Ferrarese, 2010, p. 40).

The rise of soft law in legal culture implies the abandoning of a binary view of law, according to which either it is law or it is non-law but *tertium non datur*, for a 'graduated conception of normativity', according to which 'there is no bright line between hard and soft law. Legal texts can be harder or softer' (Peters and Pagotto, 2006, p. 12; Peters, 2011, p. 23). The binary law view of Kelsenian origin holds that recognizing the status of law as non-binding ultimately obscures the meaning of law and undermines its normative power. It erodes the normative power of the international legal order as a whole. Notwithstanding the clear advantage of providing simple dichotomic structures, the first danger of this conceptualization can be an oversimplification. Instead, the graduated normativity view provides a quantitative interpretation of reality that appears more realistic and closer to the real evolution of international standards. In fact, reality leads us to a normative continuum where norms can be harder or softer but not simply black or white. Starting from a typological conception of law we can understand that there is a plurality of types of law, but we must also acknowledge that conceptual categories are merely indicative and only the practice of law can tell us whether we are facing law or non-law phenomena. A characteristic of graduated normativity is the distinction between rules, principles, and standards that nowadays scholars[19] increasingly acknowledge (Peters and Pagotto, 2006, p. 9). This variety of typological norms better explains the continuum of norms inside the law and the contemporary presence of differences and similarities among different realms of normativity (legal, ethical, technical and so forth).

2.5.2 In the perspective of regulation theory, resort to soft regulation has been justified in two ways: either with the 'reflexive regulation theory' or with the 'responsive regulation theory' (Pariotti, 2011). The first considers the law as the centre of a process through which individuals, organizations, and other social entities give rise to forms of spontaneous cooperation using the means of soft law (Teubner, 1983). According to the latter, law is a means to express and spread social values, and soft regulation can be a tool for helping to pursue this objective effectively (Ayres and Braithwaite, 1992; Marchant et al., 2008; Bowman and Hodge, 2006).

In the view of 'reflexive regulation theory', instead of pursuing the outcomes of social processes through the means of regulatory responsibility, reflexive law restricts itself to the installation, correction, and redefinition of democratic self-regulatory mechanisms (Teubner, 1983, p. 255). In this context, legal structures can reinterpret themselves in the light of external needs and demands that are selectively filtered through regulatory processes and adapted in accordance with a logic of normative development (ibid., p. 249). In this context the use of voluntary forms of arrangements is not only strategic, but the only way of realizing the 'self-reference of legal structures' by providing forms of integration and coordination of each sub-sphere within the same systemic environment (ibid., p. 274). In other words, law is only the medium for achieving the self-reflexivity of all sub-systems (politics, science, economy, moral and law) in functionally differentiated societies, thus performing a merely integrative function (ibid., p. 275).

In the perspective of 'responsive regulation theory', law has a motivational goal. Here '[p]ublic regulation can promote private

---

[19]In particular, this distinction is acknowledged by scholars of legal hermeneutics (Dworkin, 1977; Esser, 1971; Zaccaria, 1990; Pastore, 2003). On the distinction between rules/principles see also Alexy (2002).

market governance through the enlightened delegation of regulatory functions' (Ayres and Braithwaite, 1992, p. 4). As regulation responds to industry conduct, to how industry is effectively making private regulation work, law attempts to make market private actors responsible through the delegation of normative functions that should be deemed as neither wholesale nor unconditioned but motivated by given social values. In this regard, private actors are maximizers or at least satisfiers of some social values (ibid., p. 79). This flexible approach can be illustrated by the construction of an 'enforcement pyramid' which depicts the spectrum of all possible sanctions, from persuasion and warnings at the base, up through civil, licensure, and criminal penalties at the apex (Marchant et al., 2008, p. 51). This picture captures the range of regulatory strategies available, from self-regulation at the base, through supervised or enforced self-regulation and other forms of interaction among public and private actors, to standard forms of command-and-control arrangements with a range of different penalties provided for. 'The threat of regulatory intervention both deters non-compliance by potential defectors and encourages all firms to develop an attitude of social responsibility' (ibid., p. 52). Thus, according to 'responsive law theory', self-regulation is the means for expressing and implementing responsibility in all key actors. Recently an 'incremental version' of the 'responsive regulation theory' has been developed in the field of nanotechnologies. According to this version the different regulatory layers of the pyramid have to be seen through time, thus considering different regulatory strategies as sequentially ordered, beginning with softer and decentralized measures able to gain greater information, then using less flexible and more intrusive means according to a command-and-control style (ibid., pp. 52–53).

2.5.3 We can identify three functions in soft regulation: pre-law, plus-law, and para-law functions.

Soft law instruments can have a pre-law function when they are adopted with a view to the elaboration and preparation of future international treaties or Community legislation (Peters and Pagotto, 2006, p. 22; Peters, 2011, p. 34). When binding rules are unavailable or inopportune for other reasons, soft law may give an impulse to legislative processes, phenomena of building mutual confidence, and further political negotiation. The first legal effect produced by the adoption of soft law is that the subject involved is removed from the *domain reserve* of the State (de-nationalizing effect). Therefore, *ultra vires* soft law can pave the way for a formal extension of competences of the organization concerned. A second significant effect is that the promulgation of soft law declarations and conclusions can be indicative of an existing *opinion iuris* in the direction of those instruments (promoting effect). A further pre-law effect, on the other hand, could be the creation of legal uncertainty. In this regard both Article 263(3) TFEU and the European Parliament warned against the abuse of pre-legislative instruments (Peters, 2011, p. 35).

Soft law instruments can have a plus-law function within 'mixed regimes' of hard and soft law by complementing hard regulation. In this regard it is generally accepted that soft law makes hard law concrete and guides the interpretation of it (Peters and Pagotto, 2006, p. 23). Thus in the interpretation of hard legislation norms the CJEU can take into account soft texts to clarify the meaning of a general concept or some hard legislation provisions (Peters, 2011, p. 36).[20] Moreover national judges are also obliged to take into account soft law texts such as recommendations in their interpretation.[21]

Lastly, soft law regulation can constitute a surrogate of hard law. In situations where the only alternative to the resort to soft law is anarchy, soft law is more than the second-best solution (Peters, 2011, p. 37). In these instances, a soft solution can help to overcome deadlocks in the relations between States when the efforts of hard law have failed (or might fail). Powerful States might prefer recourse to soft law to retain their freedom of action while showing a cooperative attitude at the same time. Weaker States can adopt soft law instruments as the best they can politically achieve. But soft law can be the only legal instruments possible (ibid., pp. 14–18). Indeed, in international law there is a category of subjects

---

[20]Judgement of the Court of Justice, *The Queen v the Licensing Authority, established by Medicines Act 1968, (acting by The Medicines Control Agency) ex parte Generics (UK) Ltd and Others,* of 3 December 1998 (Case C-368/96) *ECR I-7967*, para. 12.

[21]Judgement of the Court of Justice, *Salvatore Grimaldi v. Fonds des maladies profesionelles,* of 13 December 1989 (Case C-322/88) *ECR I-4407*, para. 18.

such as transnational enterprises which are not endowed with legal subjectivity (Pariotti, 2007, 2013). In this case the only way to impose legal obligations, such as those concerning human rights, can, when they act abroad, be resort to soft law. In this regard, in the field of nanotechnologies we have observed the phenomenon of the 'proliferation of codes of conduct' especially among large corporations (Kernes and Rip, 2009).

2.5.4 The spread of soft instruments among private actors has led to a sort of 'privatization of law', namely to the diffusion of private or semi-private soft law such as self-regulation and co-regulation (Peters, 2011, p. 41). This phenomenon raises problems of both effectiveness and legitimation which appear to be mutually inter-connected. Excessive standardization can lead to the phenomenon of over-regulation. Moreover, unlike the case of private exchanges and private law, the base of legitimation of standard setting does not rest on private autonomy and non-state actors' consent as standards have a general scope. In fact, unlike the contract sphere, standards address and bind not only the authors of these norms, but also third–parties. In these instances, non-state actors not only regulate themselves and their future behaviour, but also others who have not participated in the processes of standard setting. Thus consent is only a partial basis of its legitimation (ibid.).

Indeed, a further basis of legitimation could be the delegation by governments or by the European Union when member States or Community authorities have delegated the standard-setting powers to private actors. Since States have the overall legitimacy and authority to produce norms, acts of delegation can provide private actors with the legitimacy lacking in standard-setting processes. In highly complex global contexts governments lack both knowledge and the capacity to regulate issues that transcend state boundaries. In this framework business actors can offer their expertise to improve standard-setting processes and at the same time concur to design norms in a more suitable fashion according to their primary interests. In this regard, at the global level 'power appears diffuse among public and private nets which, due to their structure, ways of functioning, aims, and lack of international law subjectivity', tend to elude national and international rules, such as human rights

(Pariotti and Ruggiu, 2012, p. 162). In these contexts, traditional tools of regulation can be useless or largely ineffective. Self-regulation can set constraints and standards to relevant subjects' behaviour by acting from the inside of the enterprise organization. Furthermore, it can also involve relevant private actors whose expertise may be taken advantage of when setting the rules (i.e. researchers, research centres, funding organizations). Since, in the globalized world decisional processes are increasingly decentralized and *de facto* involve non-state actors, the main criterion of the legitimacy of regulation is not given by its connection to the idea of representation as its ability to foster accountability. In other words, possibilities to improve compliance with norms of soft law and their effectiveness depend on the fact that firms' behaviour goes along with other actors' expectations. In this sense, accountability can increase only in the presence of adequate accountability practices that foster information exchange, transparency, and the justifiability of an enterprise's choices in the face of other stakeholders. In this context the voluntary resort to forms of undue regulation can be a means to consolidate stakeholders' trust. Only the acknowledgement and the assumption of responsibility for actions, decisions and an enterprise's policies can strengthen the network of relations among all actors and create a climate of confidence for supporting compliance with norms of both hard and soft regulation. In this regard soft norms all fall within the path of accountability practices since the lack of acknowledgement of one's own accountability leads to the failure of the compliance mechanisms provided. Thus, in the end, accountability is to be deemed as the source of soft law legitimation (ibid.).

2.6 Soft Law and Corporate Social Responsibility

2.6.1 According to Parker (2009, p. 207) the idea of Corporate Social Responsibility (CSR) hides an inner paradox since it tries to include compliance with business' legal responsibilities, but goes 'beyond compliance' to encompass the economic, ethical and discretionary expectations of civil society. Indeed, society needs the production of certain goods and services that are not yet on the market, the adoption by enterprises of additional forms of behaviour that are not necessarily codified into legal texts, even when members of society do not have a clear-cut message on those forms of behaviour, but nonetheless they want them to be followed by business. In these instances, CSR tends to build a sort of 'compliance beyond compliance'. While regulation works by holding people legally accountable to meet thresholds set up by legal standards through forms of liability, CSR 'internalizes standards by building them into self-conceptions, motivations and habits' of the organization, thus extending the responsibility of the business enterprise (ibid., p. 213).

Since technological innovation became a successful business strategy, the rise of emerging technologies has gone hand in hand with the development of the market, and many high-technological products have reached the store shelves. The case of nanotechnologies is in this sense emblematic. While nanotechnological research promises to revolutionize our life, by providing new cures for cancer, solving the global food scarcity, enhancing human performance, giving rise to even more wearable technologies, nanotechnologies invade our shops at every corner. Nanotechnologies are currently used in OLED technology (organic light-emitting diode) that gives rise to a multitude of high-tech products in our houses, such as television screens, computers monitors, mobile phones and gaming consoles. Yet, nanotechnologies are also largely used in cosmetics, such as face and hand creams, toothpastes, in food packaging, toys, sportswear, and so forth. Thus, although some hold that mass personalization and distribution has substituted forms of mass production, such as in the case of nanomedicine (Roco, 2006), mass production processes still represent the main landing place of emerging technologies in the global market. The problem, however, is that many people often do not know it yet (Throne-Host and Strandbakken, 2009).

The large diffusion of emerging technologies in goods used in our everyday lives underlines the relevance of firms and corporations, mainly those of a transnational size, within the global governance framework. We must not forget that the globalization of the market gives these multi- and transnational corporations, which are without any subjectivity in international law, the chance to avoid compliance with international obligations set up for the protection of human rights (Pariotti, 2007). Due to their dimension, transnational corporations can exert a quasi-political power (Pariotti, 2013, p. 140). In fact, they can influence the goals and the agenda of both politics and regulation. Then, when transnational corporations dislocate part of their production abroad, they can influence the policy choices of their host countries. They can, in any case, directly produce rules through acts of self-regulation. In sum, they can shape policies, rules and even ideas (ibid.). Moreover, since the economic context has become global, normative differences between several legal orders can encourage the 'lex and fora shopping phenomena' among enterprises which have the resources to do business in a plurality of countries and choose the best regulatory conditions in which to develop it (ibid., p. 140). When transnational corporations act abroad, outside their home State borders, they can become opaque to international law and elude the State's control.

2.6.2 Both international and national laws have relevant limits in imposing the State's responsibility for human rights violations when private actors are at stake (ibid., p. 147). Due to their structural features, human rights are able to reach the State's responsibility (their 'vertical effect'), but they tend to evaporate when non-state actors such as transnational corporations are at play. In these instances, in order to strengthen their 'horizontal effect' among private actors, self-regulation is suitable and the paradigm of CSR arises (Pariotti, 2007, 2013; Pariotti and Ruggiu, 2012).

With regard to the possibilities of protecting human rights when non-state actors are at stake, three theories structure the relationship between responsibility and international personality in different ways (Pariotti, 2013, pp. 142–143). The first conception refuses to acknowledge any legal personality to entities different from States and thus to attribute business enterprises with any responsibility for breaches of human rights. The second theory, though it does not tackle the question whether they have any personality or not, holds that it is possible to ascribe the responsibility for human rights violation to firms. The third theory, finally, tends to attribute transnational corporations with partial international subjectivity with regard to the duty to abstain from human rights violations

(Weschka, 2006, p. 659) and the possibility of considering them responsible for human rights breaches.

International law has developed several instruments to strengthen the protection of human rights when transnational corporations' behaviour runs the risk of leading to violations (the so-called 'horizontal effect' of human rights). Here we can mention several International Labour Organization (ILO) initiatives such as the Tripartite Declaration on Fundamental Principles Concerning Multinational Enterprises and Social Policy[22] of 1977, which addresses governments, workers' and enterprises' organizations and transnational corporations by offering detailed guidelines especially with regard to human rights; the Declaration on Fundamental Principles and Rights at Work[23] of 1998 which encompasses all four core labour standards, namely freedom of association and the rights to collective bargaining, the effective abolition of child labour and the elimination of discrimination in employment and occupation (Weschka, 2006, pp. 645–646). We must also mention the Organization for the Economic Co-operation and Development (OECD) Guidelines for Multinational Enterprises.[24] Finally the International Chamber of Commerce (ICC) Business Charter for Sustainable Development[25] addresses enterprises with a view to steering their behaviour according to principles which are not merely based on business (Pariotti, 2013, p. 145).

---

[22] ILO (1977) *Tripartite Declaration of Principles concerning Multinational Enterprises and Social Policy*, adopted by the Governing Body of the International Labour Office at its 204th Session (Geneva, November 1977) as amended at its 279th Session (Geneva, November 2000), (2002) 41 ILM, 186–201, at 186–187, http://www.ilo.org/wcmsp5/groups/public/—ed_emp/—emp_ent/—multi/documents/publication/wcms_094386.pdf. Accessed 2 September 2013.

[23] ILO (1998) *ILO Declaration on Fundamental Principles and Rights at Work*, 86th Session, Geneva, June 1998, http://www.ilo.org/public/english/standards/relm/ilc/ilc86/com-dtxt.htm. Accessed 2 September 2013.

[24] Organisation for Economic Co-operation and Development (1976) *The OECD Guidelines for Multinational Enterprises. 2011 Edition* OECD Publishing, 2011, http://www.oecd.org/daf/inv/mne/48004323.pdf. Accessed 2 September 2013.

[25] International Chamber of Commerce (ICC) (1991) *Business Charter for Sustainable Development: Principles for Environmental Management, Year 2000 Edition*, Second World Industry Conference April 1991, ICC, Rotterdam, http://www.iccwbo.org/Advocacy-Codes-and-Rules/Document-centre/2000/ICC-business-charter-for-sustainable-development-(2000)-(EN/FR/ES)/. Accessed 2 September 2013.

At the UN level it is worthwhile mentioning the Global Compact and the UN-Norms. The Global Compact[26] is a governance framework, an initiative of the ex-UN-Secretary General Kofi Annan, of a voluntary nature, 'open to business, which strives to promote ten principles through a variety of instruments, such as dialogue, learning and projects' (Weschka, 2006, p. 650), covering the areas of human rights, labour rights, the environment and corruption. In 2004 it was endowed with a sanction mechanism based fundamentally on public criticism, whose aim is to lead towards greater accountability of business enterprises. In August 2003 the UN-Sub Commission on the Promotion and Protection of Human Rights adopted the UN Norms on the Responsibilities of Transnational Corporation and other Business Enterprises (UN-Norms)[27] which provides a comprehensive set of global business standards including a broader range of human rights, then those protected by other instruments and bearing the UN *imprimatur* (ibid., p. 654). The UN-Norms directly envisage provisions of their implementation in the text itself (ibid., p. 655). The document encountered significant opposition from the business sector and thus the Commission on Human Rights ultimately determined in 2004 that it had no legal standing (i.e. it was annulled). Finally it was substituted in 2005 by the Protect, Respect and Remedy Framework on Transnational Corporation and Human Rights[28] also known as Ruggie's principles (from John Gerard Ruggie, the author of the guiding principles), recently endorsed in 2011 by the UN Human Rights Council.

None of these instruments are binding and rest on merely voluntary bases.

The rise of forms of self-regulation is at the base of the development of the CSR paradigm in business ethics. According

---

[26]UN Global Compact Homepage, http://www.unglobalcompact.org/AboutThe GC/TheTenPrinciples/index.html. Accessed 4 September 2013.

[27]ONU (2003, 13 August) *UN Norms on the Responsibilities of Transnational Corporation and other Business Enterprises with regard to Human Rights*, Document E/CN.4/Sub.2/2003/38/Rev.2 (26 August 2003), http://www1.umn.edu/humanrts/links/norms-Aug2003.html. Accessed 4 September 2013.

[28]ONU (2011) *Protect, Respect and Remedy Framework on Transnational Corporation and Human Rights-Guiding Principles*, http://www.business-humanrights.org/media/documents/ruggie/ruggie-guiding-principles-21-mar-2011.pdf. Accessed 4 September 2013.

to CSR business enterprise does not only hold obligations of an economic origin (i.e. making profits), nor exclusively legal obligations (to workers, suppliers, costumers) stemming from binding rules, but also kinds of moral and social obligations to all stakeholders variously affected by its activity (Pariotti, 2013, p. 145). Voluntary tools (such as codes of conduct, the promotion of social projects having value for a given community where the firm itself acts, third-party certification systems) can extend the scope of an enterprise's responsibility. The rationale of these voluntary tools rests on the progressive involvement of stakeholders in order to build an alliance among all different interests. In this sense CSR is 'a set of vague, discretionary and non-enforceable corporate responses to social expectations' (Parker, 2009, p. 208).

Within stakeholder theory, the term 'stakeholder' means either who has an interest or an eligible claim to the firm; any group or individual who can affect, or can be affected by, the firm's interests; or, finally, any group or individual whose level of well-being can be affected by decisions concerning the firm's action or inaction (Pariotti, 2013, p. 145). Thus business enterprise not only has to limit negative externalities and to respect legal rules such as those on human or fundamental rights, but also to pursue social development and community well-being (ibid., p. 146). In this context mere respect for the law is no longer enough. The reason for this is that respect for the law cannot prevent some interests that are increasingly acknowledged at the international law level, in particular human rights, from being affected by the enterprise's business both within the State borders and abroad. Thus the activity of transnational corporations is even more the target of increasing expectations of civil society and the international community (Parker, 2009, p. 2). This is the background against which a new form of responsibility in equilibrium between legal and ethical spheres is being created.

Scholars usually distinguish between internal stakeholders such as workers, suppliers, customers and creditors, and external stakeholders, such as the environment and the community (Pariotti, 2013, p. 145). Thus, according to the CSR paradigm the duties of business enterprise become larger by including not also those who have a relevant interest in the firm's activity such as stockholders and shareholders (i.e. the owners and investors), but also other categories belonging to civil society (ibid.). In this framework three dimensions emerge within the enterprise's activity which form a 'triple bottom line', namely profit, persons, and the environment in its broad sense.

2.6.3 The main instrument to develop forms of CSR is the code of conduct. Codes of conduct are increasingly used in the field of emerging technologies, especially with regard to nanotechnologies (Kearnes and Rip, 2009; Kurath et al., 2015). Thus we have observed the rise of several ethical codes on nanotechnologies: the Responsible NanoCode by the Royal Society, Insight Investment and Nanotechnology Industries Association; the 'Code of Conduct: Nanotechnology' developed by BASF[29] which addresses the responsibility of the organization to its employees, customers, supplies and other stakeholders, such as future generations; the Nanocare Initiative launched by a group of chemical companies[30]; the European Commission's Code of conduct for nanosciences and nanotechnologies research[31] which is a case of meta-regulation (Dorbeck-Jung and Shelley-Egan, 2013; Ruggiu, 2014); the 'Nano Risk Framework to Aid in Responsible Development of Nanotechnology' developed by the joint work of DuPont and the (US) Environmental Defense; and the 'Code of Conduct: Nanotechnologies' developed by the Swiss retailers association IG DHS (Kearnes and Rip, 2009, p. 15). In these instances, the lack of legal sanctions is reinforced by a synergic relationship between public and private actors that partially reinforce each other.

First of all, we need to distinguish between the code of conduct and the code of ethics (Arrigo, 2006). The first is *ruled-based*, that is, a set of rules aimed at driving the conduct of recipients in order to solve each problem of the enterprise's existence (such as harassment in the workplace, mobbing, safety etc.). The latter is

---

[29] http://www.basf.com/group/corporate/nanotechnology/en/microsites/ nanotechnology/safety/code-of-conduct. Accessed 9 September 2013.

[30] http://www.nanopartikel.info/cms. Accessed 9 September 2013.

[31] European Commission (2008) Commission recommendation of 7 February 2008 *on a code of conduct for responsible nanosciences and nanotechnologies research* C(2008) 424 final, http://ec.europa.eu/research/participants/data/ ref/fp7/89918/nanocode-recommendation_en.pdf. Accessed 19 February 2015.

value-based, that is, a set of principles (such as the protection of the environment, health, non-discrimination), without specifying how these values should be concretized (ibid., p. 93).

The code of conduct is usually made up of a formulation of the mission of the organizations that will adopt it, namely what their goal is, why they exist and why stakeholders should have relations with them; an assessment of the means deemed suitable to reach the organization's goal; finally, an explanation of the reason why to adopt the code (D'orazio, 2011, p. 474). Its aim is to make all stakeholders, the business enterprise *in primis*, accountable for their actions, thus acting according to a responsibilization rationale, i.e. the progressive distribution of responsibilities (Ruggiu, 2014).

When, through the means of meta-regulation which regulate the development of self-regulation, '[l]aw attempts to constitute corporate "consciences" getting companies "to do what they should do"' beyond mere compliance with legal norm, it seeks to hold business accountable for taking their responsibility seriously (Parker, 2009, p. 208). Meta-regulation must be aimed clearly at values or policy goals (e.g. respect for human and fundamental rights) for which business enterprise can take responsibility. Then, meta-regulation must be aimed at making sure these social values such as individual rights are built into the practices and organizational structure of the corporation. Finally, it must recognize that the main goals of the organization are still pursued within the responsibility framework of the enterprise. In this regard, the tool of the code of conduct can contribute to meeting the enterprise's main goals of producing particular goods and services, providing a return to its investors, and providing paid employment to its workers and managers within the framework of established values. In this way these values become an integral part of business practices and entrepreneurial structure (ibid., pp. 215–217). Thus, in this conceptual framework, codes of conduct may present a way to foster the accountability of business enterprises effectively.

A variety of models structure the balance of interests which are about the enterprise's business (D'orazio, 2011, p. 474). In particular, we can count the 'company codes' or 'industry associations' codes' where the corporation's interests are at the centre of all relations and only primary stakeholders (i.e. those groups without whose participation the corporation cannot exist) are concerned. There are also 'multi-stakeholder codes' where there is an alliance among all stakeholders in a joint partnership, and the interests of the business enterprise are just some among others. 'Multi-stakeholder codes' represent the last generation of ethical codes and better express the potentiality of new forms of governance as regards human rights (Pariotti and Ruggiu, 2012). In this framework stakeholders are bearers of mutual rights and duties and are equally involved in the business of the enterprise. Here the firm's interest is not at the apex of the whole process but one of the interests at stake. This framework can be deemed as the more suitable to pursue the aim of protecting human rights within the business ambit. Here also individual rights are assumed as a legitimate source of the final arrangement of the business enterprise. Basically 'multi-stakeholders codes' are based on the idea of 'stakeholder democracy' (Matten and Crane, 2005), which stresses the openness of businesses to the social system as a source of the legitimation of self-regulation. A case of a multi-stakeholder code in the field of nanotechnologies can be deemed the Responsible NanoCode by the Royal Society, Insight Investment and Nanotechnology Industries Association (D'orazio, 2011, p. 476ff.; Pariotti and Ruggiu, 2012) or the Commission code of conduct on nanotechnologies research (Ruggiu, 2014).

As said, the multi-stakeholder perspective can better cover those interests expressed by human rights. Since right to health, workers' and consumers' rights, right to a healthy environment are all at the same time human rights and interests affected by business, they need to be accordingly represented in the life of the enterprise (Pariotti and Ruggiu, 2012). This also corresponds to an interest of the enterprise itself. In this sense, the early care of these interests can be a precise strategy of corporations. As publicly recognized, '[a]n early and open examination of potential risks of a new product or technology is not just good common sense- it's good business strategy' (Krupp and Holiday, p. B2). As the recent worldwide scandal of Volkswagen taught, trust is a good which is difficult to gain, but it is more difficult to maintain. This is true also in technoscientific field. 'Though the opportunities for a technology may be literally endless, these opportunities cannot be achieved if a technology is not developed in a secure manner that maintains public confidence' (Mendel, 2009, p. 2). In this context, human rights are not only a 'negative externality' of business in technoscientific field, since they can, where violated, lead to adverse decision of national and supranational courts, by representing an instance of system failure (von Schomberg, 2013, p. 61ff.; Ruggiu, 2015, p. 229ff.). If they are considered from the outset within the CSR paradigm, they can thus proactively transform the life of business enterprises in depth. In this regard, beyond traditional mechanisms of regulation that are also provided for human rights, soft law tools, such as codes of conduct, guidelines or certification systems can be a further way of implementation in the business sphere: the necessary complement of their legal dimension. In other words, legal constraints stemming from the protection of human rights lead also to assume a proactive attitude in governance frameworks by adopting coherent tools of soft law and implementing them in the sphere of CRS.

References

Alexy, R. (2002). Diritti fondamentali, bilanciamento e razionalità, *Ars interpretandi. Annuario di ermeneutica giuridica*, **7**, pp. 131–144.

Arrigo, E. (2006). Code of Conduct and Corporate Governance, *Symphonya. Emerging Issues in Management*, **1**, pp. 93–109, http://symphonya.unimib.it/article/view/2006.1.07arrigo/8776. Accessed 18 February 2015

Ayres, I., and Braithwaite, J. (1992). *Responsive Regulation: Transcending the Deregulation Debate*, New York: Oxford University Press.

Beck, U. (1986). *Risikogesellschaft: Auf dem Weg in eine andere Moderne*, Frankfurt am Main: Suhrkamp; trad. it. (2000). *La società del rischio. Verso una seconda modernità*, Roma: Carocci.

Bowman, D.M., and Hodge, G.A. (2006). Nanotechnology: Mapping the Wilde Frontier, *Futures* **38**, pp. 1060–1073.

Brownsword, R. (2008). *Rights, Regulation and the Technological Revolution*, Oxford: Oxford University Press.

Coglianese, C., and Mendelson, E. (2010). Meta-regulation and Self-Regulation, in: Baldwin, R., Cave, M., and Lodge, M. (eds.) *The

Oxford Handbook on Regulation, Oxford: Oxford University Press, pp. 146–168.

de Búrca, G. (2006). EU Race Discrimination Law: A Hybrid Model? in: G. De Burca and J. Scott (eds.), *Law and New Governance in the EU and the US*, Oxford: Hart Publishing, pp. 97–119.

de Búrca, G., and Scott, J. (2006). *Introduction: New Governance, Law and Constitutionalism*, in: G. De Burca and J. Scott (eds.), *Law and New Governance in the EU and the US*, Oxford: Hart Publishing, pp. 1–12.

Dorbeck-Jung, B., and Shelley-Egan, C. (2013). Meta-regulation and Nanotechnologies: The Challenge of Responsibilisation within the European Commission's Code of Conduct for Responsible Nanosciences and Nanotechnologies Research, *Nanoethics*, **7**(1), pp. 55–68.

D'Orazio, E. (2011). Responsabilità degli stakeholder e approccio multi-stakeholder ai codici etici: riflessioni sul *Responsible Nano Code*, in: Guerra, G., Muratorio, M., Pariotti, E., Piccinni, M., and Ruggiu, D. (eds.), *Forme di responsabilità, regolazione e nanotecnologie*, Bologna: Il Mulino, pp. 425–507.

Dworkin, R. (1977). *Taking Rights Seriously*, Cambridge (Mass.); trad. it. (1982). *I diritti presi sul serio*, Bologna: Il Mulino.

Eberlein, B., and Kerwer, D. (2004). New Governance in the European Union: A Theoretical Perspective, *Journal of Common Market Studies*, **42**(1), pp. 121–142.

Esser, J. (1972). *Vorverständnis und Methodenwahl in der Rechtsfindung. Rationalitätsgrundlagen richterlicher Entscheidungspraxis*, 2 ed., Frankfurt am Main: Fisher Athenäum Taschenbücher; trad. it. (1983). *Precomprensione e scelta del metodo nel processo di individuazione del diritto. Fondamenti di razionalità nella prassi decisionale del giudice*, Napoli: Edizioni Scientifiche Italiane.

European Commission (2001). *European Governance: a White Paper*, 25.7.2001, COM(2001) 428 final, http://eur-lex.europa.eu/legal-content/EN/TXT/PDF/?uri= CELEX:52001DC0428&rid= 2. Accessed 18 February 2015

European Commission (2008). *Commission Recommendation of 07/02/2008 on a Code of Conduct for Responsible Nanosciences and Nanotechnologies Research*, Brussels: European Commission, http://ec.europa.eu/research/science-society/document_library/pdf_06/nanocode-apr09_en.pdf. Accessed 18 February 2015

Ferrarese, M.R. (2010). *La governance tra politica e diritto*, Bologna: Il Mulino.

Ferrarese, M.R. (2012). *Prima lezione di diritto globale*, Roma-Bari: Laterza.

Francioni, F. (2007). Genetic Resources, Biotechnology and Human Rights, in: Francioni, F. (ed.) *Biotechnologies and International Human Rights*, Oxford and Portland, Oregon: Hart Publishing, pp. 3–32.

Groves, Ch. (2015). Logic of Choice and Logic of Care? Uncertainty, Technological Mediation and Responsible Innovation, *Nanoethics*, **9**(3), pp. 321–333.

Harvey, T.K. (2006). The European Union and the Health Care, in: De Burca, G. Scott, J. (eds.), *Law and New Governance in the EU and the US*, Oxford: Hart Publishing, pp. 179–210.

Heydebrande, W. (2003). Process Rationality as Legal Governance: A Comparative Perspective, *International Sociology*, **18**(2), pp. 325–349.

Jasanoff, S. (2003). Technologies of Humility: Citizen Participation in Governance Science, *Minerva*, **41**, pp. 223–244.

Karkkainen, B.C. (2006). Information- forcing Regulation and Environmental Governance, in: De Burca, G. Scott, J. (eds.), *Law and New Governance in the EU and the US*, Oxford: Hart Publishing, pp. 293–322.

Kearnes, M.B., and Rip, A. (2009). The Emerging Governance Landscape of Nanotechnology, in: Gammel, S., Losch, S., and Nordmann, A. (eds.), *Jenseits Von Regulierung: Zum Politischen Umgang Mit Der Nanotechnologie*, Berlin: Akademische Verlagsgesellschaft, pp. 97–121.

Kilipatrick, C. (2006). New EU Employment and Constitutionalism, in: G. De Burca, and J. Scott (eds.), *Law and New Governance in the EU and the US*, Oxford: Hart Publishing, pp. 121–151.

Kooiman, J., and Van Vliet, M. (1993). Governance and Public Management, in: Eliassen, K., and Kooiman, J. (eds.), *Manging Public Organizations: Lessons from Contemporary European Experience* (2nd ed.), London: Sage.

Krupp, F., and Holliday, C. (2005). Let's get nanotech right, *Wall Street Journal*, Tuesday, June 14, 2005, Management Supplement, p. B2.

Kuhlmann, S., Stegmaier, P., Konrad, K., and Dorbeck-Jung, B. (2012). *Tentative Governance—Conceptual Reflections and Impetus for Contributors to a Planned Special Issue of Research Policy on 'Getting Hold of a Moving Target—the Tentative Governance of Emerging Science and Technology'*.

Kurath, M., Nentwich, M., Fleischer, T., and Eisenberger, I. (2014). Cultures and Strategies in the Regulation of Nanotechnology in Germany, Austria, Switzerland and the European Union, *Nanoethics*, **8**(2), pp. 121–140.

Lyall, C., and Tait, J. (2005). Shifting Policy Debates and the Implications for Governance, in Lyall, C., and Tait, J. (eds.) *New Modes of Governance. Developing an Integrated Policy Approach to Science, Technology, Risk and the Environment,* Adelshot: Ashgate, pp. 1–17.

Lobel, O. (2006). Governing Occupational Safety in the United States, in: De Burca, G., and Scott, J. (eds.), *Law and New Governance in the EU and the US,* Oxford: Hart Publishing, pp. 269–292.

Mandel, G.N. (2009). Regulating Emerging Technologies. Legal Studies Research Paper Series, Research Paper N. 2009-18, 04-08-2009, *Law, Innovation & Technology,* **1**, p. 75.

Marchant, G.E., Sylvester, D.J., and Abbott, K.W. (2008). Risk Management Principles for Nanotechnology, *Nanoethics,* **2**(1), pp. 43–60.

Matten, D., and Crane, A. (2005). What is Stakeholder Democracy? Perspectives and Issues, *Business Ethics: A European Review,* **14**(1), pp. 6–13.

Maurer, S.M., Lucas, K.V., and Terrel, S. (2006). *From Understanding to Action: Community-Based Options for Improving Safety and Security in Synthetic Biology,* Berkeley: University of California.

Meili, C., Widmer, M., Schwarzkopf, S., Mantovani, E., and Porcari, A. (2011). *NanoCode MasterPlan: Issues and options on the path forward with the European Commission Code of Conduct on Responsible N&N Research,* November 2011.

Owen, R., Stilgoe, J., Macnaghten, P., Gorman, M., Fisher, E., and Guston, D. (2013). A Framework for Responsible Innovation. in: Owen, R., Bessant, J., and Heintz, M. (eds.), *Responsible Innovation,* London: John Wiley & Sons Ltd, pp. 27–50.

Pariotti, E. (2007). 'Effetto orizzontale' dei diritti umani e imprese transnazionali nello spazio europeo, in Trujillo, I., and Viola, F. (eds.), *Identità, diritti, ragione pubblica in Europa,* Bologna: Il Mulino, pp. 171–201

Pariotti, E. (2011). Normatività giuridica e *governance* delle tecnologie emergenti, in: Guerra, G., Muratorio, A., Pariotti, E., Piccinni, M., and Ruggiu, D. (eds.), *Forme di responsabilità, regolazione e nanotecnologie,* Bologna: Il Mulino, pp. 509–549.

Pariotti, E. (2013). I diritti umani: concetto, teoria, evoluzione, Padova: CEDAM.

Pariotti, E., and Ruggiu, D. (2012). Governing Nanotechnologies in Europe: Human Rights, Soft Law, and Corporate Social Responsibility, in: Van Lente, H., Coenen, C., Konrad, K., Krabbenborg, L., Milburn, C., Seifert,

F., Thoreau, F., and Zülsdorf, T. (eds.), *Little by Little. Expansions of Nanoscience and Emerging Technologies*, Heidelberg: IOS Press/AKA-Verlag, pp. 157–168.

Parker, C. (2009). Meta-Regulation: Legal Accountability for Corporate Social Responsibility, in: McBarnet, D., Voiculescu, A., and Campbell, T. (eds.) *The New Corporate Accountability: Corporate Social Responsibility and the Law*, Cambridge: Cambridge University Press, pp. 207–241.

Pastore, B. (2003). *Per un'ermeneutica dei diritti umani*, Torino: Giappichelli Editore.

Peters, A. (2011). Soft Law as a New Mode of Governance, in: Diedrichs, U., Reiners, W., and Wessels, W. (eds.) *The Dynamics of Change in EU Governance*, Cheltenham, UK, Northampton, MA, USA: Edward Elgar, pp. 21–51.

Peters, A., and Pagotto, I. (2006). *Soft Law as a New Mode of Governance: A Legal Perspective, report of the project NEWGOV New Modes of Governance. Integrated Project. Priority 7 – Citizens and Governance in the Knowledge-Based Society*, 04, D11, http://papers.ssrn.com/sol3/papers.cfm?abstract_id=1668531&rec=1&srcabs=1876508&alg=1&pos=1. Accessed 18 February 2015.

Polanyi, M. (1962). The Republic of Science: Its Political and Economic Theory, *Minerva*, **1**, pp. 54–74.

Rip, A. (2002). *Coevolution of Science, Technology and Society: An Expert Review for the Bundesministrium Bildung und Forshung's Förderinitiative Politik, Wissenschaft und Gesellschaft* (Science Policy, Studies), http://www.google.it/url?sa=t&rct=j&q=&esrc=s&frm=1&source=web&cd=2&ved=0CEEQFjAB&url=http%3A%2F%2Fciteseerx.ist.psu.edu%2Fviewdoc%2Fdownload%3Fdoi%3D10.1.1.201.6112%26rep%3Drep1%26type%3Dpdf&ei=g7vKUfj0NoWpOrG5gbgP&usg=AFQjCNH9NKjk8E5pIoFqJsMVSPofWw-TLw&sig2=GjdoftjztHB-vqW160HMEA&bvm=bv.48340889,d.ZWU. Accessed 18 February 2015.

Roco, M. (2006). Progress in Governance of Converging Technologies Integrated from the Nano Scale, *Annals of the New York Academy of Science*, **1093**, pp. 1–2.

Rhodes, R. (1996). The New Governance: Governing without Government, *Political Studies*, **44**, pp. 652–667.

Ruggiu, D. (2012a). Synthetic Biology and Human Rights in the European Context: Health and Environment in Comparison of the EU and the

Council of Europe Regulatory Frameworks on Health and Environment, *Biotechnology Law Report*, **31**(4), pp. 337–355.

Ruggiu, D. (2012b). *Diritti e temporalità. I diritti umani nell'era delle tecnologie emergenti*, Bologna: Il Mulino.

Ruggiu, D. (2013a). A Rights-Based Model of Governance: The Case of Human Enhancement and the Role of Ethics, in: Konrad, K., Coenen, C., Dijkstra, A., Milburn, C., and van Lente, H. (eds.). *Shaping Emerging Technologies: Governance, Innovation, Discourse*, Berlin: IOS Press/AKA-Verlag, pp. 103–115.

Ruggiu, D. (2013). Temporal Perspectives of the Nanotechnological Challenge to Regulation. How Human Rights Can Contribute to the Present and Future of Nanotechnologies, *Nanoethics*, **7**(3), pp. 201–21.

Ruggiu, D. (2014). Responsibilisation Phenomena: The EC Code of Conduct for Responsible Nanosciences and Nanotechnologies Research, *European Journal of Law and Technology*, **5**(3), pp. 1–16, http://ejlt.org/article/view/338. Accessed 18 February 2015.

Ruggiu, D. (2015). Anchoring European Governance: Two versions of Responsible Research and Innovation and EU Fundamental Rights as 'Normative Anchor Points', *Nanoethics*, **9**(3), pp. 217–235.

Salter, B., and Jones, M. (2002). Human Genetic Technologies, European Governance and the Politics of Bioethics, *Nature*, **3**, pp. 808–814.

Scott, J., and Holder, J. (2006). Law and New Environmental Governance in EU, in: De Burca, G., and Scott, J. (eds.), *Law and New Governance in the EU and the US*, Oxford: Hart Publishing, pp. 211–242.

Scott, J., and Trubek, D.M. (2002). Mind the Gap: Law and New Approaches to Governance in the European Union, *European Law Journal*, **8**(1), pp. 1–18.

Simon, W.H. (2006). Toyota Jurisprudence: Legal Theory and Rolling Rules Regime, in: De Burca, G., and Scott, J. (eds.), *Law and New Governance in the EU and the US*, Oxford: Hart Publishing, pp. 37–64.

Sloat, A. (2002). Governance: Contested Perceptions of Civic Participation, *Scottish Affairs*, **39**, spring.

Smismans, S. (2008). New modes of governance and the participatory myth, *West European Politics*, **31**(5), 874–895.

Sørensen, E., and Triantafillou, P. (2009). Introduction, in: Sørensen E., and Triantafillou P. (eds.), *The Politics of Self-Governance*, Farnham: Ashgate, pp. 1–22.

Stoke, E., and Bowman, D.M. (2012). Looking Back to the Future of Regulating New Technologies: The Case of Nanotechnology and Synthetic Biology, *European Journal of Risk Regulation*, **2**, pp. 235–241.

Stoke, G. (1998). Governance as Theory: Five Propositions, *JSSJ*, **155**, pp. 17–28.

Strum, S. (2006). Gender Equity Regimes and the Architecture of Learning, in G. De Burca and J. Scott (eds.), *Law and New Governance in the EU and the US*, Oxford: Hart Publishing, pp. 323–360.

Teubner, G. (1983). Substantive and Reflexive Elements in Modern Law, *Law & Society Review*, **17**, pp. 239–285.

Throne-Host, H., and Strandbakken, P. (2009). 'Nobody Told Me I Was a Nano-consumer': How Nanotechnologies Might Challenge the Notion of Consumer Rights, *Journal of Consumer Policy*, **32**(4), pp. 393–402.

Trubeck, L.G., Cottrell, P., and Nance, M. (2006). 'Soft law', 'Hard Law', and EU Integration, in: De Burca, G., and Scott, J. (eds.), *Law and New Governance in the EU and the US*, Oxford: Hart Publishing, pp. 65–94.

Trubeck, L.G. (2006). New Governance Practices in US Health Care, in: De Burca, G., and Scott, J. (eds.), *Law and New Governance in the EU and the US*, Oxford: Hart Publishing, pp. 245–268.

von Schonberg, R. (2007). *From the Ethics of Technology Towards an Ethics of Knowledge Policy & Knowledge Assessment. A Working Document from the European Commission Services*, Brussels: European Commission, Directorate-General for Research.

von Schomberg, R. (2010). Organising Public Responsibility: On Precaution, Code of Conduct and Understanding Public Debate, in: Fiedeler, U., Coenen, Ch., Davies, S.R., and Ferrari A (eds.), *Understanding Nanotechnology: Philosophy, Policy and Publics*. Amsterdam: IOS Press, pp. 61–70.

von Schonberg, R. (2011). Prospects for Technology Assessment in a Framework of Responsible Research and Innovation. in: Dusseldorp, M., and Beecroft, R. (eds.), *Technikfolgen abschätzen lehren: Bildungspotenziale transdisziplinärer Methoden*, Wiesbaden: Vs Verlag, pp. 39–61.

von Schonberg, R. (2013). A Vision of Responsible Research and Innovation, in Owen, R., Heintz, M., and Bessant, J. (eds.) *Responsible Innovation: Managing the Responsible Emergence of Science and Innovation in Society*, London: John Wiley, pp. 51–74.

Waldron, J. (1999). *Law and Disagreement*, Oxford: Oxford University Press.

Weschka (2006). Human Rights and Multinational Enterprises: How Can Multinational Enterprises Be Held Responsible of Human Rights Violations Committed Abroad? *Zaitschrift für ausländisches öffentlisches Recht Völkerrecht*, **66**, pp. 652–661.

Zaccaria, G. (1990). *L'arte dell'interpretazione*, Padova: Cedam.

Chapter 3

The European Governance of Emerging Technologies

3.1 Introduction

3.1 For decades the national interest and the international competition between States steered the expansion of research policy in Europe. This role of the State capacity in research and innovation was increasingly eroded by the globalization of firms, market and technology (Grande, 2001, p. 905). In early 1980s this fact caused in Europe the slow shift from a scenario where the nation State was the only promoter of technological advance and industrial competitiveness to one where European Union increasingly gained skills and competencies. It's true that this process is still far from being completed, as the actual European crisis clearly shows.[1]

First attempts to launch a sort of cooperation in this field at the European level are quite faraway. The European Atomic

---

[1] A recent analysis of patent and scientific publication data (400 thousand patents, and 2 million and 600 thousand over the period 1986–2010) showed that the cross-border R&D collaborations between EU countries did not increase more than those between European and non-European countries meaning that Europe is mainly a collection of national research communities (Chessa et al., 2013, p. 651).

Human Rights and Emerging Technologies: Analysis and Perspectives in Europe
Daniele Ruggiu
Copyright © 2018 Pan Stanford Publishing Pte. Ltd.
ISBN 978-981-4774-93-2 (Hardcover), 978-0-429-49059-0 (eBook)
www.panstanford.com

Energy Community (EURATOM) founded in 1957 can be deemed the first example of the European integration in research and innovation policy (Grande, 2001, p. 912). However, after the crisis of EURATOM in mid-1960s the cooperation among European countries was prevalently outside the Community framework such as in the aviator sector with the Anglo-French collaboration in Concorde and the Franco-German one relating to Airbus. Moreover, in 1975 the European Space Agency was established in the framework of international organizations, thus, again, outside the European Community's scope. In this period, the Community efforts towards a greater internal cooperation failed. For instance, the Community cooperation in data processing (Eurodata, Unidata) in late 1960s and 1970s did not gain the expected success. In 1980s the decline of European industries in key sectors such as microelectronics and data processing and the US dominance in international markets with the surge of new key actors, such as Japan, led to a favourable climate triggered by the 'technology gap' debate (Grande, 2001, p. 912). With the First Framework Program (FP1) in 1983 the ground for a common research and innovation policy at the Community level was established. In following years, a 'new generation of programs' tried to foster the growth of high-tech industries deemed strategic in fostering the economic development and industry competitiveness (Grande, 2001, p. 912). For example, in early 1980s a new Community's program in R&D in information technology industry, called ESPRIT, became the 'flagship' in technology policy in Europe (Grande, 2001, p. 913). From 1998 with the FP5 to the 2000 Lisbon Treaty the European Union pushed to the integration between States and innovation, as well as their funding. In this cultural climate, research in emerging technologies was increasingly perceived as fundamental in pursuing the goal of making the Europe one of the most competitive economies in the world, together with US and Japan.[2] In this regard in 2000 the European Research Area (ERA) initiative was established at the European Council with the aim of overcoming 'national borders through direct funding, increased mobility, and streamlined innovation policies' (Grande,

---

[2]Lisbon European Council 23–24 March 2000: Presidency Conclusion, point 5, http://www.europarl.europa.eu/summits/lis1_en.htm.

2001, p. 913). The ERA's ambition was of creating the fifth freedom in Europe, namely the freedom of circulation of knowledge and knowledge workers in order to lead to a 'new renaissance' for Europe (de Saille, 2015). Thus ERA was partially implemented through the FP6, leading to major changes with regard to the organization of research in Europe. Not only in research funding we can count a new meaningful trend. To the detriment of the States' domestic jurisdiction, Community acquired new skills and competencies in several areas that are strategic for the regulation of technological developments and the industrial competitiveness. Guidelines for environmental protection, regulation of genetic engineering, the liberalization of communication markets and data protection transformed the European Union in one of the potential key actors on the worldwide scene (Grande, 2001, p. 913). In this framework emerging technologies were increasingly perceived by the EU authorities as an extraordinary mean to fill the gap with other economies. In front of the continuous rise of new promising research fields, such as nanotechnology and synthetic biology, past experiences with GMOs gave the European Union the chance to find a wiser way to cope with the challenges of an ever-evolving world. In this sense the experience with biotechnology sector became a warning for each subsequent policy on emerging technologies. If there are lessons to learn from it, they are the need of a greater involvement of the public (societal trust is fundamental for the success), a greater transparency (the rejection of the 'substantial equivalence' criterion, or of the artificial process/product distinction which would justify the lack of labelling), a new approach to the precautionary principle, which would not neglect it or overestimate it (Metha, 2004). In particular, two directions seem to be recognized in new technology policy in Europe: a better engagement of society at early stage and a greater anticipation of both positive and negative impacts of technology (Mandel, 2009). It is a common knowledge that in emerging technologies actors are distributed at the global level and the success or the failure of a field depends on the rate of contribution of all parties (industries, enterprises, research institutions, researchers, funding organizations, policymakers, non-governmental organizations, civil society, etc.) (Ferrarese, 2010, p. 40ff.). In this sense the trust is the main value at play, a good that needs to be cultivated through the right balance among governance tools, and, consequently, the distrust is the worst misfortune in technology policy (Mandel, 2013, p. 60). For this reason, given the current state of scientific, ethic and regulatory uncertainty related to emerging technologies, the research community shares the opinion that governance should be adaptive, proactive and flexible by escaping forms of regulation too hard influenced by command-and-control style (Mandel, 2013; Widmer et al. 2010; Kearnes and Rip, 2009; Stoke and Bowman, 2012). This governance model should be able to adapt the existing regulations and to evolve following the data gathering in order to respond to changing knowledge and information, to foster self-regulatory attitude of all parties, to widen the stakeholders' participation, thus anticipating the regulator even at early stage (Mandel, 2013, p. 59).

3.2 The EU Governance of Emerging Technologies: The Case of Biotechnology

3.2.1 The case of biotechnology is the first instance of European governance in the field of emerging technologies. It can be deemed as the first, striking case where European Union experimented the precautionary principle, by moving away, under the increasing pressure of European public, from an initially pro-GMOs position in harmony with the World Trade Organization (WTO) and the Organisation for Economic Co-operation and Development (OECD) in 1995 to a more restrictive one (Tiberghien, 2006). This outcome was reached following a long debate and a number of controversies in the European countries where the civil society requested a more involvement in decisions on policy-making (Wynne and Felt, 2007, p. 27; Mendel, 2013, p. 48). As recent studies have shown, by raising further concerns on health risks of genetically modified organisms (Séralini et al., 2014), this turbulence of the public debate is far from ending. This was especially due to the fact that issues raised by biotechnology cannot be reduced to health and safety concerns in a simplistic manner, since they involve practices, cultural values, the protection of biodiversity, as well as strategic choices regarding to what agricultural system a country want to choose (Carro-Ripalda and MacNaghten, 2014; Falkner, 2009). However, this instance has heavily influenced then the governance initiatives developed at the EU level on other emerging technologies, such as nanotechnologies yesterday (e.g. Metha, 2004) and, today, synthetic biology (Stoke and Bowman, 2012).

Genetically modified organisms (GMOs) 'can be defined as living organisms that have been modified by the manipulation of genes (usually by the introduction into organism of new gene coding for specific properties)' (Tiberghien, 2006, p. 6). '[T]his technique allows selected individual genes to be transferred from one organism into another, also between non-related species' (Plan and Van den Eede, 2010, p. 3). Biotechnology arose in late 1970s as one of the most promising fields and as one of the possible drivers of the economic and industrial growth. In the 1980s there were first biotechnological applications with bacteria such as Bacillus subtilis, which was functioned for producing useful proteins (Tiberghien, 2006, p. 6). In this period the EU started out as a liberal supporter of biotechnology in order to foster technology promotion and competitiveness (ibid., pp. 19–20). At that time, it was quite diffuse the belief that a major funding of this research field could have filled the gap with other leaders of the economic worldwide scene, such as the US and Japan. However, the initial pro-GMO attitude in harmony with the position of WTO and OECD was soon abandoned in the subsequent decade. Significantly the GMOs support in Europe decreased in parallel with the development of the sector and the placing into the market of new products. In the 1990s the first generation of GM plants incorporated new features augmenting their efficacy such as herbicide resistance, insect resistance, improved ripening. However, though the next generation focused on higher nutritional features (e.g. golden rice) and pharmaceutical features (ibid., p. 6), the surge of the precautionary principle at the international level and institutional changes in 1997–1999 occurred at the EU level, produced a substantial shift in the approach of EU authorities. First of all, with the Single European Act of 1987, then with the Maastricht Treaty of 1992 and the 1997 Amsterdam Treaty a considerable change occurred in EU policies on GMOs during the Cartagena negotiations between 1999 (Cartagena) and 2000 (Montreal final negotiations). In this period, we need to locate the 1993 European Parliament's resolution calling for a moratorium on the use of GMOs in agriculture. This event, joined with the change in mood of the European public, triggered a deep transformation of EU policies. This led to the incorporation of the precautionary principle in the EU law and, finally, in 1999, to a *de facto* moratorium on new GMO approvals (ibid., p. 20). The *de facto* moratorium regards in particular the cultivation of genetically modified organisms, but it does not involve their importation within the EU, especially animal feed which can freely circulate throughout Europe. Besides, the scope of this moratorium is conditioned by the fact that under current EU legislation while mandatory labelling requirements are provided for GM food, no mandatory labelling requirement is provided for livestock nourished with GM animal feed. Nevertheless, this framework is deemed by many as one of the strictest regulation on the world (e.g. Tiberghien, 2006, p. 14).

This shift was also due to a deep change in Europeans' perception of biotechnologies which was melt with an increasing crisis of legitimation of EU institutions. In this sense the attempt to boost the GMOs market without the needed societal support caused a sudden lack of trust of European public on EU authorities that fatally led to the failure of GMOs in Europe (Metha, 2004). For example, the 2000 Eurobarometer on Biotechnology based on data collected between 1 November and 15 December 1999 showed that Europeans became increasingly opposed to GM foods (Tiberghien, 2006, p. 22). A subsequent survey revealed that this opposition was even consolidated in 2003. Thus, while in 2000 about 80.5% of Europeans was favourable to scientific and technological progress, the opposition to GM food and crops was still high (71%) (ibid.). The support for GMOs rebounds only in few countries, e.g. UK, Sweden, Denmark, as reaction of the EU restrictive regulation. In 2005 95% of the worldwide market was represented by only five countries (the US, Argentina, Brazil, Canada, and China), though GMOs were grown in 21 countries (ibid., p. 6). Here the development of the GMOs sector was possible under the umbrella of the principle of 'substantial equivalence' according to which 'novel food products' 'have the exact same functions as traditional ones' so as they 'should not be regulated differently' (ibid., p. 3). On the other side the application of the precautionary principle in Europe led to one of the most stringent regulations in this field based on 'a more thorough approval mechanism for GMOs products (including both food safety and environmental impact) and mandatory labelling and traceability of GMOs' (ibid., p. 4). Now this legal framework is evolving towards the progressive incorporation of synthetic biology and it will probably lead to further modifications.

In this paragraph I aim at reconstructing the basis of the current governance model existing in Europe. In this regard I will follow the evolution of the EU legislation on biotechnology through the interaction of different agencies and committees such as EGE, EFSA, JRC, SCHENIR with a multitude of reports, advices, Community acts of soft and hard nature that has concurred to lead to the current regulatory framework on biotechnology. By doing this, I will provide a picture of the current EU legislation in this field, from its beginning to the recent Directive (EU) 2015/412, which has become the discussed model of subsequent experiences of governance in the ambit of emerging technologies in Europe. In the case of biotechnology, indeed, we can find out the basis of the current case-by-case approach (used in nanotechnologies and synthetic biology), as well as the origin of the strategic role of several agencies such as EFSA and EGE, that led to the centrality of ethics in the European model of governance. However, we also find the seeds of an increasing attention for novel ways of participation in governance processes, as well as the need for more anticipation, which are the premise of the current *Responsible Research and Innovation* model. However, the progressive evolution of EU legislation on biotechnology towards the incorporation of the rising field of synthetic biology could raise further concerns and won't likely be able to stop controversy in this field.

3.2.2 Genetic modification (GM) 'involves the modification of living organisms with heritable material, independent of the chemical nature of the heritable material and the way in which it has been manufactured' (SCENIHR, 2014, p. 31). In the Cartagena Protocol on Biosafety the established definition of living organisms is: 'any biological entity capable of transferring or replicating genetic material, including sterile organisms, viruses and viroids' [Art. 3(h)].[3] This definition has been embodied by Article 2,1 of Directives 2009/41/EC on contained use of genetically modified microorganisms (GMMs) and Directive 2001/18/EC on the deliberate release of genetically modified organisms (GMOs) (Art. 2,2).[4]

The EU legislation on GMOs is the result of several shakes stemming from European populations that progressively affected EU institutions. This great variability of the European legislative landscape was mainly due to the adoption of a case-by-case approach, which reflected the mood swings of the European public that culminated with the opinion of the European Parliament on 27 October 1993 (EGE, 1993, p. 8; EGE, 1995, p. 1). The Parliament's resolution was thus the apex of a move which triggered a direction change in the EU policy on biotechnology which affected, first, the EU advisory board, the EGE, then, the whole regulation in this field.

In the beginning, the EGE (previously GAIEB) adopted a prudent behaviour in front of movements stemming from the civil society against GMOs, and only in a second stage it decided to take note of the increasing public's pressure against GMOs. By doing this, it gained an increasing centrality in the EU action, leading to the current EU model of governance of emerging technologies.

In its first opinion in 1993 (GAIEB, 1993) it faced the question of the patentability of human (and non-human) genes and partial gene

---

[3] ONU (2000) *Cartagena Protocol on Biosafety to the Convention on Biological Diversity*, adopted in Montreal on 29 January 2000, entered into force on 11 September 2003, https://www.cbd.int/doc/legal/cartagena-protocol-en.pdf. Accessed 21 June 2015.

[4] 'A genetically modified organism (GMO) means an organism, with the exception of human beings, in which the genetic material has been altered in a way that does not occur naturally by mating and/or natural recombination. Techniques of genetic modification referred to in (2001/18/EC) Article 2(2)(a) are inter alia: (1) recombinant nucleic acid techniques involving the formation of new combinations of genetic material by the insertion of nucleic acid molecules produced by whatever means outside an organism, into any virus, bacterial plasmid or other vector system and their incorporation into a host organism in which they do not naturally occur but in which they are capable of continued propagation; (2) techniques involving the direct introduction into an organism of heritable material prepared outside the organism including micro-injection, macro-injection and micro-encapsulation; (3) cell fusion (including protoplast fusion) or hybridisation techniques where live cells with new combinations of heritable genetic material are formed through the fusion of two or more cells by means of methods that do not occur naturally' (SCENIHR, 2015, p. 34 nt. 16).

sequences at the centre of the proposal directive on biotechnological inventions. In this document it did not suggest to impose a complete ban on the production of transgenic animals provided that 'they are used for adequate purposes, not suffer inadequate pain or cause damage for the general public' (ibid., p. 9). Notwithstanding the increasing societal scepticism at the European level, it was not persuaded that biotechnological inventions could be per se 'a threat for biodiversity' (ibid., p. 10). In this sense, in order to recover a climate of trust, the European Commission should adopt 'measures to familiarize the public not only with the scientific and economic side of biotechnology but also with the social, legal and ethical implications' (ibid.). In this document we can see the seeds of current quest for more public engagement. In 1995 the societal resistance against GMOs did not involve the EGE (previously GAIEB) yet. With the subsequent 1995 opinion on labelling of foods derived from biotechnology (GAIEB, 1995), the EGE rightly noted that 'the modern biotechnology, as a technique, used in food production, cannot be regarded in itself as ethical or non-ethical' (GAIEB, 1995, p. 3). However, this did not necessarily lead to a greater transparency, but to state the principle of substantial equivalence between traditional and GM products. In fact, even acknowledging the centrality of consumers' information rights, 'appropriate and understandable information for consumers about food derived from the use of modern biotechnology should go hand in hand with the development of these technologies' (ibid.). This led to conclude that, since 'modern biotechnology, however, does not necessarily substantially change the composition and characteristics of foods [. . .], it is not appropriate to indicate the nature of the process' (ibid.). However, 'information networks [can] provide mechanisms by which producers and/or retailers can give individual consumers, consumer organisations, religious and special interest groups the material they need to reach informed judgements about food derived from the use of modern biotechnology' (ibid., p. 4).

Subsequently the initial prudence of the EGE was abandoned. In 1996 with the opinion on the genetic modification of animals the resistance of the civil society for GMOs was also reflected in the work of the European Union's advisory board by addressing a number of issues such as transparency, ethical acceptability, risks for human health and the environment, proportionality of means and ends, procedures, and possible alternatives at stake. In particular, according to the EGE (previously GAIEB), '[g]reat care should be taken to prevent the release into the environment of genetically modified animals capable of surviving and breeding in the wild' (ibid., p. 5). In this instance a better acknowledgement of the public's role should be gained through the promotion and implementation of ELSI studies and social dialogue that can be deemed as the beginning of the recognition of more participatory pathways in the science and technology issues. It is at this stage that the *de facto* moratorium started taking shape in Europe.

Industrial production systems linked to biotechnology and their problematic coexistence with local farming systems are focused in the 2008 EGE's opinion (EGE, 2008). In this framework the implementation of modern developments in agricultural technology is coupled with the hard task of the solution of UN millennium development goals such as 'the fight against world hunger, and the impact of changing agricultural methods on rural and urban communities' (ibid., p. 15). In a context characterized by the increasing need for food, 'new technologies are necessary for creating and encouraging new methods of agricultural production and trade with a view to developing equitable food distribution capacity and a food-secure world' (ibid., p. 54). This objective needs to be balanced with the two ethical principles of human dignity and justice of the EU Charter of fundamental rights on which responsible action in agriculture aimed at food security and sustainability must be rooted (ibid., p. 48). Both principles of dignity and justice converge on the right to food of the 1948 Universal Declaration, market equity and intergenerational justice. In this framework food security, food safety and sustainability are deemed fundamental to achieve: healthy food for the population, a better management of food resources with a minor food waste and lower food prices, the soil and water protection. As regards food safety EGE requested that EU food safety standards are based on scientific data only, in particular thanks to the support of member States and EU relevant bodies such as EFSA (ibid., p. 61). In this context it must be considered an integrated approach to impact assessment which takes into account both environmental and social implications of the use of new technologies in agriculture. Thus, implications as regards the use of industrial production systems linked to biotechnology, their impact on local agricultural systems, as well as on traditional knowledge, the dependence farmer-GM seeds, the impact of products related to the use of GMOs (e.g. pesticides) on health and the environment, information for consumers, etc. should take place at this level. The EGE's work leads to implement a given type of biotechnology as anticipatory framework of ethical, legal and social implications of science and technology. In particular, modern biotechnology can be used 'where appropriate for [achieving] sustainable use of soil [...], reduction of spray pollution, active ingredients in herbicides and CO_2 emissions, [...] plant tolerance to high salinity, [...] better water management and prevention of water pollution' (ibid., p. 63). This leads the EGE to build an ambitious programme, which should relaunch European economies in the global market. Since 'the EU has been a world player in agricultural trade, both as an importer and exporter of agricultural products', this 'urges the EU to take the identified priorities of food security, safety and sustainability as ethical principles in its role in the global economy' Thus, '[t]o achieve the goals identified, global agricultural trade needs an ethical framework'. This can be deemed the beginning of the ethics centrality in EU governance of emerging technologies. In this context '[s]olidarity, justice and free and fair trade in agricultural products and technologies are priorities. The group therefore advocate[d] that the EU promotes a market system that includes aid for trade, fair (protected) trade and free trade and emphasises the importance of aid for technological development as laid down in the UN millennium development goals' (ibid., p. 65). To achieve this objective, biotechnology can still play a central role, though within a renewed ethical framework where dignity, justice, sustainability are deemed as primary.

In this changing framework EU legislation evolved in a schizophrenic manner and now it is destined to embed also the field of synthetic biology. Since 1990s the European Union regulated biotechnology with regard to three main areas: contained use, environmental release, the placing on the market of food and feeds. Finally, the EU regulated also the movement of GMOs between the Union and non-EU countries and vice versa. The EU legislation on GMOs has been constantly updated several times during the last 25 years, especially between 2000 and 2003, losing in some aspects in clarity (Fontana, 2010, p. 1). Recently, the EU system of authorization and marketing of genetically modified food and feed has been further amended by enabling member States to have the final say on the introduction of the marketed GMOs in their national territory. Finally, as said, the EU regulatory framework on GMOs is being reshaped in view of including also recent developments of synthetic biology within the same legislation (SCENIHR, 2014, 2015). With this ever-evolving legislation two main goals are pursued: to maintain a high level of protection of health and the environment according to treaties provisions (Arts. 168, 191 TFEU) and to ensure the free movement of safe and healthy genetically modified products within the Union (Arts. 26ff., 34–36 TFEU) (Plan and Van den Eede, 2010, p. 3). However, the incorporation of synthetic biology within the same regulatory umbrella might have a considerable impact on the whole regulation of both sectors in future.

(i) Actually the contained use of genetically modified microorganisms (GMMs) and the deliberate release of genetically modified organisms (GMOs) are regulated within the European Union by Community Directives 2009/41/EC on contained use of GMMs[5] (i.e. in a closed environment, such as a laboratory) and Directive 2001/18/EC[6] on the deliberate release of GMOs (without any confinement measure). Under these directives a case-by-case risk assessment is provided for the above-mentioned activities in order to limit their possible negative impact on the

---

[5] European Union (2009) Directive 2009/41/EC of the European Parliament and of the Council of 6 May 2009 *on Contained Use of Genetically Modified Micro-organisms, Official Journal of the European Union*, L125, 21/05/2009, pp. 75–97, http://eur-lex.europa.eu/legal-content/EN/TXT/PDF/?uri=CELEX:32009L0041&from=EN. Accessed 24 March 2015.

[6] European Union (2001) Directive 2001/18/EC of the European Parliament and of the Council of 12 March 2001 *on the Deliberate Release into the Environment of Genetically Modified Organisms and Repealing Council Directive 90/220/EEC, Official Journal of the European Union*, L106, 17/04/2001, http://eur-lex.europa.eu/resource.html?uri=cellar:303dd4fa-07a8-4d20-86a8-0baaf0518d22.0004.02/DOC_1&format=PDF. Accessed 24 March 2015.

environment and human health. The two directives establish a general risk assessment procedure for conventional genetically modified organisms partially based on the comparative analysis of genetically modified organisms and non-genetically modified counterparts (Fontana, 2010, pp. 1–2). However, since the data related to non-natural organisms could be lacking, performing the comparative analysis could be harder (Pauwels et al., 2012, p. 6).

Several modifications occurred in the field of the contained use until a unique act has re-ordered it in the current Directive 2009/41/EC which is thus not a truly new rule (Fontana, 2010). The matter was initially regulated by the Council Directive 90/219/CEE and by the Council Directive 98/81/CE until the adoption of the 2009 text which has resumed the whole multitude of amendments occurred during the time (ibid., p. 2). Under this Directive entire research and even simple storage activity involving GMMs, in which the contact with the population, as well as the environment, is avoided, are regulated. The reason of provisions on the contained use is expressed in the 'whereas' of the Directive, stressing that the objective of the economic expansion through the development of biotechnology can be pursued only if the possible negative implications for health and the environment are limited (fourth and fifth whereas). To fulfil these goals appropriate and common measures for the contained use of genetically modified microorganisms (GMMs) are adopted (Art. 1).

Directive 2009/41/EC mainly adopt an approach focused on safety. According to the fifth whereas, for example, '[t]he contained use of GMMs should be such as to limit their possible negative consequences for human health and the environment, due to attention being given to the prevention of accidents and the control of waste'. Thus, risk assessment is strategic and it should be made according to a case-by-case basis (ninth whereas) and to a classification in line with international practice (tenth whereas). However, in order to meet principles of good microbiological practice and good occupational safety and hygiene, '[p]eople employed in contained used should be consulted in accordance with requirements of the relevant Community legislation' (nineteenth whereas).

For the purpose of the 2009 Directive by GMM you have to mean 'a micro-organism in which the genetic material has been altered in a way that does not occur naturally by mating and/or natural recombination' (Art. 2,1(b)). Instead, by contained use you have to mean 'any activity in which microorganisms are genetically modified or in which such GMMs are cultured, stored, transported, destroyed, disposed of or used in any other way, and for which specific containment measures are used to limit their contact with, and to provide a high level of safety for, the general population and the environment' (Art. 2,1(c)). In this framework, member States are invited to ensure that all appropriate measures are taken to avoid that adverse consequences on human health and the environment arise from the contained use (Art. 4,1). To that end any person, before undertaking the contained use of a GMM in a closed environment, must forward a notification to the competent authority in order to assess that the installation is appropriate for the purposes of carrying out in a manner that does not present hazards to human health and the environment (Art. 4,2). Thus, risk assessment mainly rests on the user's responsibility and competent authorities only have to check how it is ensured. In this regard risk assessment procedures through a risk classification from negligible risk to high risk activities are provided (Art. 4,3). While under the class 1 (no or negligible risk activity) the contained use can proceed without any notification (Art. 7), the subsequent class 2 (low risk) requires that the notification is supplied by the information listed in Annex III (Art. 8). On the basis of the classes 3 and 4 (moderate and high risk activities) the contained use requires that the compulsory notification the information as above and the prior consent of the competent authority to whom the applicant has to communicate in writing (Art. 9).

In 2000 the Commission delivered a guidance for the risk assessment outlined in the Annex III of Directive 90/219/CEE on the contained use of genetically modified microorganisms.[7] This guidance 'address[es] the magnitude of potential hazards and

---

[7] European Commission (2000) Commission Decision 2000/608/EC of 27 September 2000 *Concerning the Guidance Notes for Risk Assessment Outlined in Annex III of Directive 90/219/EEC on the Contained Use of Genetically Modified Micro-Organisms* (notified under document number C(2000) 2736), *Official Journal of the European*

adverse effects of genetic engineering on human health and the environment and on the probability that these events will lead to hazards (exposure chain)' and is still operative (both for genetic engineering and synthetic biology). The EU regulatory framework on genetically modified organisms is process-based and refers to tools and approaches relying on, among others, three elements: 1) recombinant DNA techniques, 2) the direct introduction of heritable material into an organism and 3) cell fusion or hybridization techniques (Annex I, Part A of Directive 2009/41/EC and Annex I A Part I of Directive 2001/18/EC). However, there is debate on whether this process-based approach should be applied for the regulatory oversight of given novel techniques such as new plant-breeding techniques (NPBT) producing plants that do not contain recombinant DNA in their genome, by challenging thus the current legal definition of GMO (SCENIHR, 2014, p. 16; (Lusser et al., 2011). Up to now, 'procedures in Europe include the evaluation of substantial differences between GM crops and their non-GM counterparts, molecular characterisation, toxicity and allergenicity studies and the assessment of the environmental impacts and unintended effects' that contribute in rise the registration costs (SCENIHR, 2014, 16 nt. 1).

An emergency plan to cope with possible accidents should be provided by the user 'where failure of the containment measures could lead to serious danger, whether immediate or delayed, to humans' (Art. 13,1(a)). In this regard the competent authority has to organize inspections and other control measures to ensure that the user complies with the provisions of the Directive (Art. 16). Member States through the competent authority have to ensure a wide circulation of information. Every three years the Commission shall publish a summary report on class 3 and class 4 contained uses notified (Art. 17).

Directive does not apply to GMMs placed on the market according to the Directive 2001/18/EC on the deliberate release of genetically modified organisms (Fontana, 2010, p. 4). The main and not yet solved problem is whether the Directive covers clinical trials with

---

Union, L258, 12.10.2000, pp. 43–48, http://eur-lex.europa.eu/legal-content/EN/TXT/PDF/?uri=CELEX:32000D0608 &from=EN. Accessed 1 April 2015.

GMOs or they should fall within the scope of the provisions on the environmental release (ibid.). In fact, some member States consider clinical trials with GGMs in clinical settings as deliberate release while some others consider them as contained use.

(ii) The core of the EU legislation is made up of the directives on the environmental release and on food and feed. With these two directives we are at the centre of the movement that shacked the European public opinion until today.

In the Directive 2001/18/EC more clearly emerges the EGE's influence. First 2000 Directive protects human dignity by stating that '[w]hen defining genetically modified organism for the purpose of this Directive, human beings should not be considered as organisms' (fifteenth whereas). On the basis of the Directive '[a] case-by-case environmental risk assessment should always be carried out prior to a release' (nineteenth whereas). In the case of environmental release '[t]he introduction of GMOs into the environment should be carried out according to the step by step principle. This means that the containment of GMOs is reduced and the scale of release increased gradually' (twenty-fourth whereas). In this case '[a]ny person, before undertaking a deliberate release into the environment of a GMO, or the placing on the market of GMOs, as or in products, where the intended use of the product involves its deliberate release into the environment, is to submit a notification to the national competent authority' (thirty-second whereas). Thus, also in this instance risk assessment, related costs partially weigh on who wants to release or place on the market a GMO. This means that the management of environmental risks, at least in its first stage, risks being out of the hands of public authorities. Finally, '[i]n order to ensure that the presence of GMOs in products containing, or consisting of, genetically modified organisms is appropriately identified', adequate information rights should be ensured either on a label or in an accompanying document (fortieth whereas). This is functional 'to implement a monitoring plan in order to trace and identify any direct or indirect, immediate, delayed or unforeseen effects on human health or the environment of GMOs' (forty-third whereas).

Directive 2001/18/EC on the deliberate release of genetically modified organisms (GMOs)[8] has replaced the Directive
90/220/EEC adopted in 1990 and applies when no precise
confinement measure to restrict the contact with the population
as well as the environment is provided (ibid., p. 2). Under the
current Directive 2001/18/EC on the deliberate release two types
of activities regarding to GMOs are covered: 'placing on the market
genetically modified organisms as or in products' or 'carrying out
the deliberate release of genetically modified organisms for any
other purposes than placing on the market within the Community'
(Art. 1). Thus, it regulates the environmental release of GMOs, be
it for experimental purposes (field of trials) or for commercial
purposes (placing on the market) (Plan and Van den Eede, 2010,
p. 5). It provides the procedure for authorizing the release into the
environment of genetically modified organisms. The competence
in authorizing or rejecting the application on the release into the
environment of genetically modified organisms for experimental
purposes is of the competent national authority that has received
the notification. Thus, the process has merely national nature.
The applicant, called the 'notifier', must submit an application
(notification) with which he notifies his intention of releasing GMOs
into the environment for experimental purposes which must include
the evaluation of the environmental risk carried out by the notifier.
In this document the notifier must communicate, in particular:
information relating to the conditions of release and the potential
receiving environment, information on the interactions between the
GMO(s) and the environment, a plan for monitoring in accordance
in order to identify effects of the GMO(s) on human health or the
environment, information on control, remediation methods, waste
treatment and emergency response plans, (Art. 6,2). In the case of
positive response, the authorization will apply only in the member
State where the notification has been submitted. In view of ensuring

---

[8]European Union (2001) Directive 2001/18/EC of the European Parliament and
of the Council of 12 March 2001 *on the Deliberate Release into the Environment
of Genetically Modified Organisms and Repealing Council Directive 90/220/EEC*,
Official Journal of the European Union, L106, 17/04/2001, http://eur-lex.europa.eu/
resource.html?uri=cellar:303dd4fa-07a8-4d20-86a8-0baaf0518d22.0004.02/
DOC_1&format=PDF. Accessed 24 March 2015.

the due transparency there is a public register provided for where information about deliberate field trials and placing on the market of genetically modified organisms are published.[9]

Differently from the procedure provided for the authorization for experimental purposes, the procedure of the authorization for the placing on the market genetically modified organisms has not national scope, but it has a Community one by involving all member States. This different scope can be explained by the fact that the authorization for the placing on the market implies the free movement of authorized products throughout the European Union territory (ibid., p. 6). This authorization system was amended in 2015.

The notification is submitted to the competent national authority of the concerned member State. It must include, in particular, a number of information such as detailed information on the GMO for the detection and identification of particular GMO products, the environmental risk assessment, a proposed period of consent which must not exceeding 10 years, a post-market monitoring plan, a proposal for labelling including the words 'This product contains genetically modified organisms' (Art. 13,2). Once having received the notification, the competent national authority must issue an opinion ('assessment report') within 90 days after its receipt (Art. 14). In the event of favourable opinion, the member State must inform the other member States via the European Commission. In this case the European Commission requests the European Food Safety Authority (EFSA) to provide an environmental risk assessment which includes: the identification of any characteristics of the GMO(s) which may cause adverse effects, the evaluation of the potential consequences of each adverse effect, the evaluation of the likelihood of the occurrence of each identified potential adverse effect, the estimation of the risk posed by each identified characteristic of the GMO(s), the application of management strategies for risks resulting from the deliberate release or placing on the market of GMO(s), the determination of the overall risk of the GMO(s) EU authorization (ibid.). In the case of favourable opinion of the EFSA, the Commission must present a draft decision to the

---

[9]See http://gmoinfo.jrc.ec.europa.eu/. Accessed 29 September 2017.

Regulatory Committee composed by representatives of member States for an opinion. If the Committee is favourable, the Commission adopts the decision. If not, a more complex procedure is provided. The draft decision must be submitted to the Council of Ministers for its adoption or rejection. If the Council does not act in three months, the Commission shall adopt the decision (ibid.). The consent for placing on the market is given for a maximum of 10 years and is renewable (Art. 15,4). The consent is valid throughout the territory of the whole Union. However, in the case of any objection from any member State a safeguard clause is provided for (Art. 23). The State, which recurs to the safeguard clause, can ban the placing on the market in its territory of an approved GMO on the basis of the availability of new safety information. An EU decision must be taken on regard to this provisional ban following the consultation of EFSA and member States (ibid., p. 7). During the authorization process the public is informed. A specific register is provided for[10] whose content is ruled by the Commission Decision 2004/204/EC.[11]

Risk assessment, given by agencies such as EFSA, is thus a strategic element for the public acceptance of a given technology. In this context independence and transparency are indispensable for societal trust. In this sense, conflicts of interests can undermine independence of EU agencies. Some concerns were raised, for example, as regards EFSA's work. Since experts of EFSA are all highly qualified in fields of medicine, nutrition, toxicology, biology, chemistry, namely in disciplines which request expertise that often can be developed only within the industry sector, it is not easy to ensure the due independency. In fact, 'most research is industry funded which makes it nearly impossible finding an expert who is not involved to varying degrees in projects funded by, or involving, industry'. In this regard '[u]nlike some other risk assessment bodies, EFSA relies heavily on external expertise from academia or research

---

[10]See http://gmoinfo.jrc.ec.europa.eu/. Accessed 29 September 2017.
[11]European Commission (2004) Commission Decision 2004/204/EC of 23 February 2004 *Laying Down Detailed Arrangements for the Operation of the Registers for Recording Information on Genetic Modifications in GMOs, Provided for in Directive 2001/18/EC of the European Parliament and of the Council, Official Journal of the European Union*, L65, 3/3/2004, pp. 20–22, http://eur-lex.europa.eu/legal-content/EN/TXT/PDF/?uri=CELEX:32004D0204&from=EN. Accessed 18 June 2015.

organizations (50% of the experts) and national risk assessment bodies to generate its scientific advice' (BEUC, 2014, p. 2).[12] For, 'it is at the same time evident that in practice scientists of good repute who could serve on staff committees of agencies will always be or have been involved in industry or national affairs' (Vos, 2016, p. 219). For these reasons, specific policies are being developed and implemented in view of ensuring the credibility of EU agencies such as EFSA (Vos, 2016).

In accordance with the precautionary principle Annex II of Directive identifies six steps in the environmental risk assessment of GMOs provided for by Articles 4 and 13: (1) problem formulation including hazard identification, with the identification of characteristics which may cause adverse effects; (2) hazard characterization, with the evaluation of the potential consequences (magnitude) of each adverse effects, if it occurs; (3) exposure characterization, with the evaluation of the likelihood of the occurrence of each identified potential adverse effects; (4) risk characterization which includes the estimation of the risk (magnitude x likelihood) posed by each identified characteristic of the GMO; (5) risk management strategies; (6) an overall risk evaluation (SCENIHR, 2015, pp. 21–22).

Thus far, the market authorization process primarily involves GM plants, while to a lesser extent GMM. At the international level the comparative approach for GMOs and GMMs is well accepted as baseline for the assessment of risks as regards human and environmental health. The comparative approach means to make a comparison between a genetically modified organism or microorganism and its non-modified counterpart in order to facilitate the process of risk assessment by referring to existing and comparable data. In the Annex II the comparator is defined as similar organisms 'with similar traits and their interaction with similar environments' produced without the help of genetic modification as defined in Directive 2001/18/EC and for which there is a well-established history of safe use (Annex II, C.1.). 'According to Directive 2001/18/EC, the general principle followed

---

[12]Recently, media raised a charge according to which EU report on weedkiller safety copied text from Monsanto study. See https://www.theguardian.com/environment/2017/sep/15/eu-report-on-weedkiller-safety-copied-text-from-monsanto-study. Accessed 5 October 2017.

when performing environmental risk assessment is to identify the characteristics of the GMO and its use which has the potential to cause adverse effects and should be compared to those presented by the non-modified organism from which it is derived and its use in similar situations' (ibid., p. 21).

Finally, for the deliberate release of genetically modified plants, the European Commission European Food Safety Authority (EFSA) provided some guidelines such as the EFSA Panel on Genetically Modified Organisms and the Guidance on environmental risk assessment of genetically modified plants (SCENIHR, 2014, p. 19).

As noted above, in 2015 the EU intervened again into this magmatic regulatory framework, which is continuously evolving. With Directive (EU) 2015/412[13] the Commission amended Directive 2001/18/EC by giving the final say to the member States on the introduction into their territory of genetically modified organisms regularly authorized through the single authorization process. This amendment seems thus to merely photograph the on-going lack of consensus existing in Europe on this matter, which evidently was not stopped notwithstanding multiple direction changes in legislation. This last change is likely leaving unsolved the issues at stake.

Currently Directive 2001/18/EC and Regulation 1829/2003 represent the overall legal framework under which the authorization process of genetically modified organisms is regulated within the EU. In order to obtain the authorization for being placed into the Union market in accordance with the Annex II of the Directive 2001/18/EC any GMO which is intended for cultivation has to undergo to a risk assessment process before being placed in the market by 'taking into account the direct, indirect, immediate and delayed effects, as well as the cumulative long-term effects, on human health and the environment'. In this sense, the aim of the authorization process is 'to ensure a high level of protection of

---

[13] European Union (2015) Directive (EU) 2015/412 of the European Parliament and of the Council of 11 March 2015 *Amending Directive 2001/18/EC as Regards the Possibility for the Member States to Restrict or Prohibit the Cultivation of Genetically Modified Organisms (GMOs) in Their Territory, Official Journal of the European Union*, L68, 13.3.2015, pp. 1–8, http://eur-lex.europa.eu/legal-content/EN/TXT/PDF/?uri=CELEX:32015L0412&from=EN. Accessed 18 June 2015.

human life and health, animal health and welfare, the environment and consumer interests, whilst ensuring the effective functioning of the internal market' (second whereas of Directive (EU) 2015/412). In this context the precautionary principle applies. While the placing on the market and the import of GMOS is regulated under EU legislation, the cultivation of GMOs 'require[s] more flexibility in certain instances as it is an issue with strong national, regional and local dimensions, given its link to land use, to local agricultural structures and to the protection or maintenance of habitats, ecosystems and landscapes' (sixth whereas). In the past in order to restrict or prohibit the cultivation of GMOs, member States had recourse to the safeguard clauses and emergency measures pursuant to Article 23 of Directive 2001/18/EC and Article 34 of Regulation (EC) No. 1829/2003. For this reason now the European Union, in accordance with the principle of subsidiarity, has meant, without prejudice to above-mentioned Articles 23 and 34, to grant member States more flexibility to decide whether or not they wish to cultivate GMOs on their territory without affecting the risk assessment provided in the authorization system either in the course of the authorization procedure or thereafter, and independently of the measures that States cultivating GMOs are entitled to take by application of Directive 2001/18/EC to avoid the unintended presence of GMOs in other products (eighth whereas). In this regard the Commission Recommendation 2010/C200/01[14] provides guidelines for avoiding cross-pollination between crops of member States that allow the GMOs cultivation and the neighbouring those that prohibit it (see below). In this regard the geographical scope of the notification/application must be updated in accordance with the territorial limitation within the Union requested by member States that do not want GMOs. Thus, member States can adopt reasoned measures prohibiting or restricting in all or part of their territory of GMOs crops once authorized on the

---

[14] European Commission (2010) Commission Recommendation 2010/C200/01 of 13 July 2010 on *Guidelines for the Co-existence Measures to Avoid the Unintended presence of GMOs in Conventional and Organic Crops, Official Journal of the European Union* C200/1, 22/07/2010 P. 0036–0047, http://ec.europa.eu/food/plant/docs/plant_gmo-agriculture_coexistence-new_recommendation_en.pdf. Accessed 2 July 2015.

basis of grounds distinct from those assessed under Community laws (thirteenth whereas and Article 26b Directive 2001/18/EC). Those grounds may be related, individually or in combination, to environmental or agricultural policy objectives (such as the need to protect the diversity of agricultural production and the need to ensure seed and plant propagating material purity), or town and country planning, land use, socioeconomic impacts, coexistence and public policy but cannot be those already covered by the risk assessment process of the authorization (in particular health and the environment) (Art. 26b,3). The restrictions or prohibitions adopted must refer to the cultivation, and not to the free circulation and import, of genetically modified seeds and plant propagating material as, or in, products and of the products of their harvest (sixteenth whereas and Article 26,8). This confirms the fact that the current GMOs moratorium refers to the cultivation, not to the importation of GM products in Europe. Once amended the geographical scope of the notification/application, the member State must inform the Commission which shall amend the decision of authorization accordingly (Art. 26b,6). No later than 3 April 2019 the Commission must present a report to the European Parliament and the Council of Ministers on the application of these provisions and one on the actual remediation of the environmental damages that might occur due to the GMOs cultivation. The revision of the Annexes to Directive 2001/18/EC as regards the environmental risk assessment is forecasted. In this sense, we can expect that this matter shall be further amended in the near future.

(iii) The other main regulatory act is the Regulation 1829/2003 on genetically modified food and feed.[15] Before its entry

---

[15] European Union (2003) Regulation (EC) 1829/2003 of the European Parliament and the Council of 22 September 2003 on *Genetically Modified Food and Feed*, *Official Journal of the European Union*, L268/1, 18.10.2003, pp. 1–23, http://ec.europa.eu/food/food/animalnutrition/labelling/Reg_1829_2003_en.pdf. Accessed 9 May 2015. European Union (2003) Regulation (EC) No 1830/2003 of the European Parliament and of the Council of 22 September 2003 *Concerning the Traceability and Labelling of Genetically Modified Organisms and the Traceability of Food and Feed Products Produced from Genetically Modified Organisms and Amending Directive 2001/18/EC*, *Official Journal of the European Union*, L 268, 18.10.2003, pp. 24–28.

into force which introduced a single authorization process, in accordance with Regulation 258/97 on 'novel foods and novel food ingredients' two procedures of entry for genetically modified organisms were provided: a 'simplified procedure' based on mere notification of the manufacturer and a 'formal procedure'. This prior authorization system led to several interventions of the European Court of Justice paving the way to several controversies that affected the evolution of the EU legislation (Ruggiu, 2012a).

(A) Before 2003, under Regulation 258/97 on 'novel foods' in the event of marketing of foods derived from, but not containing, genetically modified organisms the 'simplified procedure' applied.[16] According to it a new product was supposed to be 'substantially equivalent' to foods or food ingredients already on the market. In this instance the manufacturer could simply make a simple notification which merely informed the essential information about the product. (I) In this case, when it was held on the basis of new information that the product marketed with the 'simplified procedure' was endangering human health or the environment, the member States could restrict or suspend, in extremis, in their territory the trade and the use of the food in question, by using the so-called 'safeguard clause',[17] following they immediately should inform the Commission. After merely formal proceedings, the Commission decided on the validity of the unilateral safeguard measures and provided for the restoration of a common level of protection within Community borders. Eventually, if the State did not comply with the Commission decision, the Court had to intervene. (II) Instead, according to the 'formal procedure', the Commission authorized the introduction of new food products containing genetically modified organisms in the EU market by giving *ab initio* (earlier) an assessment on their safety for public

---

[16]Cfr. *Association Greenpeace France and others v. Ministère de l'Agricolture et de la Pêche and others* (Case C-6/99) judgement of 21 March 2000 *European Court Reports*, 2003, p. I-01651.

[17]Cfr. *Monsanto Agricoltura Italia SpA and others v. Presidenza del Consiglio dei Ministri and others* (Case C-236/01) judgement of 9 September 2003 *European Court Reports*, 2003, p. I-08105.

health.[18] At the stage of the marketing authorization by the Commission, the precautionary principle was and is fully implicated.

(B) Since the past double authorization procedure did not stop controversies by some member States that were against to the use of GMOs, the Regulation was finally amended. With the entry into force of the 2003 regulation on genetically modified food and feed the situation radically changed and a single authorization process was introduced. Within the new framework risk assessment equally weighs on producers and EU authorities.

Current Regulation 1829/2003 on the authorization and marketing of genetically modified food and feed applies to either food containing, consisting, or produced by GMOs or feed containing, consisting, or produced by GMOs (Art. 2). In order to ensure a high level of protection of health and the environment, as well as the effective functioning of the internal market, it provides for the Community procedure for granting consent for placing on the EU market of GMOs food and feed (Plan and Van den Eede, 2010, p. 10). The authorization procedure for GMOs food and feed provides a single risk assessment process conducted by EFSA and a single risk management process under the joined responsibility of the Commission and member States through the regulatory committee procedure. The producer who means to place on market a GMO product belonging to food and feed must to submit an application to the competent national authority which must to inform EFSA within 14 days without delay. The application must include *inter alia* the copy of safety studies carried out for demonstrating that GMO products do not have adverse effects for human health, animal health and the environment, methodologies used for detection, identification and sampling of GMO food and feed, samples of GMO food and feed and control samples (Art. 5). The application and all its attached documentation must be made available to EFSA under whose responsibility the risk assessment on human, animal safety, as well as that on environmental risk, must be carried out. Inclusive processes with regard to civil society are provided. Within 6 months

---

[18] Regulation (EC) 1829/2003 Article 4 and Directive 2001/18/EC. With regards to the deliberate release of GMOs authorization procedure see Christoforou (2008, 200 ss. e 207 ss.).

EFSA must adopt its opinion on which the public has the opportunity to make comments (Art. 6). For granting the authorization, the detection method submitted for the GM food/feed must be validated by the Community reference laboratory, which is the European Commission's Joint Research Centre (JRC) (Art. 5(f)). This is an integral part of the whole authorization process. Consequently, a GM food/feed cannot be authorized in the EU before a relevant detection method has been validated (ibid., p. 9).

Annex I of Regulation 641/2004[19] provides that detailed information on detection methods that shall be supplied by the applicant in order to verify the fitness of method (ibid., p. 14). This information includes information on the method as such and on the method testing carried out by the applicant. National reference laboratories assist the Community Reference Laboratory (i.e. JRC) in this process of method validation in accordance with Regulation 1981/2006.[20] It is possible to know the applications received by EFSA and its adopted opinions on the EFSA website.[21] Within 3 months after receiving the EFSA opinion, the Commission must submit a draft decision for approval to the Standing Committee on the Food Chain and Animal Health, which is composed by representatives of all member States (Art. 7). If the opinion is favourable the Commission adopts the decision. If it is not, or in the event of rejection by a qualified majority, the draft decision must be

---

[19]European Commission (2004) Commission Regulation (EC) No 641/2004 of 6 April 2004 *on Detailed Rules for the implementation of Regulation (EC) No 1829/2003 of the European Parliament and of the Council as Regards the Application for the Authorisation of New Genetically Modified Food and Feed, the Notification of Existing Products and Adventitious or technically unavoidable presence of genetically modified material which has benefited from a Favourable Risk Evaluation, Official Journal of the European Union*, L102, 7.4.2004, pp. 14–25, http://eur-lex.europa.eu/legal-content/EN/TXT/PDF/?uri=CELEX:32004R0641&from=EN. Accessed 18 June 2015. On this see also http://gmo-crl.jrc.ec.europa.eu/default.htm. Accessed 30 June 2015.

[20]European Commission (2006) Commission Regulation (EC) No 1981/2006 of 22 December 2006 *on Detailed Rules for the Implementation of Article 32 of Regulation (EC) No 1829/2003 of the European Parliament and of the Council as Regards the Community Reference Laboratory for Genetically Modified Organisms, Official Journal of the European Union*, L368, 23.12.2006, pp. 99–109, http://eur-lex.europa.eu/legal-content/EN/TXT/PDF/?uri=CELEX:32006R1981&qid=1434625654447&from=EN. Accessed 18 June 2015.

[21]See http://www.efsa.europa.eu/en/panels/gmo.htm. Accessed 26 June 2015.

submitted to the Council of Ministers for its adoption or rejection. In this case the decision is taken with a qualified majority. If the Council does not take any decision within three months or does not obtain the qualified majority, the Commission shall adopt the decision. Once authorized, the authorization is valid throughout the whole Union and can be renewed for 10 years. In this case the GMO food and feed must respect the labelling requirements including the words 'genetically modified' (Art. 13).

Also in this case there is a register of genetically modified food and feed.[22]

Under Regulation 1830/2003 traceability and labelling require-ments are provided for products containing, or consisting in, or derived from GMOS.[23] Under Regulation 1830/2003 'traceability' is defined as 'the ability to trace GMOs and products produced from GMOs at all stages of their placing on the market through the production and distribution chains' (Art. 3). Traceability is mandatory in order to facilitate the control and verification labelling claims, the targeted monitoring of potential effects on health and the environment and thus the prompt withdrawal of products where unforeseen adverse effects on human, animal health, as well the environment occur, the implementation of risk management measures in accordance with the precautionary principle (third and fourth whereas). Traceability provisions address all who produce and place on the market or receive a product placed on the market within the EU in order 'to identify their suppliers and the companies to which the products have been supplied' (ibid., p. 11).

Traceability requirements vary from products containing or consisting in GMOs or deriving from GMOs. In the first instance the producer must provide in writing the operator who receive the product the following information: an indication that the product

---

[22] See http://ec.europa.eu/food/dyna/gm_register/index_en.cfm. Accessed 27 June 2015.

[23] European Union (2003) Regulation (EC) No 1830/2003 of the European Parliament and of the Council of 22 September 2003 *Concerning the Traceability and Labelling of Genetically Modified Organisms and the Traceability of Food and Feed Products Produced from Genetically Modified Organisms and Amending Directive 2001/18/EC, Official Journal of the European Union*, L268/24, 18.10.2003, http://eur-lex.europa.eu/LexUriServ/LexUriServ.do?uri=OJ:L:2003:268:0024:0028:EN:PDF. Accessed 18 June 2015.

contains or consists in GMOs and the unique identifier assigned to those GMOs (Art. 4). In the latter the producer must provide in writing the operator who receive the product the following information: an indication of each of food ingredients which is produced from GMOs and an indication of each of feed materials which is produced from GMOs (Art. 5). In both instances the operator must hold all information for a period of five years (Arts. 4,4; 5,2). Details of the system of assignment of a unique identifier are ruled by Regulation 65/2004[24] (ibid.).

Products authorized under Directive 2001/18/EC and under Regulation 1829/2003 are subjected to mandatory GM labelling requirements according to Regulation 1829/2003 and Regulation 1830/2003 (ibid., p. 11). The aim of this regulation is to inform consumers and users of the product and to make an informed choice. This choice of EU authorities' responses to pressing pressure of European civil society and is in accordance with objectives of founding treaties (i.e. consumer and health protection). Regulation 1830/2003 provides that products containing or consisting in genetically modified organisms must have indicated in label the genetically modified origin through the words 'This product contains genetically modified organism(s)' or the equivalent indication 'This product contains genetically modified [name of organism(s)]' (Art. 4,6). Analogously Regulation 1829/2003 lays down specific requirements for GM food and feed. Articles 12 and 13 provides mandatory labelling for genetically modified food whose product must contain the words 'genetically modified' or 'produced from genetically modified [name of the ingredient(s)]' 'irrespective of the detectability of DNA or protein resulting from genetic modification' (twenty-first whereas Regulation 1829/2003). '[B]efore 2003 GM labelling requirements were based on the detection of DNA or protein resulting from genetic modification' (ibid.). Thus, there was a considerable change that caused a foreseeable reaction by producers requesting no labelling requirements for new plant-

---

[24]European Commission (2004) Commission Regulation (EC) No 65/2004 of 14 January 2004 *Establishing a System for the Development and Assignment of Unique Identifiers for Genetically Modified Organisms, Official Journal of the European Union*, L10, 16/1/2004, pp. 5–10, http://eur-lex.europa.eu/legal-content/EN/TXT/PDF/?uri= CELEX:32004R0065&from= EN. Accessed 18 June 2015.

breeding techniques that do not lead to plants which are free of transgenes. These latter concerns related to higher costs for plants classified as GMOs have led to an official report which could open up further changes in EU regulation on GM labelling in future (Lusser et al., 2011, p. 10).

Same labelling provisions of Regulation 1829/2003 apply to animal feed, including any compound feed containing or being produced from genetically modified feed (e.g. soya or maize), so as to give right and complete information to livestock farmers on the composition and properties of feed (Arts. 24 and 25). No information to consumers with regard to the livestock nourished with this type of products is requested under the EU law. Thus, while mandatory labelling is established for animal feed, no labelling requirement is provided for livestock nourished with GM animal feed. This fact not only leaves open concerns stemming from consumer organizations, but also puts the question of the difficult coexistence between this agricultural production system with different agriculture models. Also in this instance, before 2003 no requirements for labelling of feed was requested (Plan and Van den Eede, 2010, p. 11).

Cross-pollination during cultivation, or adventitious or techni-cally unavoidable mix of genetically modified and non-genetically modified organisms during harvesting, storage, transport or pro-cessing may unintentionally contaminate conventional products (ibid., p. 12). For this reason, Articles 12 and 24 of Regulation 1829/2003 provide a threshold of 0.9% of the presence of GM material in food and feed below which provisions on traceability and labelling do not apply, provided that its presence is adventitious or technically unavoidable. This means that GMO crops can easily contaminate nearby farms also within the current legislation with inevitable economic harms. Notwithstanding the current *de facto* moratorium, the EU legislation on GMOs is not able to avoid the circulation of genetically modified organisms throughout Europe since they circulate both due to the above-mentioned threshold of 0.9% and due to the fact that GM animal feed is often used to nourish livestock which we eat without any labelling requirement.[25] This

---

[25]For example, in 2014 the majority of the maize (90%) and soy (85%) in Italy are used for producing animal feed. In most cases this animal feed is made up

inconsistency between the situation regarding to the cultivation of GMOs in Europe (the *de facto* moratorium) and the importation of GMOs (animal feed) causes uncertainty and is unable to mitigate the conflicts between opposite positions that request on the one hand the extension of the moratorium to the importation too and on the other its overtaking for the cultivation.

In this regard in order to establish that the presence of genetically modified material is adventitious or technically unavoidable operators must supply evidence to let to competent authority verifies that they have taken all steps to avoid the unintentional contamination of conventional products. 'In principle, farmers should be able to cultivate the types of agricultural crops they choose – be it GM crops, conventional or organic crops' (ibid., p. 17). In this sense the framers' and operators' wish to have their corps with the lowest possible presence of genetically modified organisms should be ensured. In order to avoid the unintended presence of GMOs (regularly authorized) in crops and other products co-existence measures can be provided. As said, this point was addressed by EGE. However, since measures of co-existence involve already authorized GM products, on which a comprehensive assessment of health and environmental risks has been done, they cannot concern health-related and environmental risks, but only economic risks such as the potential economic loss and impact of the admixture of organic and genetically modified crops (ibid.). However, health issues can be raised by the use of products related to GM crops such as pesticides that can be more intense (e.g. powerful weedkiller).[26] In this regard, it is for member States to implement management measures at national level. At the EU level some efforts are made to achieve a peaceful coexistence between different crops. The EU, with its Commission, provided for a set of guidelines for member States in this matter with the

---

by genetically modified which comes into the country through the importation mainly from the US (244 thousand tons) and the Brazil (517 thousand tons). See http://ec.europa.eu/eurostat/data/database. Accessed 16 July 2015.

[26] For example, recently the Court of appeal of Lyon found the US Monsanto legally responsible for poisoning a French farmer who inhaled Lasso, a powerful weedkiller which was used to accompany Monsanto corn cultivation (Ruggiu, 2015, p. 230). See http://www.bbc.com/news/world-europe-17024494.

Recommendation 2003/556/EC.[27] This text was replaced in 2010 by the Commission Recommendation 2010/C200/01.[28] According to these guidelines '[t]hey are intended to provide general principles for the development of national measures to avoid the unintended presence of GMOs in conventional and organic crops'.[29] This could also be influenced by local, regional, as well as national conditions. Furthermore, according to the 2010 guidelines '[t]he adventitious presence of GMOs above the tolerance threshold set out in EU legislation triggers the need for a crop that was intended to be a non-GMO crop, to be labelled as containing GMOs. This could cause a loss of income, due to a lower market price of the GM crop or difficulties in selling it. Moreover, additional costs might incur to farmers if they have to adopt monitoring systems and measures to minimize the admixture of GM and non-GM crops'.[30] The potential loss of income of organic products is not necessarily limited to exceeding the above-mentioned threshold of 0.9%, as set out by the EU legislation. In certain circumstances, depending on market demand, as well as on the national regulation, the presence of traces of GM materials in particular food crops can be excluded even at a lower level than 0.9% (ibid., p. 18). The European Commission has set up the European Coexistence Bureau (ECoB), located at the Institute for Prospective Technological Studies (IPTS) of the Commission's Joint Research Centre with the purpose of developing technical reference documents for best practices to achieve coexistence through non-binding guidelines addressed to member States. These documents are drawn on a crop-by-crop basis and the first one in 2008 focused on maize crop cultivation (ibid.).

---

[27] European Commission (2003) Commission Recommendation 2003/556/EC of 23 July 2003 on *Guidelines for the Development of National Strategies and Best Practices to Ensure the Coexistence of Genetically Modified Crops with Conventional and Organic Farming, Official Journal of the European Union* L189, 29/07/2003 P. 0036–0047, http://eur-lex.europa.eu/LexUriServ/LexUriServ.do?uri=CELEX:32003 H0556:EN:HTML. Accessed 2 July 2015.

[28] European Commission (2010) Commission Recommendation 2010/C200/01 of 13 July 2010 on *Guidelines for the Co-existence Measures to Avoid the Unintended presence of GMOs in Conventional and Organic Crops, Official Journal of the European Union* C200/1, 22/07/2010 P. 0036–0047, http://ec.europa.eu/food/plant/docs/plant_ gmo-agriculture_coexistence-new_recommendation_en.pdf. Accessed 2 July 2015.

[29] Ibidem, §1.4.

[30] Ibidem, §1.1.

(iv) As regards the transboundary circulation of GMOs the EU set out a framework of rules, which also is in the middle of an evolution process.

The European Union is part of the Cartagena Protocol on Biosafety entered into force in 2003.[31] This agreement of the United Nations aims at establishing common rules in transboundary movement of genetically modified organisms in order to ensure the protection of biodiversity as well as human health at the global level (ibid., p. 19). The Cartagena Protocol established a central database, the Biosafety Clearing-House, which has the aim of providing and exchanging information for implementing the Protocol as regards the competent national authorities on GMOs, the GMO national legislation, regulatory decisions on GMOs (approval or prohibition) including the related risk assessment (ibid.). The incorporation of the Cartagena Protocol into the EU legislation relies on a regulatory framework that has its cornerstone in Directive 2001/18/EC on the deliberate release of genetically modified organisms which regulates the imports of GMOs into the Union (ibid.). With regard to the exchange between the Union and abroad (and vice versa) regarding to GMOs there is the Regulation No. 1946/2003,[32] which was derived by Directive 2001/18. It represents a translation of the Cartagena Protocol on Biosafety and creates a framework, on the basis of the precautionary principle for the safe, transfer, handling and use of living modified organisms resulting from biotechnology which may cause adverse consequences to the conservation and sustainable use of biological diversity or to human health (Fontana, 2010, pp. 2–3). As recalled in its preamble: '[i]t is important to organise the supervision and control of transboundary movements of GMOs in order to contribute to ensuring the conservation and sustainable use of biological diversity, taking also into account risks to human health, and so as to enable citizens to make a free and

---

[31] On this see also Francioni (2007).
[32] European Union (2003) Regulation (EC) No. 1946/2003 of the European Parliament and of the Council of 15 July 2003 *on Transboundary Movements of Genetically Modified Organisms, Official Journal of the European Union,* L287, 5/11/2003, http://eur-lex.europa.eu/legal-content/EN/TXT/PDF/?uri= CELEX:32003R1946&from= EN. Accessed 24 March 2015.

informed choice in regard to GMOs' (forth whereas). The Regulation applies on any movement of GMOs from the EU to third countries (and vice versa) 'that may have adverse effects on the conservation and sustainable use of biological diversity' (Art. 2). It provides: the obligation for the exporter to notify (to authority of Party and non-Party of import) exports of GMOs intended for deliberate release into the environment and to secure express consent prior to a first transboundary movement (Art. 4); the obligation for the Commission to inform the public and international partners on EU practices, legislation and decisions on GMOs (Art. 15,2); a number of rules for the export of GMOs intended to be used as food, feed or for processing (Section 2); the information to be provided to the Biosafety Clearing House (BCH) on behalf of the Union (Art. 12) (Plan and Van den, Eede, 2010, p. 19).

Export is not immediate since '[i]t is necessary to ensure the identification of GMOs being exported from or imported into the Community. With regard to traceability, labelling and identification of imports into the Community, such GMOs are subject to rules in Community legislation' (tenth whereas). Thus, '[e]xporters [has to] await the prior written express consent of the Party or non-Party of import before proceeding' (tenth whereas). There are also confidentiality provisions. In particular: '[t]he Commission and the Member States shall not divulge to third parties any confidential information received' (Art. 16,1). Furthermore, '[t]he exporter may indicate the information in the notification submitted under Article 4 which should be treated as confidential' (Art. 16,2). However, in no case may be kept confidential information as regards: name and address of the exporter and importer; general description of the GMO; a summary of the risk assessment taking also into account risks to human health; methods and plans for emergency response. Within the EU the main import way of GMOs refers to animal feed. As said above as regards labelling of animal feed containing or being produced from genetically modified feed there is Regulation 1829/2003 apply to animal feed. This information to livestock farmers regards to the composition and properties of feed. However, as regards information of the livestock nourished with animal feed containing or derived from GMOs no obligation is set out. This could represent an element of weakness and a potential source of future conflicts.

(v) As noted above, the EU legislation on GMOS is being reshaped in order to include new developments in the field of synthetic biology under the same legislative umbrella (see below). The rise of synthetic biology as a fast-evolving field that differs from previous gene modification techniques (SCENIHR, 2014, p. 8) opened up to two possible paths for regulating this new sector: either by establishing *ad hoc* provisions or by adapting the existing regulation on GMOs. The EU chose the latter on the basis of the conviction that it should be deemed as just an evolution of the 'old' genetic engineering (see below). The chosen way is probably simpler but puts some difficulties that the EU faced with great prudence by carefully preparing this fundamental step through the involvement of several agencies and committees. Accordingly, this fact will lead, soon or later, to the gradual fusion of both governance frameworks by considering genetic modification and synthetic biology as a single research field with significant repercussion from the legal standpoint. In fact, the coexistence of genetic engineering and synthetic biology under the same conceptual umbrella is highly debated within the community of scientists (e.g. Arkin et al., 2009). In this regard, scholars are equally divided: some argue that SynBio products and GMOs are substantially equivalent, some do not. Some purposes that can be achieved through synthetic biology, also thanks to new techniques such as CRISPER-CAS 9, which enormously facilitates research in this ambit, surely overpass traditional genetic engineering which aims at modifying existing organisms. However, it is a fact that some aspects of synthetic biology cannot be covered by current legislation on GMOs by rising new and unforeseen risks. For this reason, the EU prudently moved all its agencies to prepare the future docking of synthetic biology within the GMO legislation. Thus, first in 2008 the European Commission established a EU Member State expert Working Group to analyse a list of new techniques which results in GMOs as defined under Directive 2001/18/EC on the deliberate release of GMOs and Directive 2009/41/EC on contained use of GM microorganisms (GMMs). This report also considered the field of synthetic biology constituting the logical extension of genetic modification techniques (SCENIHR, 2015a, p. 9). Consequently, this Working Group, known as New Techniques Working Group (NTWG), delivered the report in January 2012 (NTWG, 2012). The report analysed eight different breeding techniques: oligonucleotide directed mutagenesis; zinc finger nuclease; RNA-dependent DNA methylation; cisgenesis; grafting on a GMO rootstock; reverse breeding; agro-infiltration; and, as said, synthetic genomics. It is meaningful that this report came to the conclusion that the legal definition of a GMO does not apply to most of the new breeding techniques and that these techniques fall under the exemptions already established by the legislation or should be exempted as they are not different from plants obtained by traditional breeding (Schiemann and Hartung, 2014, p. 202). Then, as we will better see later, the Commission asked the Scientific Committee on Emerging and Newly Identified Health Risks (SCENIHR) to deliver a set of reports aimed at unifying both fields under the existing regulatory framework on GMOs. One laid down the conceptual premises of this action (a unique definition able to cover also synthetic biology), (SCENIHR, 2014), the second one focused the related risk assessment issues (SCENIHR, 2015), the third deals with safety aspects and research priorities. Also in this case the premise of the operational definition of synthetic biology upon which the process of adaptation of the GMO legislation is built, is the conceptual difference between the two fields. This revision process of the EU legislation should prepare the docking of synthetic biology under the existing regulatory framework by an adapting phenomenon of current provisions on GMOs. However, the consequences caused by the inclusion of synthetic biology under the same legal framework of biotechnology cannot be well assessed at this stage, especially if we consider that the wider and different concerns raised by this field. In other words, there is the risk to add new unpredictable risks of a rising field to a framework, which denotes some structural weakness and was not able to stop conflicts.

3.3 The EU Governance of Emerging Technologies: The Case of Nanotechnologies[33]

3.3.1 A clear attempt of using the new governance paradigm can be seen in the field of nanotechnologies in Europe. Like for biotechnology, this governance framework is a concrete application of the case-by-case approach. It arose as an adaptive, experimental, flexible framework, with the initial agreement of all the three leading bodies (Commission, Parliament, Council), but subsequently, following the initiative of the European Parliament, a reform process started. Thus it can be noticed a progressive process of consolidation leading towards more traditional forms of regulation which is still in course. However, forms inspired by the Responsible, Research and Innovation model are being experimented within this hybrid framework.

The European Group on Ethics in Science and New Technologies (EGE) defines nanotechnology as the study (nanoscience) and the manipulation (nanotechnology) of the matter at the nanoscale, namely at the atomic, molecular and macromolecular scales (usually objects with dimensions included between 0 and 100 nanometres, though there are nanomaterials which can reach even 200 or 300 nm) (EGE, 2007, p. 11). As known, one nanometre is one billionth of a meter or about one eighty thousandth the width of a human hair. Since at this level the matter enjoys uncommon properties depending on size, shape, chemical bonds, polarity, etc., engineered nanomaterials are all different with different physical and ethical implications with regard to risks and benefits. This is the reason why according to the Scientific Committee on Emerging and Newly Identified Health Risks (SCENIHR) 'nanomaterials are similar to normal chemicals/substances in that some may be toxic and some may not' (SCENIHR, 2009, p. 56). This conclusion should lead to recommend a case-by-case approach in this field. This also means that it is not possible having a general paradigm of risk assessment

---

[33]This paragraph partially retakes and deepens the analysis of Ruggiu (2014b, 2013b, 2015b).

applicable in the case of nanomaterials (i.e. nanomaterials are not generally harmful or good), but we need to distinguish in case. In other words, there is no substantial equivalence with other common materials existing in nature, but we have not to generalize as regards nanotechnologies. For this reason, it is more appropriate to talk of nanotechnologies in the plural, instead of simply nanotechnology (Ruggiu, 2013b, p. 203).

However, the uncommon features of nanotechnologies leave some open questions representing a limit for the regulator. In general terms, there is a lack of data regarding to implied risks of nanomaterials (and the way to overcome them); there is no shared definition; there is a great uncertainty with regard to terminology, classification and metrology (there is the need to common standards to measure them); although nanotechnologies do not act in a regulatory vacuum (as some noted, the existing regulations apply in this sector (van Calster, 2006), given their novelty, there is uncertainty with regard to both the adequacy and efficacy these regulations. Moreover, there is the need for further legal, ethical, and sociological studies on their implications (Rip, 2002).

Without generalizing or even doing scaremongering, it is truth that together with their undoubted advantages, nanotechnologies imply some risks we need to consider. In fact, there are some studies revealing the existence of a criticality regarding to the toxicology and the ecotoxicology of some engineered nanomaterials. For example, there are studies addressing that carbon nanotubes (both single and multi-walled) can cause lung inflammations (Shvedova et al., 2005), as well as granulomas once insert in abdomen of mice (Poland et al., 2008)[34]; there are some others addressing potentials for health and the environment with regard to silver nanospheres (Mwilu et al., 2013), or regarding to titanium dioxide nanoparticles included in several sun creams (Zhang et al., 2007), or the use of nickel nanoparticles in the workplace which would cause allergic reactions

---

[34]With regard to the carbon nanotubes toxicology, there is also a promising study in nanomedicine which identified an enzyme produced by some kinds of white blood cells (myeloperoxidase) that can biodegrade them by decomposing the carbon nanotubes into two innocuous elements, water and carbon dioxide (Kagan et al., 2010).

in a setting without any specific respiratory protection or control measures (Journeay and Goldman, 2014).

Nanotechnologies are not only a concrete promise for our future in the hand of more advanced research. Notwithstanding there is a multitude of current studies on nanotechnologies which promise to revolutionize our life in medicine (e.g. the cure against cancer), electronics (micro- and nanochips, ultrapowerful processors), contrast of environmental pollution (water purification systems, nanoremediation of oil spill), energy generation and storage (fuel cells, batteries, photovoltaic cells), nanotechnologies are already widely commercialized in cosmetics, electronics, automotive industry, food and agricultural field (ingredients, food storage systems, feed), paints, biocidal products (such as pesticides, insect repellents, disinfectants, etc.) and so on[35] (Throne-Host and Strandbakken, 2009). If there is any risk, it is truth that it could be, in some degree, already present (Ruggiu, 2013b). In this regard it is quite difficult to encompass with the same regulation the entire range of the nanotechnologies world, which is made up of research and market with pretty different needs.

This paragraph means to analyse the legal framework on nanotechnologies from its beginning to recent regulatory initiatives in the field of cosmetics, labelling on food, biocides. Main goal is to show how governance has shifted from the initial approach of adaptive, experimental and flexible nature, relying on soft law tools that had in the Commission code of conduct their major result, to a more traditional approach which resorted to instruments of hard legislation such as directives and regulation. While the case-by-case approach remained the key common feature during its whole evolutionary course, the current model of governance is a hybrid which mixes forms of new governance such as agency, networking, comitology, consultation processes with tools of traditional style (regulations, directives under revision). In this sense what emerges is the incapacity of the current approach to establish a coherent and anticipatory framework able to evolve in accordance with the

---

[35]There are also nanotechnological drugs used in anti-cancer therapy in nanomedicine (e.g. Myocet) that are commercialized within the EU (Hafner et al., 2014).

development of the sector. Abrupt shifts seem to be the unavoidable outcome of this governance course with a gradual stiffening of the overall legal framework.

3.3.2 Taking into account the lesson learnt from the case of biotechnology (Metha, 2004), in 2000s the Community strategy in the field of nanotechnologies was inspired by a great prudence. At that time, although discovered about fifteen years ago, nanotechnologies were at their infancy.[36] The uncertainty was so diffused at any level (scientific, ethical, regulatory) so as to discourage any try of regulating this field by resorting to classic hard legislation measures (Mendel, 2009; Pariotti and Ruggiu, 2012b). The strategy adopted by the European Commission was initially to wait and see: (i) whether there were risks and what they were; (ii) whether the existing regulation was adequate for this rising field; (iii) whether and how principal actors were developing spontaneous forms of self-regulation. Then, Community authorities would have adopted the needed measures in case. On the basis of this mix of 'wait and see' and 'case-by-case' logic, initially the European Commission went along with the European Parliament. However, this initial agreement did not last for long.

Behind this strategy there was the hope of feeding a general climate of responsibility and cooperation among stakeholders, most of all enterprises. At the core there was the belief that the market would be able to quickly develop and, if possible, self-regulate, at least until the moment when scientific data would be complete or sufficiently robust for developing a more stringent regulatory framework in this promising field. It is an adaptive, flexible and inclusive approach of experimentalist nature that would have had to accompany the growth of the entire sector till its ripeness (Roco, 2006; Kearnes and Rip, 2009; Widmar et al., 2010).

This route starts with the 2004 Commission communication *Towards a European Strategy for Nanotechnology* where the Community authorities put the basis of the 'safe, integrated and responsible

---

[36]Scanning tunnelling microscope with which matter was visible at the atomic scale since 1981 was firstly used for manipulating single atoms just in 1989.

approach'[37] (European Commission, 2004). This approach was thus aimed at fostering a responsibilization process by enlarging the involvement of all parties in order to build the premises of a general self-regulatory attitude (Ruggiu, 2014b).

With the subsequent 2005 communication,[38] the Commission elaborated 'a series of articulated and integrated actions for the immediate implementation of a safe, integrated and responsible strategy for' nanotechnologies (European Commission, 2005, p. 3). In particular, it proposed: to double the FP7 budget compared to FP6 especially in the nanoelectronics sector; to boost toxicological and ecotoxicological studies on the impact of nanoparticles on health and the environment; to ensure that EU funded projects in nanotechnologies are subjected to ethical review so as to respect 'fundamental ethical principles' in view of helping to build confidence in decision-making; to develop simultaneously ethical, legal, sociological studies (ELSI studies) in this matter in order to anticipate at earlier stage their impact; to ask the EGE to carry out an ethical analysis of nanomedicine; to promote measures to minimize the exposure of workers, consumers and the environment; to support existing infrastructures and transnational networks among universities, research organizations and industry in order to assembly critical mass through distributed poles of excellence; to foster the exchange of best practices in industry and the increase of industry involvement in EU R&D projects on nanotechnologies; to establish a monitoring system of patents in this sector; to coordinate and strengthen actions in standardization process and the development with member States, international organizations, European agencies, and industry of terminology, guidelines, models, and standards for risk assessment throughout the whole lifecycle of nanoproducts; to boost an inclusive, aware, public dialogue on

---

[37]European Commission (2004) Communication from the Commission of 12/05/2004 *Towards a European Strategy for Nanotechnology*, COM(2004) 338 final. Luxembourg: Commission of the European Communities, http://ec.europa.eu/research/industrial_technologies/pdf nanotechnology_communication_en.pdf.

[38]European Commission (2005) Communication from the Commission of 07/06/2005 *Nanosciences and Nanotechnologies. An Action Plan 2005–2009*, COM(2005) 423 final. Brussels: Commission, http://ec.europa.eu/research/industrial_technologies/pdf/nano_action_plan_en.pdf. Accessed 23 February 2015.

the impacts of nanotechnologies; to use the pattern of the Open Method of Coordination as best way for the information exchange; to increase the dialogue at the international level, especially with the industry, for making a code of conduct on the use and development of nanotechnologies; to review and, where appropriate, propose adaptations of the existing regulation at the both Community and national level. In sum, the main directions of the safe, integrated and responsible approach were: (1) to boost networking infrastructures for research and development of the sector, (2) to develop a better integration of research with its ethical dimension, (3) to involve all stakeholders into the development of nanotechnologies and their regulation, and in the meantime to apply the current regulatory framework by checking its adequacy. In this latter regard, public engagement was explicitly addressed as one of the cardinal ways to follow.

Then, with the 2006 Parliament resolution[39] the approach proposed by the Commission was substantially accepted. In particular, according to the European Parliament for a responsible strategy on nanotechnologies it was needed to integrate knowledge on social, health and safety aspects into the technological development. For this reason, it was better to engage the European Commission, member States and industry in an effective dialogue with all stakeholders in order to steer nanotechnology developments along a sustainable path (inclusive approach). In this regard, industry needed to take into account risks posed to human health, consumers, workers, the environment during the whole lifecycle of nanoproducts and contribute to disseminate information concerning their use and risks (integrated approach).

The same year the European Union launched the Seventh Framework Programme (FP7) which identified the nanotechnologies as one of the key areas to be widely encouraged in order to build a 'strong industrial base' in Europe and improve the industry competitiveness (European Union, 2006, p. 17). While during FP6

---

[39]European Parliament (2006) Resolution of the European Parliament of 28/09/2006 *Nanosciences and Nanotechnologies. An Action Plan 2005–2009* (2006/2004 (INI)). Strasbourg: European Parliament, http://www.europarl. europa.eu/sides/getDoc.do?pubRef=-//EP//TEXT%2BTA%2BP6-TA-2006-0392 %2B0%2BDOC%2BXML%2BV0//EN. Accessed 23 February 2015.

about €1.36 billion were invested in nanotechnologies, with the FP7 €3.5 billion were dedicated to this research field (EGE, 2007, p. 5). For this purpose, the FP7 built a broad ethics framework wherein to develop the European research and innovation. Thus, it tried to foster for key enabling technologies, such as nanotechnologies, a better integration among disciplines and different areas, by 'broadening the engagement of researchers and the public at large [...] with scientific related questions, to anticipate and clarify political and societal issues, including ethical issues' (European Union, 2006, p. 34). In this regard any supported research should respect the 'fundamental ethical principles including those reflected in the Charter of Fundamental Rights of the European Union', as well as the EGE's opinions and the Protocol on the protection and welfare of animals (European Union, 2006, p. 41). This marks the centrality of ethics within the European governance framework. In conformity with the direction initially adopted by the European Union, the anticipatory dimension of the EU policy rested thus mainly on a process of wide inclusion of society in order to build the premises of a general dialogue open to societal and ethical aspects of technological development. Accordingly, in 2007 the EGE delivered its opinion on nanomedicine (EGE, 2007). In this it underlined 'the need to establish measures to verify the safety of nanomedical products' at both national and Union level (EGE, 2007, p. 5), it expressed the necessity of launching a wide public participation also about uncertainties and knowledge gap, and it addressed the opportunity of fostering interdisciplinary research in this field by including also ethical, legal and sociological studies according to the pattern developed by the European Commission in 2000s. In this regard it assessed as adequate the existing regulatory structures by suggesting the Commission to consider to elaborate changes within this framework, but addressing the risk of overlapping of different regulations (EGE, 2007, p. 6).

3.2.3 The high point of this approach should have been the nanotechnologies code of conduct. The clear aim was to trigger a self-regulatory attitude in stakeholders in accordance with the inclusive trend of this turn of governance. However, instead of dealing with the entire innovation cycle (research, production, commercialization, recycling), it dealt with only research. The reason of this choice will be understandable later.

The Commission code of conduct (EC CoC) is the result of a governance experimentation which adopts the NGM in the field of emerging technologies for the first time, since it was reached through a consultation process launched by the Commission in 2007 (Ruggiu, 2014b). In a landscape where risks are scattered (Beck, 1986) and the relevant actors are distribute at the global sphere (Stoke, 1998, p. 21; Pariotti and Ruggiu, 2012), also responsibility needs to be distributed (von Schomberg, 2010, p. 56), especially in the face of the current process of erosion of the State capacity of regulating the whole world (Grande, 2001). In this regard instances of self-regulation, such as codes of conduct, can be interpreted as belonging to the responsibilization phenomena, since they are aimed at distributing the responsibility among stakeholders (Parker, 2007; Dorbeck-Jung and Shelley-Egan, 2013; Ruggiu, 2014). Under this perspective, self-regulatory tools, such as codes, are a form of meta-regulation (Coglianese and Mendelson, 2010). By meta-regulation I mean 'a process of regulating the regulators, whether they are public agencies, private corporate self-regulators or third party gatekeepers' (Dorbeck-Jung and Shelley-Egan, 2013, p. 56). In other words, meta-regulatory tools enact norms regulating the process of regulation, that is, in the case of code of conduct, self-regulation (Parker, 2007, p. 211). Any tool that is aimed at steering the law-making process (regulation, self-regulation such as codes of conduct), either public (statute law) or private (social dialogue, networking), belongs to the dimension of meta-regulation. In this sense, consultation processes can represent an efficacious meta-regulatory tool (Ruggiu, 2014b).

When the European Commission had to elaborate a new instrument able to trigger forms of self-regulation, it resorted mainly to consultations. Between 2007 and 2010 two consultations were launched: one aimed at drafting a code of conduct on nano-technologies in 2007[40] and one aimed at detecting stakeholders'

---

[40]See European Commission (2007b) *'Code of Conduct for Responsible Nanosciences and Nanotechnologies Research'—Detailed Analysis of Results from the Consultation*, Brussels: European Commission, http://ec.europa.eu/research/

opinions on this self-regulatory experience between 2009/2010.[41] The consultation processes launched both directly by the European Commission (through its Directorate-General for Research) and, indirectly, outside it (e.g. the NanoCode project through the FP7)[42] represent an attempt of governing the emerging field of nanotechnologies by triggering spontaneous paths of co-responsibility among stakeholders. In this sense, the Commission code of conduct should be interpreted within the larger framework of the rising NGM in the field of emerging technologies, which tends to foster spontaneous forms of self-regulation among stakeholders (Ruggiu, 2014b).

The first 2007 consultation was launched on the basis of a draft that would have represent the starting point of the process of drafting the code called 'consultation paper' (European Commission, 2007a). The consultation paper is thus the trigger for the process of the code formulation, by being the basis of the subsequent 2007 consultation that led to the setting of the code principles. In this document the Commission drew reference points for a future consultation '[i]n order to promote safe and responsible nanotechnology research and pave the way to its safe and responsible application and use' (ibid., p. 1). In this text the Commission explained the reason for limiting the scope of the code on research: '[o]n the one hand it develops new technologies for application in industry [...] on the other hand it investigates the potential risks and establishes the appropriate measures to take' (ibid., p. 1). In other words, research has an across-the-board nature, by being at the basis of both industrial advance and university inquiry.

This choice probably reflects a more concrete and realistic approach, by avoiding too ambitious attempts that could have led to

---

science-society/document_library/pdf_06/consultation-nano-sinapse-feedback_en.pdf. Accessed 23 February 2015.

[41]European Commission (2010) Recommendation *on a Code of Conduct for Responsible Nanosciences and Nanotechnologies Research: 1st Revision. Analysis of Results from the Public Consultation*, http://ec.europa.eu/research/consultations/nano-code/results_en.pdf. Accessed 23 February 2015.

[42]Between 2010 and 2011 a survey, as a part of a EU funded FP7 project, held. Since it was also aimed at implementing the Commission code of conduct, it can be deemed as a part of responsibilization phenomena triggered at the Community level (Meili et al., 2011).

a foreseeable failure (for example, by providing a code of all emerg-ing technologies or a code for overall research and innovation). However, it came at some (remediable) costs: the perception of an unfair distribution of responsibilities (Ruggiu, 2014b, p. 7). As noted on emerging technologies, the development of nanotechnologies stands on the shoulders of several actors: public authorities (EU and member States), enterprises and industry, research organizations (both public and private), funding organizations (both public, e.g. the EU, States, etc., and private, e.g. banks, foundations), civil society organizations (trade unions, consumers', patients', environmental organizations, animal rights organizations), the public at large (Ferrarese, 2010). In this sense, researchers are only a small part of the entire audience of stakeholders in the field of emerging technologies.

It is worthwhile to follow variations of the presence of EU fundamental rights within this process. In the consultation paper, the role of fundamental rights and the precautionary approach emerge at the core of the European Commission's action since 'confidence in its safety' and 'public acceptance are preconditions for the application and commercialization of nanotechnology-based products' (European Commission, 2007a, p. 1). This centrality of fundamental rights was progressively weakened in the 2008 recommendation on a code of conduct and in the subsequent Council conclusions in 2009. In this framework the Commission code of conduct and the first consultation promoted by the EU in 2007 appear coherent with the 2004 communication of the Commission (European Commission, 2004) which outlined the basis of a safe, integrated and responsible approach, and with the nanotechnologies Action Plan 2005/2009 (European Union, 2005) proposing the adoption of a code of conduct (European Commission, 2007a, p. 2).

The consultation process launched in 2007 was anything but ineffective. While the proposed principles of the consultation paper were three ('precaution', 'inclusiveness' and 'integrity'), they became seven in the final version (i.e. 'meaning', 'sustainability', 'precaution', 'inclusiveness', 'excellence', 'innovation', 'accountability') thanks to some suggestions stemming from the consultation participants (European Commission, 2007a, p. 3; European Commission, 2008a;

Ruggiu, 2014b, p. 8). Moreover, some suggestions regarding to the limitation of those applications aimed at enhancing human performances, nanofood and feed were also embraced, meaning that the 2007 consultation reached its goals (Ruggiu, 2014b).

At the beginning of 2008 the code of conduct was adopted with a recommendation.[43] The nature of the code was mainly practical. While we can distinguish between code of conduct (a set of rules) and code of ethics (a set of values) (Arrigo, 2006), we need to conclude that it belongs to the first type. In fact, although it was made up of a set of principle and an ensemble of guidelines, the Commission code of conduct was mainly aimed at regulating the existence of the organization in detail. It had thus a clear practical aim.

The responsible action in nanotechnologies research rests on a set of principles which should clarify the goals of the Commission code of conduct rules (i.e. the guidelines), what they aim to. First of all, research has to pursue the aim of being comprehensible for the lay public ('meaning'). Research needs to be safe and ethical and contribute to the sustainable development of the EU, by avoiding of representing either a danger for human health and the environment or 'a biological, physical or *moral* threat' (European Commission, 2008a), p. 6 – italics mine) ('sustainability'). It must respect the precautionary principle, that is while there are risks, it is to anticipate the impact at earlier stage by adopting the needed measures related to the level of protection and the benefits of scientific research ('precaution'). Research needs to ensure opening to all parties, transparency and information exchange ('inclusiveness'). It must pursue highest scientific standards by avoiding data falsification, plagiarism, self-plagiarism ('excellence'). It needs to grant the maximal creativity ('innovation'). Researchers and research institutions needs to be accountable as regard to implications on health and the environment in face of the future generations ('accountability'). This latter principle was contested

---

[43]European Commission (2008) Commission Recommendation of 07/02/2008 *on a Code of Conduct for Responsible Nanosciences and Nanotechnologies Research*, C(2008) 424 final. Brussels: European Commission. http://ec.europa.eu/research/participants/data/ref/fp7/89918/nanocode-recommendation_en.pdf. Accessed 23 February 2015.

during subsequent surveys (Ruggiu, 2014). While all principles were generally well-accepted, the 'accountability' was criticized with regard to the wide reference to future generations. For example, during the NanoCode survey problems regarding to the translation of the word 'accountability' became apparent. In fact 'the French and the German translations of the "accountability" principle as "responsibility" earned mistrust as they were interpreted with a connotation of *implying legal liabilities* as well as suggesting that scientists are held responsible for what is done with their work by decision outside their control or by other actors in the future' (Meili et al., 2011, p. 6 – italics mine).[44]

Since it is not a mere container of values (Arrigo, 2006), the code is also accompanied by an ensemble of guidelines that should steer the organization in all its existence by telling what to do and what to do not and who is responsible of their realization. Unfortunately, there is a lack of correlation between the principles and the guidelines that weakens the dimension of their practicability. In this regard the authors of the NanoCode survey complained that each principle is not reflected in all the guidelines (Meili et al., 2011, p. 26). As attested by the NanoCode survey, there is 'an unambiguous demand for increasing its specificity and practicability' (ibid., p. 23). Furthermore, some guidelines are not well formulated and seem to lead to unintended consequences or not to the intended ones. For example, guideline 4.1.17[45] was criticized since it risks *de facto* leading to 'a moratorium on certain types of research in nanomedicine and nano-enabled personal care'. It was also criticized for the absence of 'criteria and indicators to clarify how to apply it' (ibid., p. 28). Moreover guideline 4.2.6[46] was also criticized since

---

[44]Even in the 2007 consultation some respondents pointed out that the 'issue of *liability* should be clarified' (European Commission, 2007b, p. 3 – italics mine).

[45]'As long as risk assessment studies on long-term safety is not available, research involving deliberate intrusion of nano-objects into the human body, their inclusion in food (especially in food for babies), feed, toys, cosmetics and other products that may lead to exposure to humans and the environment, should be avoided' (European Commission, 2008a, p. 9).

[46]'N&N research organisations and researchers should launch and coordinate specific N&N research activities in order to gain a better understanding of fundamental biological processes involved in the toxicology and ecotoxicology of nano-objects man-made or naturally occurring. They should widely publicise, when duly validated, data and findings on their biological effects, be they positive, negative or null' (European Commission, 2008a, p. 10).

'it seems unrealistic to require all N&N researchers to "launch and coordinate" nanotoxicology research' (ibid., p. 29).

These difficulties concerning mainly the wording of the code are also reflected by other parts of the code. In particular, the language of the Commission limited the comprehensibility of the code especially in the preamble and the text of the 2008 recommendation, namely the document containing the code itself (ibid., p. 10). This result was amplified by the fact that the recommendation containing the code was made up of several parts where the code was only a part, even more important and just the entire recommendation was summited to stakeholders. In particular, the 2008 recommendation is made up of a preamble where the Commission illustrates how the code has been framed within EU goals, followed by a part where the Commission recommends a set of actions mainly to member States, and finally the annex constituted by the code itself (in turn, made up of principles and guidelines). It is clear that concerns regarding the language of the Commission involve those parts of the 2008 recommendation, such as the preamble, where the Commission refers bureaucratically to EU goals that evidently were perceived quite far from its core, namely the code (Ruggiu, 2014, p. 9). However, with regard to the code, the Commission code of conduct seems to lack of the typical code structure. 'There is no introduction, outlining who should be addressed and what the benefits of using the EU-CoC are' (Meili et al., 2011, p. 10). Thus it seems that the *forma codicis*, the mere code writing itself, can be deemed as an element which can influence the acceptance and the efficacious dissemination of this instrument.[47] In other words, the process of

---

[47] In the second Commission consultation some pointed out the risk of the code being 'inapplicable considering the "present writing"' for the code and complained about the 'un-specificity of principles' (European Commission, 2010, 5). In this sense a phenomenon of semantic confusion affected the code drafting. For example, in the preamble of the 2008 recommendation fundamental rights are implicitly referred to when the Commission deals with '*ethical* aspects of nanomedicine', '*ethical* and sustainable nanosciences and nanotechnologies research in the European Union' (sixth, thirteenth *whereas*) (European Commission, 2008, p. 3 – italics mine). However, there was no specification on what are these 'ethical aspects' that need to be considered. In this sense, the unambiguous demand for increasing its specificity and practicability' emerged during all surveys is quite sharable (Meili et al., 2011a, p. 23).

stakeholders' responsibilization (i.e. the adoption of the code) is influenced by how the values at the core of the meta-regulatory instrument (i.e. the code itself) are proposed to stakeholders (Ruggiu, 2014b). It is not coincidence that subsequent surveys revealed difficulties in compliance process. From the NanoCode survey emerges that up to 2011 only The Netherlands provided a set of measures implementing the Commission code of conduct (Mantovani et al., 2010, p. 17). This data seems to be consistent with the fact that at that time only 21% of NanoCode participants (more than 400 people) had adopted the Commission code of conduct (Grobe et al., 2011, p. 12).[48]

This unexpected outcome can be deemed due to a set of causes. (i) First of all, the engagement of the EU and member States appeared quite low, as the interviewees' report stated by referring, for example, that at that time there was no official platform providing information about the code and helping stakeholders in complying with code principles and guidelines (Meili et al., 2011a, p. 7). Furthermore, the code could have been accompanied by a system of incentives and disincentives (such as white list, black list, funding distribution linked to the compliance with the code, a set of practical criteria for monitoring, assessing, and verifying the compliance degree, etc.) (Dorbeck-Jung and Shelley-Egan, 2013; Ruggiu, 2014b). (ii) Then, there were difficulties with the process of the responsibility distribution. Since the scope of the Commission code of conduct was limited on research, the weight of the whole responsibility seemed to stand on the sole research shoulders. This outcome seems to be confirmed by the detected concerns about the accountability principle that involved mainly the researchers' group (Meili et al., 2011, p. 6). This limit could be overcome, as subsequently observed in 2009 resolution (European Parliament, 2009), by replicating the experience of the code of conduct on nanotechnologies research in the other sector of the innovation chain, I mean production, commercialization and recycling. In this

---

[48]This outcome is also consistent with the data provided by Kjølberg and Strand (2011, p. 107): 'The first [i.e. obstacle for the EC CoC] is that it is dependent upon distribution (through national states, research councils, university administration, etc.) which in the case of the nanoresearch community at our university (i.e. University of Bergen) seem to have failed'.

regard the process triggered by the Commission code of conduct should be deemed as partial. (iii) Finally, what emerges from official and non-official surveys is a weak communication dimension in its broad sense, which would need to be strengthened by amending the accountability principle and realizing a press release of the code outside the framework of the 2008 recommendation. There are reasons to argue, in fact, that the perception of the code among addressees could have been weakened by its inclusion in the body of the Commission recommendation, which covered not only the code of conduct, but also some other parts that could have confused the recipients (Ruggiu, 2014b, pp. 9–10).

3.3.4 After this partially positive experience, suddenly the initial agreement between the Commission and the European Parliament broke down. While in 2008 the Commission in its first review of regulatory aspects of nanomaterials still stated that 'that current legislation covers to a large extent risks in relation to nanomaterials and that risks can be dealt with under the current legislative framework' (European Commission, 2008b, p. 3), the development of nanotechnologies seemed to reveal a different image. Immediately after the discovery of the first case of death of two female workers in a paint factory in Chine was reported in 2008, presumably due to the exposure to nanoparticles (Song Young et al., 2008), the Parliament adopted a resolution with which it asked the Commission for a strategy change in EU policy.[49] In this act the Parliament noted that while there are multiple expected benefits, so as the FP7 allocated €3.5 billion in this field, nanomaterials 'present significant new risks due to their minute size' (European Parliament, 2009, point D), accompanied by 'a significant lack of knowledge and information' concerning definition, size, properties and 'the actual use of nanomaterials in consumer products' (ibid., points F and H) so as it is to deem 'current funding for research

[49]European Parliament (2009) Resolution of the European Parliament of 24 April 2009 *on regulatory aspects of nanomaterials* (2008/2208(INI)), Brussels: European Parliament http://www. europarl.europa.eu/sides/getDoc.do?pubRef= -//EP//NONSGML+ TA+P6-TA-2009-0328+ 0+ DOC+ PDF+ V0//EN. Accessed 14 March 2015.

into environmental, health and safety aspects' being 'far too low' (ibid., point M). In this regard it was noted that 'SCENIHR[50] identified some specific health hazards as well as toxic effects for environmental organisms for some nanomaterials' (point L). For these reasons it called for the adoption of a specific regulation that applies the precautionary principle, the principle of producer responsibility and the 'polluter-pays' principle 'to ensure the safe production, use and disposal of nanomaterials before the technology is put on the market' (ibid., point Q). In addition, it asked for the legislation on chemicals (REACH),[51] the main regulatory sector involved by nanotechnologies development, which 'reveals several further deficiencies to deal with nanomaterials' to be reviewed (ibid., point S), as well as the Community legislation in food, workers' conditions, air quality and waste, need to also address nanotechnologies as well (ibid., point V). Furthermore, it highlighted the need for a new EGE's opinion on the convergence of nanotechnology with biotechnology, biology, cognitive sciences and information technology, by assessing as insufficient the scope of its previous 2007 opinion on nanomedicine (ibid., point Y) and called for the adoption of a code of conduct by 'all producers intending to manufacture or place goods on the market' (ibid., point Z). According to the Parliament's view, the whole innovation cycle (production, commercialization and recycling) should have been involved by a process of further responsibilization, by resorting to, again, the tools of self-regulation.

The consequence of this Parliament's act was an immediate and significant change of direction in the EU policy on nanotechnologies and the adoption of a set of legislative initiatives of hard law nature in this field (Ruggiu, 2013b).

---

[50] It refers to opinions of the Scientific Committee on Emerging and Newly Identified Health Risks (SCENIHR) on definitions and risk assessment of nanomaterials (SCENIHR, 2009).

[51] European Union (2006) Regulation (EC) No 1907/2006 of the European Parliament and of the Council of 18 December 2006 *concerning Registration, Evaluation, Authorisation and Restriction of Chemicals* (REACH), *Official Journal of the European Union*, L396, 30 December 2006, http://eur-lex.europa.eu/legal-content/EN/European Union (2006) TXT/PDF/?uri=CELEX:32006R1907&from=EN.

First of all, a regulation on cosmetic products containing several provisions specifically referring to nanomaterials[52] was adopted (Bowman, van Calster and Friedrichs, 2010). According to this new act anyone who want to place new cosmetics product containing nanomaterials into the market has to provide safety information to the Commission six months earlier (Art. 16.3). Besides, safety information needs to be notified to the European Commission for those products containing nanomaterials which are already in the market. Furthermore, the regulation provides a specific mandatory labelling, according to which the names of the ingredients present in the form of nanomaterials shall be followed by the word 'nano' in brackets (Art. 19.1(g)). Finally, the European Commission was requested to create a publicly available catalogue of 'all nanomaterials used in cosmetic products placed on the market . . . and the reasonably foreseeable exposure conditions' (Art. 16.10(a)). Then, on this basis Cosmetics Europe, an industry's trade association, including more than 4000 cosmetics companies and national associations in this sector, laid down a set of guidelines integrating the EU regulation on cosmetic products. In particular, the 2011 Colipa guidelines on cosmetic products integrate information requested in labelling when nanomaterials are present in cosmetics, by helping in this way enterprises in complying with Community provisions (Cosmetics Europe, 2011). Besides, the 2012 guidelines provide further indications for identifying a subject responsible of the actuation of labelling commitments, by strengthening thus the companies' process of compliance with the EU regulation (Cosmetics Europe, 2012). Given the overlapping between the action of EU authorities and private actors, this fact realized an interesting integration between self-governance patterns and Community governance arrangements.

Accordingly, in 2011 the EU introduced a first provision regarding nanomaterials in electronic equipment[53] requiring, with a quite

---

[52] European Union (2009) Regulation (EC) No 1223/2009 of the European Parliament and of the Council of 30 November 2009 *on Cosmetics Products, Official Journal of the European Union*, L349/59.

[53] European Union (2011) Directive 2011/65/EU of the European Parliament and of the Council of 08/06/2011 *on the Restriction of the Use of Certain Hazardous Substances in Electrical and Electronic Equipment, Official Journal of the European Union*,

questionable formulation, as soon as the data would be increased, the substitution of nanomaterials 'by more environmental friendly alternatives' and the modification of the list of restricted substances (Art. 16). It also launched a consultation involving all stakeholders on the impact on small medium-sized enterprises. However, this sudden change in a key sector of the European economy ran the risk of stopping the development of nanoelectronics in Europe, a sector where the probable benefits of nanotechnologies are enormous both in terms of economic competitiveness and technological advance, without gaining, instead, any specific advantage in terms of health and safety in workplace or waste recycling (since electronic articles would have come anyway in Europe from abroad).

In the same year, the European Commission faced the unsolved terminological question regarding to nanomaterials and it issued the first definition of nanomaterials in the EU legislation to be used by member States, EU agency and companies.[54] According to it 'nanomaterial' means a natural or manufactured material containing particles, in an unbound state or as an aggregate or as an agglomerate and where, for 50% [...] one or more external dimensions is in the size range" between 1 nm to 100 nm. Wisely this definition is structured with a certain degree of flexibility since there is the possibility to assess case-by-case the inclusion of other materials which do not respect its criteria. In fact, the threshold of 50% may be replaced by a lower one for concerns involving the environment, health, safety or competitiveness. The definition set out by the recommendation should be applied, in particular, in the case of legislation on chemicals (REACH) and in the regulation on classification, labelling and packaging (CLP) of hazardous substances and mixtures.[55] With regard to the

---

L174/88, http://eur-lex.europa.eu/legal-content/EN/TXT/PDF/?uri=CELEX: 32011L0065&from= EN. Accessed 14 March 2015.

[54] European Commission (2011) Recommendation of the European Commission of the 18 October 2011 *on the definition of nanomaterial*, 2011/696/EU, 2011, *Official Journal of the European Union*, L275/38.

[55] European Union (2008) Regulation (EC) No 1272/2008 of the European Parliament and of the Council of 16 December 2008 *on Classification, Labelling and Packaging (CLP) of Hazardous Substances and Mixtures*, amending and repealing Directives 67/548/EEC and 1999/45/EC, and amending Regulation (EC) No 1907/2006 *Official Journal of the European Union*, L353/1, 31.12.2008.

Community legislation on consumer products, the Commission was committed to implement its definition of nanomaterial.

Subsequently, the Union adopted a new regulation on food labelling explicitly considering nanomaterials.[56] Food sector is a field where nanotechnologies represent a significant factor of development, especially in mass production and large consumption goods but where information is insufficient (Marrani, 2013). However, this regulation does not deal with the entire food sector, but only the mere food labelling. In this regard, probably, a more attention to this delicate matter involving directly consumers' health and safety could have been paid, since risks, associated with nanoparticles use, are determined by exposure (Khulbush, 2011, p. 1). Like in the cosmetics sector, the regulation contains a provision providing a specific mandatory labelling, according to which the names of the ingredients present in the form of nanomaterials shall be followed by the word 'nano' in brackets. However, this provision, fundamental in preserving the consumers' freedom of choice, as well that of information, entered into force only in December 2014, after a (quite long) period when the enterprises had the necessary time to reach the conditions for complying with it (and, consequently, consumers were not informed) (Art. 18.3).

It is also to note that the European Commission started an action of data collection in this sector. Indeed, at the request of the Commission, the European Food Safety Authority (EFSA) adopted a guidance document clarifying the data to be provided when submitting an application dossier for nanomaterials to be incorporated in food and feed (EFSA Scientific Committee 2011). Furthermore, since 2006 it is committed in an annual reporting activity on the risk assessment on nanotechnologies in food and feed (e.g. European Food Safety Authority, 2015). In the current

---

[56] European Union (2011) Regulation (EC) No 1169/2011 of the European Parliament and of the Council of 25 October 2011 *on the provision of food information to consumers*, amending Regulations (EC) No 1924/2006 and (EC) No 1925/2006 of the European Parliament and of the Council, and repealing Commission Directive 87/250/EEC, Council Directive 90/496/EEC, Commission Directive 1999/10/EC, Directive 2000/13/EC of the European Parliament and of the Council, Commission Directives 2002/67/EC and 2008/5/EC and Commission Regulation (EC) No 608/2004, *Official Journal of the European Union*, L304/18.

regulatory arrangement, food safety risk assessment is performed by EFSA. Where nanomaterials are at stake, according to EFSA the general risk assessment methods do apply, but 'the assessment on a case by case basis is performed if no "common risks" are identified', fact this of not easily determination (Marrani, 2013, p. 180).

At the beginning of 2012, another resolution of the European Parliament on biocidal products[57] also considering nanomaterials (pesticides, insect repellents, disinfectants, spermicides and substances which can be also used in food contact materials), triggered a revision process in this matter, asking for a nano-specific mandatory labelling (Art. 58.3(d)). This new intervention of the Parliament opened up to the gradual replacement of entire legislation in matter of biocidal products in 2012.[58] The new regulation stands on the principle that '[t]reated articles should not be placed on the EU market unless all active substances contained in the biocidal products with which they were treated or which they incorporate are approved' for this use (second whereas). Then, in view of the authorization for the commercialization of a biocidal product 'where nanomaterials are used in that product, the risk for human health, animal health and the environment has to be assessed separately' (Art. 19.1(f)). In this regard a simplified authorization procedure is provided, unless the biocidal product contains nanomaterials (Art. 25(c)). Also in the case of biocides a specific mandatory labelling[59] is, in the case of nanomaterials, provided accordingly to the specific request of the Parliament. In this case the label must indicate

---

[57] European Parliament (2012) Resolution of the European Parliament of 19 January *2012 on the Council's position at first reading with a view to the adoption of a regulation of the European Parliament and of the Council, concerning the making available on the market and use of biocidal products* (05032/2/2011 – C7-0251/2011-2009/0076 (COD)), P7_TA-PROV(2012)0010.

[58] European Union (2012) Regulation (EU) No. 528/2012 of the European Parliament and the Council of 22/05/2012 *Concerning the Making Available on the Market and Use of Biocidal Products, Official Journal of the European Union*, L167/1, 27/06/2012, pp. 1–123, http://eur-lex.europa.eu/legal-content/EN/TXT/PDF/ ?uri=CELEX: 32012R0528&from= EN. Accessed 20 February 2015.
European Union (2014) Regulation (EU) No. 334/2014 Amending Regulation (EU) No. 528/2012 of the European Parliament and the Council of 22/05/2012 *Concerning the Making Available on the Market and Use of Biocidal Products, Official Journal of the European Union*, L103, 05/04/2014, pp. 22–32.

[59] On the limits of mandatory labelling see Throne-Host and Rip (2011).

'the name of all nanomaterials contained in the biocidal products followed by the word "nano" in brackets' (Art. 58). Furthermore, member States must monitor the biocidal products and the treated articles which have been placed in the market by constituting a special documentation including the information on the use of nanomaterials and their potential risks (Art. 65.2(d)).

In the meantime the Union went back to deal with electronic devices with a 2012 directive on waste electrical and electronic equipment (WEEE)[60] which provides the principle according to which all waste needs to return to the distributors to be collected and treated. In this instance the EU asked the Commission for assessing whether 'a specific treatment may be necessary' for waste containing nanomaterials due to health concerns (Art. 8.4).

The attention drew by the EGE on nanomedicine in 2006, triggered a process of adaptation of the current EU regulation. In 2012 the European Commission delivered a proposal for a regulation on medical devices where proposes to adopt a special care 'when [medical] devices contain or consist of nanomaterials that can be realised into the patient's or user's body' (Art. 3).[61] A special labelling indicating 'where devises contain or consist of nanomaterials' should be provided, unless they are encapsulated or bound in a manner that they cannot be released into the patient's or user's body (Art. 19.2(f)). In this case they are addressed by a specific classification (class III), becoming subjects of most severe conformity assessment procedure, together with more invasive devices, devices in direct contact with heart, implantable devices, breast implants, spinal disc replacement implants and so on (Rule 19). Several medicines based on nanotechnology have already

---

[60]European Union (2012) Directive 2012/19/EU of the European Parliament and the Council of 04/07/2012 on Waste Electrical and Electronic Equipment (WEEE), *Official Journal of the European Union*, L197/38, 24/07/2012, http://eur-lex.europa.eu/legal-content/ EN/TXT/PDF/?uri= CELEX:32012L0019&from= EN. Accessed 14 March 2015.

[61]European Union (2012) *Proposal of 26/09/2012 for a Regulation of the European Parliament and of the Council on medical devices, and amending Directive 2001/83/EC, Regulation (EC) No 178/2002 and Regulation (EC) No 1223/2009, COM(2012) 542 final*, Brussels: European Commission, http://ec.europa.eu/health/medical-devices/files/revision_docs/proposal_2012_542_en.pdf Accessed 3 March 2015.

led to the approval by the European Medicines Agency (EMA)[62] and subsequently were commercialized within the EU market (e.g. Caelex containing doxorubicin, Mepact containing mifamurtide, Myocet containing doxorubicin, Abraxane containing paclitaxel, Emend containing aprepitant, Rapamune containing sirolimus).[63]

No consideration is, instead, shown regarding to the growing sector of health self-monitoring systems which many private companies are developing nowadays. For example, large companies are researching on systems of self-screening which use nanoparticles (e.g. tiny iron oxide particles) into the human body that can be monitored through a wearable device such as a watch (Barr and Wilson, 2014). This search rises concerns not only with regard to health, the use of personal data (which can be easily hacked), but mainly with regard to the patient/physician relation (which can be completely eluded where a device is freely sold without any medical consultation). This aspect was also neglected by EGE's opinion on nanomedicine, but today it acquires a significant relevance due to the fast development of the market.

After the first regulatory review and the experimentation of the Commission code of conduct, at the end of 2012 the European Commission concluded the second regulatory review on nanomaterials.[64] It followed a report, the Commission Working Staff Paper,[65] providing the necessary information on the definition, nanomaterial market, uses, benefits, health and safety aspects, risk assessment, and information and databases on nanomaterials.

At this time also the broad sector of chemicals was addressed for being amended. As pointed out by the European Parliament in

---

[62]The commercialization of medicines based on nanotechnologies came after the publication of a report of the EMA Committee for Medicinal Product for Human Use (CHMP) in 2006. See EMA (2006).

[63]See http://www.ema.europa.eu/ema/index.jsp?curl=pages/special_topics/general/general_content_000345.jsp&. Accessed 15 March 2015.

[64]European Commission (2012) Commission Communication of 3/10/2012 *Second Regulatory Review on Nanomaterial*, COM (2012) 572 final. Brussels: European Commission, http://eur-lex.europa.eu/legal-content/EN/TXT/PDF/?uri=CELEX: 52012DC0572&from= EN. Accessed 20 February 2015.

[65]Commission Staff Working Paper (2012) *on Types and Uses of Nanomaterials, Including Safety Aspects*, SWD(2012) 288 final, 3.10.2012, Brussels: European Commission, http://eur-lex.europa.eu/LexUriServ/LexUriServ.do?uri=SWD:2012: 0288:FIN:EN:PDF. Accessed 1 March 2015.

2009 there were doubts on the suitability of the REACH simplified registration for nanomaterials manufactured or imported below one tonne, on whether they can be deemed as new substances that lead to consider the need of a chemical safety report with exposure assessment for all registered nanomaterials, as well as the need of setting out notification requirements for all nanomaterials paced in the market on their own, in preparations or in articles (European Parliament, 2009, p. 7). As recognized by the Commission, '[m]any registrations for substances known to have nanomaterial forms do not mention clearly which forms are covered or how information relates to the nanoform. Only little information is specifically addressing safe use of the specific nanomaterials supposed to be covered by the registration dossiers' (European Commission, 2012b, p. 6). In this regard the Commission promoted a Commission Working Staff Paper (European Commission, 2012a) inquiring on aspects connected to the definition, the use and the diffusion of nanomaterials into the market and two reports, RIPoN 2 (Hankin et al., 2011) and RIPon 3 (Aitiken et al., 2011), dealing with fulfilling information requirements and exposure assessment and risk/hazards characterization for nanomaterials under REACH. These two reports should have been followed by specific consultation processes involving members of the REACH Competent Authorities Sub-Group on Nanomaterials (CASG-Nano) and the relevant experts from Member States, industry and non-governmental organizations (NGOs) nominated by the REACH and Classification, Labelling & Packaging (CLP) Competent Authorities (CARACAL).

During 2013 it promoted a consultation among all interested stakeholders (in particular enterprises) on the modification on REACH annex on nanomaterials[66] which shall be accompanied by an impact assessment. For ensuring transparency of participants' interest to the consultation a public register is provided. Furthermore, the Commission encouraged the European Chemicals Agency (ECHA) to develop a new guidance for registration since 'REACH sets the best possible framework for the risk management of nanomate-

---

[66]See http://ec.europa.eu/environment/consultations/nanomaterials_2013_en.htm. Accessed 4 March 2015.

rials' although 'more specific requirements for nanomaterials within the framework have proven necessary'. This conclusion was reached according to the conviction that 'nanomaterials are similar to normal chemicals/substances in that some may be toxic and some may not', by being '[p]ossible risks [. . .] related to specific nanomaterials and specific uses' (SCENIHR, 2009, p. 56; European Commission, 2012b, p. 11). In this regard the Commission is still persuaded that 'nanomaterials require a risk assessment, which should be performed on a case-by-case basis, using pertinent information' (ibid., p. 11). In the Commission's view current risk assessment methods are thus still applicable.

The monitoring activity of the Commission triggered by the 2009 resolution invested also other key fields some years later. As regards health at work when nanomaterials are at stake, the Advisory Committee on Safety and Health at Work started working on a draft opinion on risk assessment and management of nanomaterials at the workplace, to be subsequently endorsed by the Advisory Committee. Similarly, to that of EFSA, a guidance has been provided also by the Scientific Committee on Consumer Safety for cosmetics products (Scientific Committee on Consumer Safety, 2012).

At the end of 2013 the Union launched the Horizon 2020 Framework Programme where nanotechnologies research and innovation represent one of the key areas of development within the EU 2020 strategy (European Union, 2013). The expressed priority is to fill the gap between knowledge and the market since without an excellent research can be neither any progress nor any development of the economy in Europe (Berger, 2013). In order to boost both the economic growth and the occupational development, the EU has proposed €80 billion for the period 2014–2020, by harnessing research and innovation at this aim. This budget includes about €24.6 billion for science, €17.9 billion for industrial innovation and €31.75 'targeted at most pressing issues facing Europe such as climate change, sustainable transport, renewable energy and the medical care requirements of an ageing population' (European Commission, 2013, p. 40). The safe, integrated and responsible approach is now subsumed under the Responsible Research and Innovation framework (RRI) (von Schomeberg, 2011, 2103; Owen et al., 2012; Owen et al., 2013; Owen, 2014), defined as 'a transparent, interactive process by which societal actors and innovators become mutually responsive to each other with a view to the (ethical) acceptability, sustainability and societal desirability of the innovation process and its marketable products (in order to allow a proper embedding of scientific and technological advances in our society)' (von Schomberg, 2011, p. 54). RRI should thus be considered as the subsequent step of the previous Community approach launched in 2000s with the aim of re-modulating the whole EU regulatory framework. For this purpose, also Horizon 2020 has built a strong ethics framework as basis for both research and innovation, as well as paths for integrating 'society in science and innovation issues, policies, and activities in order to integrate citizens' interest and values' (European Union, 2013, p. 167). Accordingly, they have to respect the 'fundamental ethical principles', the EGE's opinions and take into account the objectives of reducing animal testing and 'ensuring a high level of human health protection in accordance with Article 168 TFEU' (ibid., p. 107). 'Particular attention shall be paid to the principle of proportionality, the right to privacy, the right to the protection of personal data, the right to the physical and mental integrity of a person, the right to non-discrimination and the need to ensure high levels of human health protection' (ibid., Art. 19). In particular, stem cells research, being both adult and embryonic, is 'subject to stringent ethical review' and '[n]o project involving the use of human embryonic stem cells research should be funded that does not obtain the necessary approval by Member States' (ibid., p. 107). In sum, a large part of the proactive dimension of the Community model tends to rest on the work of this ethical framework in the process of allocation of EU funds.

3.3.5 An overview of the arising regulatory framework at the EU level shows a considerable effort in coping with the challenge thrown down by nanotechnologies. There is no doubt that provided by EU authorities is the most regulated landscape at the global level. This is the most articulated and complex system of rules in this field. However, we have to abstain from drawing any conclusion from this fact and try to point out some limits of this model, beyond its merits.

(i) First of all, what emerges in this approach is the difficulty in maintaining the consistency of the Community choices in the EU policy on nanotechnologies. In its initial stage the safe, integrated and responsible approach seemed to foster the self-regulatory capacity of the market in order to boost the entire sector of nanotechnologies and make the EU economy one of the more competitive in the world. Then, under the impulse of the Parliament it abruptly changed direction and started to regulate several sectors in detail (e.g. cosmetics, food labelling, biocides and so on). Also the wide activity of the EU agencies is leading towards a slow consolidation process of the regulation on nanotechnologies, in particular as regards chemicals. In this regard, this arrangement is neither a case of NGM or of a CCM traditionally pursued by the Union through regulations and directives. (ii) Then, this arrangement of regulation seems lacking of homogeneity due to the case-by-case approach. Measures seem growing up without an overall vision. There is an increasing attention to some sectors, such as electronics, nanomedicine, cosmetics, biocides with regard to information, labelling and risk-assessment, but the focus on others, wherein human exposure[67] is anyway high, appears low (e.g. food and feed, safety at workplace). In some sectors, which are strategic for economies of European countries, the attention appears to be even excessive (e.g. nanoelectronics). Some measures appear quite tardive, such as those on the definition of nanomaterial, or those on labelling in cosmetics and food, which could have been adopted immediately, or the re-modulation of legislation on chemicals which was foreseeable already in 2000s. In this regard, the ethics framework provided by research funding framework programmes does not seem to affect policy choices at the EU level yet. (iii) Finally, although the EU recognized that nanomaterials are all different so as it is to avoid any generalization, no specific measure was adopted until now. For example, although since 2005 toxicological studies on carbon nanotubes are known in scientific community no measure is taken into account on their use at the

---

[67] In fact, as regards nanotechnologies the level of risk does not depend on the mere toxicological form of a given substance, but on the possibilities of coming in contact (human exposure) with that potentially toxic substance (Kuhlbash, 2011, p. 11).

workplace. But a similar discourse can be done with other materials (e.g. silver nanoparticles). In this regard the anticipatory dimension of the current model can be somehow questionable. In this sense, we could wonder whether the adoption of some upstream criteria at earliest stage could have led to a more coherent, homogeneous and timely approach to nanotechnologies.

3.4 The EU Governance of Emerging Technologies: The Case of Synthetic Biology

3.4.1 If governance of nanotechnologies is evolving towards a progressive hybridization of the regulatory framework with the simultaneous adoption of flexible tools in view of their next consolidation in more traditional forms, synthetic biology and traditional biotechnology seem to be destined to be unified under the same legislation. In this regard, the EU has devised an ambitious plan which will lead to regulate all biotechnologies in the plural, meaning that there will be a single legislation on synthetic biology and genetic engineering in Europe one day.

After genetic engineering and nanotechnologies, in recent years some of the most revolutionary breakthroughs are being produced by synthetic biology (Venter, 2013). Thanks to the design of more or less complex fragments of DNA, it is possible to reprogram viruses and bacteria to produce biovaccines, antibiotics, biodrugs, biological devices able to contrast or reduce environmental pollution, or even produce new lethal bioweapons. These are inventions that can have several applications in medical, pharmaceutical, environmental, and military fields. Thanks to synthetic biology, life can now be reprogrammed in the laboratory, acquiring unexpected functions. This fact does not only open up new possibilities for the scientific and technological development, but also leads to scenarios whose impact now has to be assessed from an ethical, legal and sociological viewpoint (Guerra, Muratorio, Ruggiu, 2014; Ruggiu 2014b). Furthermore, in 2010 a team of scientists led by Craig Venter announced the creation of the first self-replicating artificial bacterium (Gibson et al., 2010). In other words, the world faced synthetic life for the first time. Although in this case only the DNA was artificial, and not the whole cell, it was a true revolution for life sciences.

There are several definitions of synthetic biology in circulation today. In the absence of an official definition in Europe, the European Union's ethical advisory board (the EGE) introduced one which includes the following elements: (i) the design of minimal cells or organisms, including genomic sequences; (ii) the identification and use of biological parts (toolkit); (iii) the construction of totally or partially artificial biological systems (EGE, 2009, p. 14). In this sense, synthetic biology includes 'the engineering of biological components and systems that do not exist in nature and the re-engineering of existing elements' (ibid.). Its objective is the 'mechanization of life' in order to perform new useful functions that do not exist in nature. To this end, it could seem that, like nanotechnologies, applied research represents the main dimension characterizing synthetic biology. However, the scientific community tends to emphasize that thanks to synthetic biology it is also possible to achieve a better understanding of the functioning of biological mechanisms and this fact-finding dimension cannot be neglected in synthetic biology. Thus, according to the EGE, synthetic biology is the new research field where scientists and engineers seek to modify existing living organisms through the design and synthesis of artificial genes, proteins, metabolic processes or evolutionary and biological systems 'in order to understand the basic molecular mechanisms of biological organisms and to perform new and useful functions' (ibid., p. 11).

Synthetic biology permits the 'transfiguration' of genetic data into digital data, the transformation of biological information into digital information (Venter, 2013; Deibel, 2014). 'The "informatization" of the biological world may have a disenchanting effect on our view of life. Life itself is increasingly understood in terms of "information processing" or "computation" and cells and organisms are seen as computers that can be easily (re)programmed according to our wishes. Rather than evolving naturally, living beings become the product of deliberate design' (Van den Belt, 2009, p. 259). The possibility of translating biological information into a digital format blurs the borders between nature and artefact, organic and inorganic, matter and information and opens up the field of life sciences to still unexplored possibilities. The de-materialization of the informational datum into pure information, which can travel through the net to be subsequently re-materialized as food, energy, a drug, a vaccine, or an antibiotic, radically extends the audience of possible users and threatens the proprietary model of the protection of copyrights in this field. However, beyond the dialectics between property rights and open source models, this fact has led to the proliferation of libraries and databases where the mere knowledge of programming languages, the use of search engines, or managing databases, can introduce nearly anyone to the manipulation of life. In this sense, the case of BioBricks Foundation and many other instances where biotechnology companies specialized in putting together long and complex pieces of synthetic DNA (such as Blue Heron, Coda Genomics, DNA 2.0 Inc., GeneArt) clearly highlights the potential of synthetic biology (Deibel, 2014; Turker and Zilinskas, 2006). It also illustrates potential risks opened up by the diffusion of synthetic biology such as garage biology, do-it-yourself biology (DIYbio), biohacking, or garage terrorism, that has led the regulator to again face the too early/too late dilemma (Kearnes and Rip, 2009, p. 4). 'In addition to the benefits of SynBio, there are scientific uncertainties associated with the development of synthetic life, cells or genomes and their potential impact on the environment, the conservation and sustainable use of biological diversity and human health' (SCENIHR et al., 2014a, p. 8). In this regard, there are significant concerns as to its compatibility with biodiversity. As the case of invasive alien species,[68] namely non-native species, into the environment showed, that the accidental or deliberate release can lead to unpredictable consequences for other plants and animals, as well as for the environment itself, which can be difficult to contain. This means that the environmental release of organisms derived from synthetic biology is conditioned by deep uncertainty. As recognized at the international level, 'the potential

---

[68]For example, according to Money (2005, p. 3) 'invasive alien species have altered evolutionary trajectories, can disrupt community and ecosystem processes, are causing large economic losses, and threaten human health and welfare'. In this regard recently a study argued that the eradication of alien species is the only means for contrasting the biodiversity loss and the extinction of autochthon species (Doherty et al., 20016).

for invasiveness of the organism which may lead to an adverse effect on native species through the destruction of habitat or a disruption of the trophic cascade' (Secretariat of the Convention on Biological Diversity, 2015, p. 10). What is the right balance, therefore, between a too early regulation of a given research field and a too late regulation, which performs when the risks are completely manifested and can no longer be stopped?

This paragraph means to analyse the growing governance of synthetic biology at the European level and to frame it within what is rising at the global level. The picture that we find shows the strict interrelation of different governance frameworks (transnational, international, national, European) with differences as regards the approach and mutual influences. In particular, at the European Union level there is an ambitious plan of embedding synthetic biology within the same framework of the GMO legislation. This implied the use of a new governance model where institutional actors attempt to adapt existing regulations thanks to multiple agencies and committees and by involving several key stakeholders. In this context, there is a broad process of adapting the current GMO legislation through the work of several agencies and committees (e.g. EGE, EFSA, SCENIHR) that have the task of analysing, preparing and implementing this shift. This strategy aims at developing: (i) an operational definition able to cover both synthetic biology and genetic engineering, (ii) the adaptation of current risk assessment methodologies to the SynBio field, and (iii) the study of possible pathways to cover future developments. Furthermore, in order to fill the gap of legitimation implicit in these agencies and committees, the pursuing of this goal also needed the preliminary involvement, via consultations, of several sectoral operators such as DNA providers, industry, public institutions such as laboratories and research centres. At this stage the lay public at large has not involved yet in the setting of priorities in the field of synthetic biology (SCENIHR et al., 2015c). However, this strategy based on the operational definition of synthetic biology can be deemed scientifically debated, since several aspects of synthetic biology will foreseeably fail to be covered by current risk assessment methodologies that were originally developed for genetically modified organisms. In this sense, notwithstanding the wide engagement of agencies and committees at any level, as well as stakeholders (with the exclusion of the civil society at large), the model which is being developed at the EU level run the risk of finding its anticipatory dimension quite weakened. This is due to the fact that in the natural environment the behaviour of completely artificial organisms with novel and unnatural functions and properties might not be compared to that of natural or genetically modified organisms resulting thus largely unforeseeable. Furthermore, the substitution of the notion of biodiversity with one that entails the increasing presence of artificial organisms could result problematic and raise concerns in the civil society that can be hardly managed.

3.4.2 Up to now, a specific regulatory framework relating to synthetic biology has not been developed yet, both in Europe and outside. At the international level, in the meantime, governance of synthetic biology is inspired by two competing models: one is growing at the institutional level and is partially inspired by the new governance model (de Búrca and Scott, 2006); the other, which is arising at the level of market actors, is slowly evolving towards forms of more or less effective self-governance (Sørensen and Triantafillou, 2009). Both models share forms of flexible, participatory governance that tend to spread when a given sector is at its infancy. Common premise is the adequacy of the current GMO regulations and the start of processes of adapting.

The surge of this novel field stimulated first attempts of governance.

First, it is worth noting that initiatives belonging to the self-governance paradigm have been developed in the field of synthetic biology at the global level. Initially subjects of the private sector took the initiative by coordinating their growing businesses with a special focus on the biosecurity field. At this stage public authorities remained waiting. Only at a subsequent stage they intervened with specific measures of a soft and flexible nature aimed at further responsibilizing stakeholders with regards biohacking and bioterrorism. In this instance, self-governance is the forefront of a process of new governance, becoming a fluid phenomenon of the self-ordering of a field at its infancy.

In 2006 an international conference was held at the University of California, Berkley, involving a multitude of stakeholders engaged in synthetic biology research (in particular, company administrators, researchers and government representatives) (Leone, 2014; Turker and Zilinskas, 2006). On this occasion, the conference produced a final declaration which recommended 'the organization of an open working group that will undertake the coordinated development of improved software tools that can be used to check DNA synthesis orders for DNA sequences encoding hazardous biological systems' to be made freely available; 'the adoption of best-practice sequence checking technology, including customer and order validation, by all commercial DNA synthesis companies' avoiding, therefore, 'patronizing companies that do not systematically check their DNA synthesis orders'; the 'support [of] ongoing and future discussions within international science and engineering research communities for the purpose of developing creative solutions and frameworks that directly address challenges arising from the ongoing advances in biological technology, in particular, challenges to biological security and biological justice'; the 'support [of] ongoing and future discussions with all stakeholders for the purpose of developing and analysing inclusive governance options, including self-governance, that can be considered by policymakers and others such that the development and application of biological technology remains overwhelmingly constructive' (Synthetic biology 2.0., 2006). For this purpose, the open working group undertook 'the coordinated improvement of software tools for checking DNA sequences' and triggered a multi-stakeholder study 'to develop policy options that might be used to govern synthesis technology' (ibid.), such as a global code of conduct for DNA synthesis companies (EASAC, 2010, p. 2), whose materials should be presented to government representatives.

At the end of 2009 the International Association of Synthetic Biology, an organization involving companies that provide DNA synthesis services and products, adopted a code of conduct in Cambridge (MA.) for 'the safe and responsible use of synthetic DNA' within the framework of 'all regulations and international standards designed to safeguard against the intentional or unintentional abuse of Synthetic DNA' (IASB, 2009, p. 1). The code is mainly aimed at 'protect[ing] the field against misuse' by providing 'guidelines for safe, secure, and responsible commercial or non-commercial DNA synthesis' (ibid., p. 2). In this regard, its goal is expressly limited to biosecurity and biosafety issues (such as the risk of unintended release, biohacking, bioterrorism), triggering a timely and spontaneous process of responsibilization of stakeholders. Besides, it also triggered an action of weak co-ordering with public authorities in the biosecurity field. This also means that many issues related to the protection of individual rights of multiple stakeholders (such as safety at work, patients' and consumers' health, environmental protection, data protection) remain uncovered. These issues belong to an area of competence of public authorities that, though, in front of a private initiative waited for gaining further data before acting. Thus, inaction (of the public sector) seems to be the risk behind these initiatives of the private.

Against the risk of the abuse of synthetic genes used in hazardous applications, the code provides for 'the screening of DNA sequences for genes which can be intentionally abused, for example, in terrorist activities' and 'the restriction of access to synthetic DNA to legitimate users' (ibid., p. 3). Furthermore, records of suspicious inquiries and positive screening hits need to be kept for 8 years. In this regard an effort has been made to collect data in order to achieve needed information. Thus, gene synthesis providers have to promptly inform government authorities 'each time they encounter evidence which clearly suggests possible illegal activities' (ibid., p. 5). 'Gene synthesis companies should always take reasonable steps to determine the relationship of the requested sequences to risk-associated sequences before sending them to customers' (ibid., p. 6). In this regard the code provides a specific procedure establishing 'a benchmark capability for detecting threat sequences' (ibid.). As a threat is identified, a response is provided by the involvement of molecular biologists or similar experts in order to assess it and national authorities must be informed (ibid., p. 7). Gene synthesis companies have to take each step to confirm the authenticity of their customers (customer screening). The code also specifies the formation of a Technical Expert Group of Biosecurity (TEGB), which has the task of reviewing 'current design and implementations of biosafety and biosecurity measures' (ibid., p. 10). In this sense, the biosafety issue appears as the leitmotif of the whole code. In this framework other relevant issues (e.g. the protection of individual rights) are just on the background and unexpressed.

3.4.3   After these attempts to arrange coordinated action on the rising market of synthetic biology, public subjects intervened at the international level under provisions of the existing treaties. Here the governance framework is explicitly inspired by the 'new governance' paradigm. This implied the participation of public authorities with a leading role in order to guide States towards a shared regulation of novel breakthroughs of synthetic biology. However, since the power is distributed among actors, especially thanks to the market, within the international community there is an increasing awareness that no outcome can be achieved without the collaboration of private actors. In this sense, it appears crucial to create governance structures and networks where these subjects are actively involved from the outset, at least in order to validate policy decisions adopted at both national and international level.

This model is the current reference point of the international community. Recently, the debate on the implications of synthetic biology also took place at the UN level, in order to address the framework of responsibility for governments and private actors in this field able to ensure the best environment for the safe development of the sector by the States. In this regard, an international forum on risk assessment, policy and governance of synthetic biology was established under the United Nation Convention on Biological Diversity. In March 2015, following the mandate of the Executive Secretary to the Convention, the Subsidiary Body on Scientific, Technical and Technological Advice (SBSTTA) delivered a report made up of two parts: 'Potential impacts of synthetic biology on biodiversity' and 'Gaps and overlaps with the provisions of the Convention, and other agreements' (Secretariat of the Convention on Biological Diversity, 2015). The draft documents constituting the report were available for peer-review and under a process of deep revision[69] according to a style inspired by new governance.

---

[69] See http://www.cbd.int/emerging/. Accessed 30 March 2015.

The consultation process mainly affected the part I of the report which was largely amended. In general, the report 'aims to support the international debate, and bridge gaps between the science-policy interface, by providing technical information on the potential positive and negative impacts on biodiversity that synthetic biology might entail' (ibid., p. 3). The basic idea is that synthetic biology equally entails opportunities and risks for biodiversity. This implies the adoption of a pragmatic approach able to start from existing regulations in order to adapt their structures to the rising field of synthetic biology and its foreseeable commercial applications. However, in the meantime it is necessary to prepare the revision of the current regulatory framework. Therefore, the strategy emerging at the UN level confirms the 'wait and see' approach in this field and the line which tends to identify synthetic biology and genetic engineering as regards their applications, their risk assessment and, most of all, their regulatory framework. Notwithstanding '[m]ost of the current and near-term applications of synthetic biology involve living organisms that are intended for contained use in research laboratories and industrial settings raising biosafety concerns different from conventional genetic engineering' (ibid., p. 9), '[e]xisting biosafety risk assessment frameworks are likely to be sufficient to assess the risks of current and near-term applications of synthetic biology on the conservation and sustainable use of biodiversity'. However, '[a]s synthetic biology develops, this assessment may need to be revisited' (ibid., p. 10). In this sense, '[c]urrent provisions and procedures established under the Cartagena Protocol on Biosafety, at the international level, and in many existing national biosafety legislations, at the national level, can effectively cover these areas of biosafety concerns' (ibid.). Thus, while waiting for the adaptation of existing provisions, it is urgent to revise risk assessment procedures. This belief was substantially shared at the EU level.

Being an overview drawn within the ambit of the international community, the report mainly considers the position of States under the international law, instead of that of individuals.

In this regard, the second part notes that 'States are under a general obligation to ensure that activities within their jurisdiction or control respect the environment of other States or of areas beyond national jurisdiction or control. This duty to respect the environment does not mean, however, that *any* environmental harm, pollution, degradation or impact is generally prohibited. The duty prohibits a State from causing *significant transboundary* harm and obliges a State of origin to take adequate measures to control and regulate in advance sources of such potential harm' (ibid., p. 68). In this framework there should be room for the application of the precautionary principle, but, surprisingly, according to SBSTTA 'its legal status and content in customary international law has not been clearly established, and the implications of its application to synthetic biology techniques are unclear' (ibid., p. 69). In this regard, the main concern of the report, especially of the part I, seems to be that actually only negative impacts are stressed in this field, and positive ones are neglected. In this sense, the debate needs to be counterbalanced. In fact, while '[m]any components, organisms and some products resulting from synthetic biology techniques may be considered as "living modified organisms resulting from biotechnology" as defined by the Convention on Biological Diversity and, as such, would be subject to its biosafety provisions as per Articles 8(g) and 19', the Convention does not address only potential negative impacts, but it 'also recognizes potential positive effects of modern biotechnology and provides for the access to and transfer of technologies, including biotechnology, that are relevant to the conservation and sustainable use of biological diversity' (ibid., p. 69). This clearly puts in question our notion of biodiversity and whether it is destined to evolve due to the possible presence of Syn-Bio organism. In this regard, '[o]rganisms resulting from synthetic biology techniques may fall under the definition of "living modified organisms" under the Cartagena Protocol for Biosafety' which was devised, though, by referring to genetically modified organisms. In fact, according to the Convention ' "[b]iotechnology" is defined in Article 2 of the Convention as any technological application that uses biological systems, living organisms, or derivatives thereof, to make or modify products or processes for specific use (Art. 2)' (ibid., p. 82). In this regard the IUCN Guide to the Convention on Biological Diversity states that 'this definition was "designed to include both present and future technologies and processes"' (ibid.). But it is worthwhile to note that the application of biosafety provisions of the Convention to synthetic biology depends on the interpretation of expressions such as 'living modified organisms resulting from biotechnology' and The Convention does not define 'biological systems,' 'living organisms,' or 'derivatives thereof' (ibid.). Nor the expression 'living modified organisms' (ibid.). However, since 'it is possible that living modified organisms resulting from synthetic biology techniques could cause adverse effects on the conservation and sustainable use of biological diversity', a subsequent integration to the Convention should explicitly address concerns arose by synthetic biology (ibid., p. 90). Following the 2013 decision XI/11 of The Conference of the Parties, which explicitly addressed the matter of synthetic biology, '[o]nce entered into force, the Nagoya – Kuala Lumpur Supplementary Protocol on Liability and Redress to the Cartagena Protocol on Biosafety will require Parties to provide at the national level for rules and procedures that address damage from living modified organisms resulting from synthetic biology techniques' (ibid., p. 69). The objective of the Supplementary Protocol 'is to contribute to the conservation and sustainable use of biological diversity, taking also into account risks to human health, by providing international rules and procedures in the field of liability and redress relating to living modified organisms' (ibid., p. 89). However, the Supplementary Protocol only regulate damages and State's liability as regards 'adverse effect[s] on the conservation and sustainable use of biological diversity, taking also into account risks to human health' that are of 'significant' scale and where a causal link is clearly established (ibid., p. 90). Since under the Cartagena Protocol, provisions for Advanced Informed Agreement (AIA) do not apply to the transboundary movement of living modified organisms intended for direct use as food or feed, or for processing, the control over SynBio organisms will likely rest on current risk assessment procedures defined for genetically modified organisms (ibid., p. 88).

Once the State's domestic jurisdiction is at stake in the technoscientific advance, what tends to be apparent are mainly issues related to States' biosecurity. The international regulatory framework involving the development of synthetic biology also includes the Biological Weapons Convention which prohibits the development, acquisition and transfer of microbial or other biological agents for non-peaceful purposes (ibid., p. 90). Contrary to other regulations, this tool appears to be well equipped as regards challenges of synthetic biology. It sets out legally binding rights and obligations when 'microbial or other biological agents or toxins, including those which are components, organisms and products resulting from synthetic biology techniques' are produced (ibid., p. 70). Furthermore, in accordance with The Agreement on the Application of Sanitary and Phytosanitary Measures (SPS Agreement) the WTO could intervene with *ad hoc* sanitary and phytosanitary measures when 'organisms resulting from synthetic biology could, depending on the specific case, be considered as causing risks to animal or plant life or health arising from the entry, establishment or spread of pests, diseases, disease-carrying organisms or disease-causing organisms' (ibid., p. 95) or as risks to human or animal life or health 'arising from additives, contaminants, toxins or disease-causing organisms in foods, beverages or feedstuffs', (ibid., p. 93).

With the exception of the Biological Weapons Convention, according to the report it seems that '[m]ost regulatory mechanisms discussed in the present document were developed before the term synthetic biology became widely used and therefore they were not intended to cope with the scope and scale that some of the potential impacts of synthetic biology may have, including those with low and very low probability, but very high impacts', (ibid., p. 112). Accordingly, it concludes that: 'the components, organisms and products resulting from synthetic biology would fall under the scope of a number of regulatory mechanisms. While some instruments are sufficiently broad to address some of the current issues related to synthetic biology, gaps still exist relating to the practical implementation of these instruments to ensure the conservation and sustainable use of biodiversity, and the fair and equitable sharing of the benefits arising from the utilization of genetic resources' (ibid., p. 113). Thus, until actions are taken at the national and international level in order to amend the existing regulatory framework, all rests on current risk assessment methodologies developed before synthetic biology was devised.

3.4.4 Initiatives are being developed with a view to accompanying the increasing development of synthetic biology not only at the global level. At the national level we find a proliferation of voluntary acts of governance. In many countries measures aimed at fostering flexible and adaptive governance models are being developed through the more or less spontaneous engagement of private actors and the intervention of public agencies (SCENIHR et al., 2014a, p. 17ff.). These initiatives have spread all over Europe and outside it, and have delivered statements, recommendations, guidelines and codes of conduct (from governmental bodies, national academies, international networks), that show the existence of a fluid and evolving governance landscape. They have arisen in Germany,[70] The Netherlands,[71] Italy,[72] the

---

[70]The Zentrale Kommission für die Biologische Sicherheit (ZKBS, Central Committee on Biological Safety, Germany (2012). http://www.bvl.bund.de/Shared Docs/Downloads/06_Gentechnik/ZKBS/01_Allgemeine_Stellungnahmen_deutsch/ 01_allgemeine_Themen/Synthetische_Biologie.pdf?__blob=publicationFile&v= 3. Accessed 28 March 2015; Deutsche Forschungsgemeinschaft (2009). *Synthetische Biologie: Stellungnahme*. ed. Deutsche Forschungsgemeinschaft.

[71]Royal Netherlands Academy of Arts and Sciences (2007) *A Code of Conduct for Biosecurity. Report by the Biosecurity Working Group.* pp. 1–44. https://www. knaw.nl/nl/actueel/publicaties/a-code-of-conduct-forbiosecurity.   Accessed 28 March 2015; The Netherlands Commission on Genetic Modification (COGEM) (2013) *Synthetic Biology – Update 2013. Anticipating Developments in Synthetic Biology.* COGEM Topic Report CGM/130117-01, http://www.google.it/url?sa= t&rct= j&q= &esrc= s&source= web&cd= 1&ved= 0CCMQFjAA&url= http% 3A%2F%2Fwww.cogem.net%2Fshowdownload.cfm%3FobjectId%3DAB3BB083- 1517-64D9-CC40A06E6D75337C%26objectType%3Dmark.hive.contentobjects. download.pdf&ei=ytwWVe_9IIbfywPvhIGgBA&usg=AFQjCNEWA_0B1OqnbkCzHx 0-n6A2NSczDQ&sig2= YQcKetoDwjmHFAXW239Fxg &bvm= bv.89381419,d.bGQ. Accessed 28 March 2015.

[72]In 2010 Italy also adopted a set of guidelines concerning synthetic biology. See Comitato nazionale per la biosicurezza, le biotecnologie e le scienze della vita (2010) *Codice di condotta per la biosicurezza*, http://www.governo.it/ biotecnologie/documenti/Codici_condotta_biosicurezza.pdf. Accessed 28 March 2015.

United Kingdom,[73] Switzerland,[74] Belgium[75] (with the Biosafety and Biotechnology Unit [SBB]), and the United States. In this latter instance (i.e. the US), after the International Congress on Recombinant DNA molecules held at Asilomar (California) in 1975 (Berg, 2008), which tried to limit the biological risk related to the commercial diffusion of biotechnologies under quite stringent guidelines, in 2010 the American Presidential Commission for the Study of Bioethical Issues (PCSBI) dealt with synthetic biology (PCSBI, 2010; Leone, 2014, p. 152). This intervention is in line with the model of flexible and self-coordinated governance aimed at fostering the spontaneous engagement of actors within a mainly adaptive pattern. In this regard, the PCSBI eluded unlikely forms of hard legislation for less undemanding forms of self-coordination that better suit a landscape where research, as well the market, are dominated by the propulsive role of private enterprises. Thus, '[a]s a corollary to the principle of intellectual freedom and responsibility, the Commission endorses a principle of *regulatory parsimony*' (PCSBI, 2010, p. 5). In other words, the resort to forms of traditional regulation needs to be exceptional and not customary. The Presidential Commission also suggested that the government promote 'a culture of individual and corporate responsibility and self-regulation by the research community' (ibid., p. 13). Accordingly, in November 2013 the National Institute of Health (NIH) adapted its Guidelines for research involving recombinant DNA molecules to cover synthetic nucleic acid molecules too (SCENIHR et al., 2014a, p. 17ff.). The NIH

---

[73]Royal Academy of Engineering (2009) *Synthetic Biology: Scope, Applications and Implications.* https://www.raeng.org.uk/societygov/policy/current_issues/synthetic_biology/default.htm. Accessed 28 March 2015; Health and Safety Executive (HSE) (2012) *Synthetic Biology. A Review of the Technology, and Current and Future Needs from the Regulatory Framework in Great Britain.* Rr944 Research Reports, http://www.hse.gov.uk/research/rrhtm/rr944.htm. Accessed 28 March 2015.

[74]Platform of the Swiss Academy of Science, *Forum for Genetic Research, Synthetic Biology*, http://www.geneticresearch.ch/f/themen/Synthetic_Biology/index.php. Accessed 28 March 2015.

[75]Biosafety and Biotechnology Unit (SBB) (2012) *Synthetic Biology: Latest Developments, Biosafety Considerations and Regulatory Challenges*, http://www.biosafety.be/PDF/120911_Doc_Synbio_SBB_FINAL.pdf. Accessed 30 March 2015.

Guidelines for research involving recombinant or synthetic nucleic acid molecules[76] provides detailed safety practices and containment procedures in the case of synthetic biology. According to these, synthetic DNA segments are regulated in the same way as their natural counterparts, thus adopting the principle of substantial equivalence which is applied to genetically modified organisms. However, while in Europe the principle establishes an equivalence between organisms derived from synthetic biology and genetically modified organisms, here the equivalence is between the organisms derived from synthetic biology and natural organisms. Besides, if synthetic DNA segments are not expressed *in vivo*, and, thus, they can neither replicate nor generate nucleic acids that cannot replicate in any living cell, they are exempted from the NIH Guidelines (ibid., p. 20).

In this sense, as was understandable, the existing regulatory framework provided for genetic modification techniques was deemed suitable for short-term developments in synthetic biology, albeit with the necessary adaptations. This approach is shared by several countries, including, to some degree, Europe.

In the US there is a coordinated framework for the assessment and the regulation of biological products, including their intended release into the environment, drawn up by the Environmental Protection Agency (EPA), the US Department of Agriculture's Animal and Plants Health Inspection Service (APHIS), and the Food and Drug Administration (FDA). 'This coordinated framework is considered appropriate for regulating most of the organisms obtained by near-term SynBio applications' (ibid.). However, some adjustments might become necessary in the future due to the extreme novelty of this research field. Indeed, there is the risk that, 'unlike plants obtained by older genetic modification techniques, the engineering of organisms without the use of a (component of a) plant pest would shift them out of the regulatory review of APHIS' (ibid.). Besides, the EPA's risk assessment could be further challenged by the increased influx of genetically modified microbes intended for commercial use (JCVI, 2014, p. 5). This fact could lead to

---

[76]See http://osp.od.nih.gov/sites/default/files/NIH_Guidelines_0.pdf. Accessed 28 March 2015.

'regulatory delays for microbial products, inadequate review and/or legal challenges' (ibid.).

In sum, at the international, transnational and national level two models overlap and have the common characteristics of being flexible and, to some degree, proactive. One belongs to the model of new governance. It was developed by UN institutions and is aimed at fostering States' investments in the strategic sector of synthetic biology within a stable framework which is under monitoring. Accordingly, at the national level there is a process of arrangement of governance structures, both within the institutional and non-institutional dimension, which aims to create the premises for its growth. The other belongs to the self-governance model. It was developed at the level of market actors and moves the presence of institutional actors into the background. It is characterized by a clear process of self-arrangement of governance and presupposes the subsequent action of policymakers. Both models mainly focus on biosecurity issues with a limited range of issues dealt with. Furthermore, they share the belief that the current regulatory framework provided for genetically modified organisms is the necessary starting point in this process of governance and thus there is mainly the need of updating current risk assessment methodologies. Behind them lies the idea of the substantial equivalence between synthetic biology and genetic engineering applications, namely the idea that synthetic biology is nothing else but a new version of genetic modification that allows us to perform the same processes and products as the old genetic engineering. This idea is also common within the EU governance and has some consequences on its anticipatory dimension.

3.4.5 Given this set of initiatives at the global level, up to now Europe showed to be quite syntonic with the on-going worldwide trend. At the EU level, the response given by Community authorities to the challenge of nanotechnologies pointed the way to taking up the case of this latest evolution of biotechnology too. Thus, as for nanotechnologies, the European Union has adopted a 'wait and see' strategy in order to: (i) understand which direction synthetic biology research was going to take; (ii) to develop a public debate on its ethical and societal implications; and (iii) to assess whether the current regulatory framework was appropriate for this new challenge. As a corollary to this approach, in this instance too, EU authorities have stood by to adopt all the necessary measures needed to adjust and amend the current GMO legislation according to a case-by-case approach. In the case of synthetic biology too, one of the first measures was to involve the ethical advisory board of the European Union (EGE) in delivering a report on the ethical implications of synthetic biology. Thus, once requested, the EGE mediated among a multitude of interests: the stance of EU institutions towards boosting the European economy, the market interest in investing in the best technologies at stake, and the interest of the public at large in safe and environmentally friendly products. However, unlike nanotechnologies, the EGE did not have a mandate for dealing with only a specific field (e.g. medical applications), but was able to consider the entire range of synthetic biology applications. Besides, a number of different competing agencies and committees were involved at the early stage, making difficult to understand what the driving factor in this case was. These latter preliminary measures (in particular EU agency and advise) were aimed at triggering a process of the gradual adaptation of the current regulatory framework existing for genetically modified organisms on the basis of the conceptual premise that synthetic biology and genetic modification are equivalent.

As a first step, in 2009 the ethical advisory board of the European Union pointed the way. According to the EGE, as generally recognized, at that time the general consensus was that SynBio applications should fall within the existing regulation, in particular concerning GMOs, since synthetic biology should be deemed as an 'extension of genetic modified technologies' (EGE, 2009, p. 27). In this regard, the safety and regulatory aspects are considered to be covered by the existing EU GMO regulatory framework, including, in particular, Directive 2001/18/EC on the deliberate release of GMOs,[77] Regulation (EC) No. 1946/2003 on transboundary movements of genetically modified organisms

---

[77]European Union (2001) Directive 2001/18/EC of the European Parliament and of the Council of 12 March 2001 *on the Deliberate Release into the environment of genetically modified organisms* and repealing Council Directive 90/220/EEC, *Official Journal of the European Union*, L106, 17/04/2001,

that implemented the provisions of the Cartagena Protocol within the European Union[78] and Council Directive 2009/41/EC on the contained use of genetically modified microorganisms[79] (ibid.). Furthermore, Directive 2000/54/EC on the protection of workers from the risks of biological agents at work[80] also needs to be considered (SCENIHR et al., 2015, p. 18). According to the ethical advisory board of the European Union, most of synthetic biology applications would fall within the main area of the contained use of genetically modified microorganisms. The matter, previously regulated by directives 90/219/CEE and 98/81/CE, flowed into the 2009 document, which has taken up a series of amendments delivered over the years (Fontana, 2010). Since Directive 2001/18/EC regulates the deliberate release into the environment of genetically modified microorganisms, it pursues environmental and health purposes (EGE, 2009, p. 28). When a genetically modified organism is introduced into the environment, without any specific confinement measure being taken to restrict contact between it and the population, as well as the environment, Directive 2001/18/EC applies (Fontana, 2010, p. 2). Its objective is to approximate the laws, the regulations and administrative provisions of member States and to protect human health and the environment when: α) there is a deliberate release into the environment of genetically

---

http://eur-lex.europa.eu/resource.html?uri=cellar:303dd4fa-07a8-4d20-86a8-0baaf0518d22.0004.02/DOC_1&format= PDF. Accessed 24 March 2015.

[78] European Union (2003) Regulation (EC) No. 1946/2003 of the European Parliament and of the Council of 15 July 2003 *on Transboundary Movements of Genetically Modified Organisms, Official Journal of the European Union*, L287, 5/11/2003, http://eur-lex.europa.eu/legal-content/EN/TXT/PDF/?uri=CELEX: 32003R1946&from= EN. Accessed 24 March 2015.

[79] European Union (2009) Directive 2009/41/EC of the European Parliament and of the Council of 6 May 2009 *on Contained Use of Genetically Modified Micro-organisms, Official Journal of the European Union*, L125, 21/05/2009, pp. 75–97, http://eur-lex.europa.eu/legal-content/EN/TXT/PDF/?uri=CELEX:32009L0041&from=EN. Accessed 24 March 2015.

[80] European Union (2000) Directive 2000/54/EC of the European Parliament and of the Council of 18 September 2000 *on the Protection of Workers from Risks Related to Exposure to Biological Agents at Work* (seventh individual directive within the meaning of Article 16(1) of Directive 89/391/EEC), *Official Journal of the European Union*, L262, 17/10/2000, pp. 21–45, http://eur-lex.europa.eu/legal-content/EN/TXT/PDF/?uri= CELEX:32000L0054&from= EN. Accessed 4 April 2015.

modified organisms for any other purposes than placing them on the Community market; and *β*) the placing on the market of genetically modified organisms as or in any product within the European Union (EGE, 2009, p. 28). It encompasses a classification of contained uses or activities involving GM microorganisms into four classes depending on their potential risks for health or the environment (i.e. negligible, low, medium, high). The Directive presents normal minimum requirements and *measures* necessary for each level of containment. Directive 2001/18/EC on the deliberate release of GMOs should regulate, in particular, the field of trials, cultivation and the commercial release of genetically modified organisms, although it is predominately applied to regulate the field of trials, cultivation and the commercial release of only genetically modified *plants* (SCENIHR et al., 2014a, p. 16).

Directive 2009/41/EC on the contained use of genetically modified microorganisms creates a classification of contained uses or activities involving genetically modified microorganisms in 4 classes depending on their level of risk for human health and the environment (i.e. again: negligible, low, medium, high). Belgium, among other member States, implemented it into its national legislation by including also genetically modified organisms and pathogens for human, animals and plants. Belgium is not the only case. The Directive was used, for example, as a basis for national legislation covering work with biological agents in Switzerland (ibid.).

In this regard, since the European Union mainly resorts to pre-existing regulation, the current regulatory framework that synthetic biology faces in Europe is generally deemed as one of the widest and most stringent landscapes in the world (Tiberghien, 2006). However, some small adjustments are being already considered for the future.

Some limits can impede the application of the GMO regulatory framework to the case of synthetic biology. For example, 'GMO Directives apply to "any biological entity capable of replication or transferring genetic material". Synbio systems such as protocells are therefore not (yet) considered as living organisms', since they lack of the capability to self-replicate (SCENIHR et al., 2014a, p. 19).

This fact implies that protocells and nanobiotechnologies, without the capability to self-reproduce, should fall within the domain of nanotechnologies or chemistry, as long as they do not produce living organisms (SCENIHR et al., 2014b, p. 27). It is clear that a grey zone is produced in this way. Furthermore, some concerns about the intentional, accidental and unintentional release of synthetic organisms into the environment, the unexpected outcome of any deliberate release, the possible 'dual use', the diffusion of bioinformation and the risks of biohacking and bioterrorism persist (Stoke and Bowman, 2012, p. 239; Bennet et al., 2009). For example, there are concerns with regard to the interaction of xenobiological systems with the environment, humans and other natural organisms (SCENIHR et al., 2015a, p. 35). Moreover, organisms produced through genetic part libraries and methods may lead to more intense changes to the genomes of living organisms than traditional techniques of genetic modification (ibid., p. 47). These effects may also be widened by the general accessibility of these techniques. In this sense, do-it-yourself biology (DIYbio) although it does not pose any new hazard for humans and the environment might further increase the probability of unintentional harm (see below). Finally, issues on the human (and environmental) exposure of workers, users and consumers run the risk of not being adequately covered by the current regulation on GMOs (Ruggiu, 2012).

The surge of synthetic biology falls not only under the existing framework of legally binding norms. There is also a dimension of a moral nature shared by the international community that will have an impact on synthetic biology. In its opinion, the EGE recalled the ethics framework made up of basic human rights established around the concept of human dignity and other ethical principles set out at the level of international law (the UN, UNESCO, Council of Europe, EU) (EGE, 2009, p. 39). This ethics framework covers documents of non-binding nature such as the Oviedo Convention,[81] the UNESCO

---

[81]Council of Europe (1997) *Convention for the Protection of Human rights and Dignity of the Human Being with regard to the Application of Biology and Medicine* (Convention on Human rights and Biomedicine or the Oviedo Convention) (CETS n. 164), adopted in Oviedo on 4 April 1997 (entered into force on 1 December 1999).

Universal Declaration on Human Genome and Human Rights,[82] the UNESCO Universal Declaration on Bioethics and Human Rights,[83] the Declaration of Helsinki[84] of the World Medical Association (WMA) and its Statement on Nuclear Weapons,[85] and the EU Charter on Fundamental Rights[86] (ibid., p. 43).

However, some of these norms have either become legally binding soon afterwards, like in the case of the Nice Charter at the end of 2009, or are already legally binding in several broad geographical areas, such as in the case of human rights law in Europe and the American continent through the existing treaties on human rights, such as the European Convention on Human Rights.[87] Furthermore, even though they cannot be applied by supranational courts, the above-mentioned documents belong to soft law and can exert legal effects even in the field of synthetic biology (Ruggiu, 2012, p. 156 nt. 2).

The EGE focused on this ethical framework in some detail. In particular, starting from the centrality acquired by human dignity in Article II 6,1[88] of the Lisbon Treaty,[89] the EGE analysed the notion of dignity from the ethical view point, identifying two main meanings: as a restrictive principle preventing human beings from

---

[82] UNESCO (1997) *Universal Declaration on the Human Genome and Human rights*, adopted by the UNESCO General Conference in 11 November 1997 and subsequently endorsed by the United Nations General Assembly in 1998.

[83] UNESCO (2005) *Universal Declaration on Bioethics and Human rights*, adopted by acclamation by the 33rd session of the General Conference of the UNESCO in 19 October 2005.

[84] World Medical Association (WMA) (2008) *Declaration of Helsinki, Ethical Principles for Medical Research Involving Human Subjects*, adopted in Seoul in October 2008.

[85] World Medical Association (WMA) (2008) *Statement on Nuclear Weapons*, adopted in Ottawa on October 1998 and amended in Seoul on October 2008.

[86] European Union (2000) *Charter of Fundamental Rights of the European Union*, adopted in Nice on 7 December 2000.

[87] Council of Europe (1950) *Convention for the Protection on Human Rights and Fundamental Freedoms* (ECHR) (CETS n. 5), adopted in Rome on 11 November 1950 (entered into force on 3 September 1953).

[88] It is meaningful that, when it had to refer to human dignity, the EGE did not recall Article 1 of the EU Charter, which in 2009 was not yet legally binding, but Article II 6,1 of the Lisbon Treaty.

[89] European Union (2007) *Treaty on the European Union and Treaty on the Functioning of the European Union* (consolidated versions 2010/C 83/01), adopted in Lisbon on 13 December 2007 and entered into force on 1 December 2009.

being treated as mere objects, or as an enabling principle which ensures individual freedom of action and autonomy in decision making (Brownsword, 2008; EGE, 2009, p. 39). This underlining of the role of personal dignity seems to set out quite a stringent limit for those SynBio applications on human beings. However, outside this limit, applications of synthetic biology seem to find promising spaces for a further development.

Beyond the discussion on the barrier between natural and artificial, which is questioned by synthetic biology, and the related question of the environment we want to live in, its 'potential applications in fields of biomedicine, biopharmaceuticals, chemicals, environment and energy' are addressed. Another question which arises is the specific issue of the '[u]nexpected interactions between synthetic microorganisms and the environment or other organisms' and that of 'risks to the environment and public health' (ibid., p. 42). The EGE seems thus to think that the field of synthetic biology cannot be reduced to the simple, though important, issue of biosecurity. In this regard, the EGE recalls that 'the protection of health is a key condition for the marketing of products resulting from synthetic biology' (ibid., p. 42). Procedures and methods have been established for ensuring the due precautions, 'but long-term health-related risks associated with the ecological effects of synthetic biology are hard to predict' (ibid.). Analogously to what the EGE concluded regarding nanotechnologies, 'risk assessments used for synthetic biology are designed not only as a technical tool [but also] for the safe governance of synthetic biology in order to protect human dignity and the autonomy of persons directly (medical applications) or indirectly (exposure to synthetic biology products if released into the environment)' (ibid.).

Unlike the conclusion to the UN draft reports, the role of the precautionary principle is stressed at the European level. In fact, according to the EGE, in the environmental field, 'the precautionary principle is a dynamic tool to follow developments in a sector and continuously verify that the conditions for the acceptability of a given innovation are fulfilled' (ibid.). As regards 'the dangers of potentially harmful organisms being inadvertently released during the experimental phase', according to the EU ethical advisory board '[f]reedom of research cannot be invoked if serious or irreversible risks to human health or the environment may occur' (ibid., p. 43). In this regard, the rising EU approach is rather different from the one that is taking shape at UN level. However, just as they are at the international level, so biosecurity issues are also being addressed by the EGE, which points out that, although '[g]iven the present state of knowledge, the design and production of entirely novel pathogens for terrorist and/or maleficent uses may seem unlikely', there are databases of toxic and infective sequences available (ibid.).

Unlike the UN approach, in the number of ethical principles to be taken into account, the EGE also mentions a principle of justice meant as distributive justice à la Rawls, involving issues related to technology divide, common heritage and intergenerational justice (ibid., p. 45). As regard the issue of intellectual property, the EGE mentions the debate on 'whether genetic information is the common heritage of mankind, making gene patenting inappropriate', which includes that on open source, as well as the discussion on the danger that some uses such as the production of chimeras from germ cells or the inclusion of artificial chromosomes into the human cells could breach human dignity, infringing thus Article 53(a) of the European Patent Convention (inventions in which commercial exploitation would be contrary to morality) (ibid., pp. 45–46).

Particular attention is paid by the EGE to transparency and public participation, which is a pillar of recent Community strategies on emerging technologies. In this regard, the EGE reported the diffidence expressed in some surveys by groups from civil society on the potential of soft regulation in the context of synthetic biology, as well as industry self-regulation (ibid., p. 38). For this reason, public dialogue is deemed to be fundamental to create trust and confidence within civil society around this emerging field.

Given this ethical analysis of the impact of synthetic biology, the EGE has drawn up an articulated set of recommendations (ibid., p. 48ff.). (A) First of all, the EGE laid down the preconditions of the subsequent work of other EU agencies by addressing the need for a clear definition of synthetic biology shared at the international level (ibid., p. 48). (B) Then, in order to implement safety, the European Commission should initiate a study of risk assessment procedures at the EU level by: (a) monitoring relevant biosafety procedures; (b) identifying gaps in regulation; and (c) elaborating mechanisms for filling those gaps. When the above safety rules have been defined, the EU should launch an international debate on them and integrate them with tools 'for the monitoring of the implementation of such provisions' (ibid.). (C) As in the case of nanotechnologies, a code of conduct for research into synthetic biology should be provided too. In particular, this code should ensure that 'synthetic biology organisms are manufactured in a way that they cannot autonomously survive if accidental release into the environment would take place' (ibid., p. 48). However, a code on only research runs the risk of reiterating the error made with the 2008 Commission code of conduct (Ruggiu, 2014). On that occasion, the decision to create a code of conduct was perceived by the researchers' group as form of concentration of the whole weight of responsibility onto a restricted part of stakeholders, by making them responsible for the consequences, both intended and unintended, of the whole innovation cycle (production, retail, recycle) (Meili et al., 2011, p. 6). This fact probably contributed to the limited success of this initiative (Mantovani et al., 2010, p. 17; Grobe et al., 2011, p. 12). (D) In the case of environmental applications such as bioremediation, a threefold set of interests needs to be taken in account: the protection of workers and citizens, freedom of consumers and the protection of animals and plants with a responsible use of synthetic organisms. Thus, before a synthetic organism is released into the environment, 'long term impact assessment studies must be carried out'. (E) The collected data should, then, be evaluated 'taking into account the precautionary principle and measures foreseen in the EU legislation (Directive on the deliberate release into the environment of genetically modified organisms)' (ibid., p. 49). (F) In the case of the use of synthetic organisms for providing new forms of compounds aimed at substituting petrol and other fossil fuel energy or for the production of raw materials in the chemical industry, the EGE recommended monitoring 'the authorization procedures for the production of synthetic biology-derived chemicals and materials, if not identical to equivalent substances, by taking into consideration (a) risk assessment factors and (b) the safety of workers exposed to synthetic biology chemical agents and (c) environment protection' (ibid., p. 50). (G) While in the case of the use of synthetic biology derived organisms in food and feed there is the mandatory labelling provided for genetically modified organisms, the EGE requested special protection for consumers' rights, whether they are used in cosmetics or in the textile industry. However, this recommendation raises the question of the suitability of the equivalence principle between synthetic biology derived organisms and GMOs and, connected to this, the question of specific labelling for all synthetic biology derived products, not only in cosmetics or the textile industry. (H) In the instance of its applications in the medical field, it also recommends, in a rather vague formulation, taking into account 'specific ethics considerations', when specific measures to implement risk assessment procedures could have been suggested, because of the high degree of uncertainty connected to this field (ibid., p. 51). However, the ample terms of this advice pave the way to an overuse of ethics, by depriving it of sense and force through an excess of vagueness and bureaucratization (MASIS, 2009, pp. 32, 38; Tallacchini, 2009). (I) The use of synthetic biology for military purposes can create the possibility of misuse, biohacking and bioterrorism, and this needs the right balance between transparency and security. However, instead of creating a database on the current synthetic biology applications in order to map the whole phenomenon, the EGE recommended embedding ethics into the curricula of biosecurity scientists, ensuring that existing databases are available to all those who use them, creating legal ways for companies to report to competent authorities when asked, 'identify[ing] the chain of responsibility for placing particular sequences in the database(s) and identify them as potentially harmful' (ibid., p. 52). (J) In this framework, the EGE suggested the use of some flexible tools by fostering the integration of different disciplines and public engagement according to a typical new governance style. Thus, at the governance level the EGE suggested developing public dialogue by integrating studies on the ethical and social implications of synthetic biology according to a case-by-case logic, finding new means for encouraging 'transparency without creating risks of misuse', 'identify[ing] areas where soft law will provide sufficient protection and areas where hard law is deemed necessarily', 'review[ing] the legislation applicable', "address[ing] the relevant stakeholders (scientists, industries, military agents, and political and administrative agents)', and 'establish[ing] ethical, preferably global, guidelines' (ibid., p. 53). (K) In order to enlarge the participatory dimension of Community governance on specific issues of patenting too, the EU ethical advisory board recommended launching a public debate on ways of ensuring open access to results of research where the use of copyright does not appear to be appropriate. (L) Furthermore, it asked the European Patent Office (EPO) and the National Patent Office 'to refer contentious ethical issues of a general relevance to the EGE for consideration' to identify a 'class of inventions that ought not to be directly exploited commercially', according to the EU Patent Directive 98/44/EC[90] which 'defines the EGE as the Body to assess the ethical implications related to patents' (ibid., p. 54). (M) Wisely the EGE pointed out that '[i]f trials involving synthetic biology products are being conducted in developing and emerging countries the same ethical standards as are required within the EU must be implemented' (ibid.). This measure aims to avoid the creation of different levels of protection depending on the provenience of research in the same field. In this regard, the Group also recommends the implementation of UN Millennium goals and the adoption of action aimed at avoiding new gaps between the EU and developing and emerging countries. (N) Since in new governance participation has a key role, EGE tackled the issue of public dialogue with regard to the development of this sector. Thus, in order to avoid the diffusion of unjustified fears, the engagement of civil society is recommended. In this regard, the EGE suggests the adoption of wide-ranging action aimed at promoting public dialogue and the engagement of stakeholders 'in order to identify main societal concerns in the different areas covered by synthetic biology' and it urges the media to disseminate the information in a responsible manner (ibid., p. 55). At the centre of this inclusive action triggered by the European Commission there should be research which has to develop an interdisciplinary approach by means of research

---

[90] European Union (1998) Directive 98/44/EC of the European Parliament and of the Council of 6 July 1998 *on the Legal Protection of Biotechnological Inventions, Official Journal of the European Union*, L 213, 30/07/1998, pp. 0013–0021, http://eur-lex.europa.eu/LexUriServ/LexUriServ.do?uri= CELEX:31998L0044:EN:HTML. Accessed 26 March 2015.

Framework Programmes and their allocation of funding for the following research areas: risk assessment and safety; security uses of synthetic biology; ELSI studies; governance; and science and society. Since 'synthetic biology could lead, in future, to a paradigm shift in understanding concepts of life', it suggested opening up an intercultural forum to address the issues from philosophical and religious viewpoints as well (ibid., p. 56).

In conclusion, the EGE's focus appears to be of a wide scope, proposing a broad range of solutions. This wide effort was aimed at preparing the conceptual preconditions of the subsequent work of implementation by other Community bodies within the current framework of the GMO legislation. By adopting a case-by-case approach it requested for a revision process by starting from the current legislation on GMOs in order to evaluate its adequacy in the face of the used risk assessment methodologies. However, soon the focus on the adequacy of the GMO legislation shifted from the update of rules to the update of risk assessment procedures. Thus, until new evidence suggests adopting a different regulatory framework, the idea of the extension of the GMO legislation to synthetic biology risks of being questionable since it presupposes a principle, namely the principle of substantial equivalence, which is debated from the scientific standpoint. This raises the question of the adequacy of current GMO legislation again.

3.4.6 After the EGE's opinion, several reports were delivered from different agencies. This runs the risk of a weakening of democratic legitimation of the EU action. Against this shortcoming new governance tools should be meant as the remedy. Moreover, there is a risk of an overlap between different bodies working on the same research field at the EU level. In particular, it might not be enough clear whether one agency integrates the other, or one is joined to substitute the other. However, a general common objective seems to emerge from these documents and reports: the goal of building the foundations for the efficacious application of the regulatory framework provided for genetic modification to synthetic biology. In this sense, the picture of this great governance initiative suggests that the EGE drafted the overall strategy of the European Union, while other agencies were charged with implementing it in each sector. This process of reorganization of the existing governance framework needs to be developed at both the EU and national level.

At the national level, for instance, there is the need for a stable EU framework and the implementation of the necessary conditions for the development of the sector. In this context participatory means can easily take shape. Thus, the European Academies Science Advisory Council (EASAC), an organization formed by the national science academies of the EU member States, recently issued a special report in 2010 on scientific opportunities and good governance in the field of synthetic biology, which sets out a number of recommendations at the EU member State level mainly aimed at augmenting public dialogue (EASAC, 2010). The aim of the EASAC recommendations is to support the activity of member States that are involved in the development of synthetic biology through strengthening structures which can foster the rapid growth of the sector. They focused on six hotspots, namely: research capacity, training of future scientists, EU competitiveness, research governance, product regulation and societal engagement (ibid., pp. 1–2). (1) First of all, there is the need to support research by strengthening disciplines, developing centres of excellence that work in interdisciplinary ways, funding initiatives that network small laboratories across Europe and fostering transnational research and standardization platforms and tools. (2) Then, the training of the next generation of scientists in bridging strategic disciplines such as biology, engineering, and incorporating skills from chemistry, physics and informatics. (3) The increase of international competition means that the European Union must implement EU Structural Funds for innovation with a view to strengthening industry with high quality research, accordingly with implications for smaller companies. (4) Since, according to the EASAC, 'regulation should neither stifle research nor impede transparency in communication', the 'safety requirements [...] in place for other research' (i.e. biotechnology) need to be adapted to synthetic organisms. Furthermore, the European Union has to avoid a situation where patents rules 'unreasonably deter competition and slow down the translation of research advance into products' (ibid., p. 2). (5) Dissenting from the SCENIHR opinion (see below), according to the EASAC '[t]he EU control of approval of novel products [. . .] should generally be subjected to the same regulatory framework as exists for products from other sources' (ibid.). (6) In the case of synthetic biology too the importance of social dialogue is deemed to be strategic since the report recommends making 'provision of accessible and accurate information about synthetic biology [. . .] pro-actively, not simply as a reaction to emotive media reports' (ibid.).

To understand the process of adaptation of the existing regulatory framework on GMOs, there is the need of the preliminary analysis of a set of guidelines provided for genetically modified organisms, first of all in the food and agricultural sector. In 2012, at the request of the European Commission, the European Food Safety Authority (EFSA) delivered an opinion regarding the safety assessment of genetically modified organisms, which also dealt with issues related to synthetic biology (EFSA, 2012). The opinion evaluated eight different breeding techniques including synthetic biology. After the opinion on these eight genetic modification techniques ranging from cisgenetic and intragenetic modification to synthetic biology, the EFSA concluded that both its guidance for risk assessment of food and feed from GMOs (EFSA, 2011) and the environmental risk of GMOs (EFSA, 2010) 'are applicable for the evaluation of food and feed products derived from cisgenic and intragenic plants and for performing an environmental risk assessment and do not need to be developed further' (ibid., p. 2). This is because some of these techniques, such as cisgensis and trangenesis, can be performed thanks, in part, to synthetic biology. For example, '[t]ransgenics plants can contain genetic elements, e.g. coding and regulatory sequences, from any organism (eukaryotic, prokaryotic) as well as novel sequences synthesised *de novo*' (ibid., p. 9). The same is applicable to cisgenesis.[91] However, a lesser amount of event-specific data is needed for risk assessment on a case-by-case basis. In this regard, '[t]he risks to human and animal health and the environment will depend on exposure factors

---

[91]'*To produce cisgenic plants any suitable technique used for the production of transgenic organisms may be used. Genes must be isolated, cloned or synthesized and transferred back into a recipient where stably integrated and expressed*' (EFSA, 2010, p. 5)

such as the extent to which the plant is cultivated and consumed' (ibid., p. 2). In this way, issues related to risk assessment and labelling of synthetic biology derived products used in the food and agricultural sector can easily be absorbed within the existing regulation provided for GMOs.

3.4.7 The premises, both conceptual and technical, of the process of adaptation of the GMO regulatory framework were subsequently addressed by the Scientific Committee on Emerging and Newly Identified Health Risks (SCENIHR). This work represents the heart of the whole process of adaptation of the GMO legislation. This process was triggered by the Commission. At the beginning of 2014, through its bodies (Directorates Health and Consumers (SANCO), Research and Innovation (RTD), Enterprise and Environment), the European Commission[92] also asked the SCENIHR for three further opinions: one on the operational definition of synthetic biology (in particular with regard to its relationship and differences with genetic engineering), one on risk assessment methodology and one on safety aspects and research priorities. However, as said, at the end the process of adaptation of the existing legislation mainly addressed risk assessment procedures.

(i) The first opinion on the conceptual aspects of synthetic biology (definition) was delivered by the SCENIHR in association with the Scientific Committee on Consumer Safety (SCCS) and the Scientific Committee on Health and Environmental Risks (SCHER) in 2014. According to the Scientific Committees, synthetic biology 'is the application of science, technology and engineering to facilitate and accelerate the design, manufacture and/or modification of genetic materials in living organisms' (SCENIHR et al., 2014a, p. 5, 27). '[T]he operational definition[93] offered [. . .] addresses the need for a definition that enables risk assessment and is sufficiently broad to include new developments in the field' (ibid., p. 5). In fact, '[t]he above definition includes

---

[92]See http://ec.europa.eu/health/scientific_committees/docs/synthetic_biology_mandate_en.pdf. Accessed 6 December 2016.

[93]'The term "operational definition" is understood as a definition that makes it possible to unequivocally decide whether an activity is or is not SynBio based on present knowledge and understanding of the field' (SCENIHR et al., 2014a, p. 30).

genetic modification as presently defined, as well as current and expected future developments in SynBio' (ibid., p. 27). In other words, synthetic biology uses 'all available technologies for genetic modification, but in particular aims at the acceleration and facilitation of the process, which includes increasing its predictability' (ibid., p. 31). Thus, the simplification of old techniques absorbs the whole teleological profile of synthetic biology, while the goal of creating organisms with new functions that can be used for human purposes is neglected. In this way, the definition proposed by the Scientific Committees includes 'any activity that aims to modify the genetic material of living organisms as defined in the Cartagena Protocol on Biosafety. This does not exclude the consideration of non-viable, non-reproducing goods and materials generated by or through the use of such living genetically modified organisms (GMOs)' (ibid., p. 5). The main concern of the Scientific Committees was, in fact, to use existing risk assessment methodologies provided for genetically modified organisms for synthetic biology organisms as well. In this way, the definition should sound sufficiently broad to also encompass future genetic modification techniques, as well as old ones. Thus, '[t]his definition has the advantage that it does not exclude the relevant and large body of risk assessment and safety guidelines developed over the past 40 years for GM work and extensions of that work, if needed, to account for recent technological advances in SynBio' (ibid., p. 6). Basically it is built on the basis of the consideration that 'new technologies for DNA synthesis and genome editing accelerate the process of genetic modification considerably, and increase the range and number of modifications that are easily possible', although the whole range of synthetic biology techniques cannot only be reduced to these (SCENIHR et al., 2015a, p. 37). However, it worthwhile to notice that the premise is not the substantial equivalence between synthetic biology and genetic engineering, but precisely the opposite. Otherwise a conceptual definition should have been used, instead of a working definition. In fact, by recalling the 2012 opinion of the New Techniques Working Group (NTWG), according to the Scientific Committees synthetic biology should be deemed 'a fast-evolving field *that differs* from previous gene modification techniques', thus eluding the legal definition of a GMO[94] (SCENIHR et al., 2014a, p. 8 – emphasis added). In this context the thesis of the substantial equivalence between the two fields appears further weakened.[95] The functioning of this definition would fall under the scope of the case-by-case approach. In this sense, this operational definition should be deemed as temporary at least until there are the conditions to amend it, but most of all to amend the GMO regulation. In the meantime, synthetic biology could further develop, but also cause uncertainties over the regulatory status of plants developed by new plants breeding techniques (NPBTs). For example, as regards the application of Directive 2001/18/EC on the deliberate release of GMOs and Directive 2009/41/EC on the contained use of GM microorganisms '[t]he uncertainty of the regulatory status of plants developed by NPBTs could have an impact on innovation, because it is difficult for a plant-breeder to decide if he/she should invest his/her efforts in a project using one of these techniques' (ibid., p. 16 nt. 2). Moreover,

---

[94] This opinion was also shared by the New Techniques Working Group (NTWG) which in its 2012 final report reached the main conclusion that synthetic genomics, as well as synthetic biology, 'is a fast-evolving field that differs from previous genetic modification techniques' (SCENIHR et al., 2015a, p. 9; NTWG, 2012).

[95] On the definitional issue of synthetic biology, as well as on the difference between it and traditional genetic engineering, see Arkin et al. (2009). In this regard there is no agreement on the fact synthetic biology may cover or not also genetic engineering. On the distinction between the two fields see, e.g. Berry ('the difference between metabolic engineering and synthetic biotechnology is that only with the latter can you design cells that accomplish a task that is independent from what the cell normally does'), Gold ('what seems critical to me is that any resulting organism can be described in words without having to refer, directly or indirectly, to any living organism'), Greenwood ('[w]hereas systems biology studies complex natural biological systems using modelling and simulation comparison to experiment, synthetic biology studies how to build artificial biological systems and synthesize industrial products'), Poste ('[t]he boundary between synthetic biology and systems biology should reside in a single criterion: has the engineered process, product or organism been fabricated from natural materials (systems biology) or from components not adopted in natural evolution (synthetic biology)?'), Smolke ('[synthetic biology] is the focus on the development of new engineering principles and formalism for the substrate of biology that sets it apart from the more mature fields upon which it builds, such as genetic engineering') (Arkin et al., 2009). As regards the opinion favorable to interpret synthetic biology as a mere evolution of genetic engineering having only a blurred distinction see Venter (2013, p. 114).

this raises a further risk of impacting innovation in this field since 'conventional breeding techniques present relatively low registration costs, transgenic plants regulated under the GMO jurisdiction were associated with high registration costs and extensive risk assessment procedures' (ibid.; Lusser et al., 2012, p. 10). Thus, 'it is difficult for a plant-breeder to decide if he/she should invest his/her efforts in a project using one of these techniques" (SCENIHR et al., 2014a, p. 8). In this regard, the process of adaptation of Community legislation on GMOs can encounter difficulties.

As regards the substantial equivalence principle, while many alternative exclusion criteria have been considered by the Scientific Committees, they have all been discarded since they could not 'deliver clear cut-off points or thresholds that scientifically discriminate' synthetic biology from genetic engineering (ibid., p. 28). For example, the criterion according to which a 'considerable/substantial proportion of the resultant genetic material has been chemically synthesised' was rejected since, although a 'quantification of the amount or proportion of chemically synthesized genetic material is possible, any threshold would be arbitrary' (ibid., p. 28). Again: the criterion according to which a 'significant proportion of the genetic material has been intentionally removed to develop a minimal functioning genome' was rejected since, although a 'quantification of the amount of removed material is possible, any threshold would be arbitrary' (ibid.). In other words, any quantitative criterion was discarded as arbitrary, but – you could object – an operational definition is by definition arbitrary, by being essentially political. It is a matter of choice. However, one could also object that an operational definition is quantitatively more arbitrary than the establishment of a threshold. In the case of nanotechnologies the European Commission used, more convincingly, a quantitative criterion for defining the concept of nanomaterial, while providing at the same time for the possibility of some exceptions.[96]

---

[96]That is: "'nanomaterial' means a natural or manufactured material containing particles, in an unbound state or as an aggregate or as an agglomerate and where, for 50% [...] one or more external dimensions is in the size range' between 1 nm to 100 nm. See Recommendation of the European Commission of the 18 October

After the delivery of the first opinion, on 14 June 2014, the European Commission and the three Scientific Committees (SCCS, SCENIHR and SCHER) launched a public consultation on the preliminary version of the first opinion on synthetic biology, laying down its operational definition.[97] Consultation processes are strategically adopted by the European Union to ensure the necessary flexibility of governance tools and to strengthen the participatory dimension of its policies (e.g. Ruggiu, 2014). They could be meant as a means to fill the original lack of democratic legitimation of agencies and committees within a new governance framework. Results of this consultation,[98] which involved 21 contributors coming from organizations (European national institutions, universities and agencies) and individuals, triggered a process of revision which flowed into the final version of the opinion. However, what emerges from some participants' comments is the weakness of the proposed definition from the scientific standpoint.[99]

(ii) In May of 2015 there was the second opinion on synthetic biology, that on risk assessment (SCENIHR et al., 2015a). The opinion delivered by the same Scientific Committees as the first one (i.e. SCENIHR, SCCS and SCHER) 'focused on the implications of likely developments in SynBio on human and animal health and the environment and on determining whether existing health and environmental risk assessment practices of the European Union for genetically Modified

---

2011 *on the definition of nanomaterial*, 2011/696/EU, 2011, *Official Journal of the European Union*, L275/38.

[97] See http://ec.europa.eu/health/scientific_committees/consultations/public_ consultations/scenihr_consultation_21_en.htm. Accessed 29 March 2015.

[98] See http://ec.europa.eu/health/scientific_committees/emerging/docs/followup_ cons_synbio_en.pdf. Accessed 29 March 2015.

[99] For example, one stated: 'In my opinion SynBio is not in itself a "gene modification technique". This definition conflates SynBio with the tools used in it; SynBio utilises existing and new gene modification techniques with the goal of creating DNA constructs with modular, standardised parts, and it is these tools which should be the main focus of regulation rather than the concept of SynBio' (SCENIHR et al., 2014b, p. 4). Again: 'Another aspect is that it fails to convey the essential information that SynBio is by necessity a cross-disciplinary domain. In this Opinion SynBio is presented as extreme genetic engineering. How about the computer scientists who present it as extreme computer science?' (ibid., p. 5).

Organisms (GMOs) are also adequate for SynBio' (SCENIHR et al., 2015a, p. 5). Accordingly, the premise was one unique definition able to cover both synthetic biology and genetic engineering. The Scientific Committees 'confined the scope of [their] analysis to the foreseeable future (up to 10 years)', but outside it they also dealt with issues related to the social, ethical, governance and security implications of synthetic biology, which can give us an idea of the actual trajectory adopted in Europe (ibid., p. 45).

It assesses six developments in the field of synthetic biology: (1) genetic part libraries and methods; (2) minimal cells and designer chassis; (3) protocells and artificial cells; (4) xenobiology (XB); (5) DNA synthesis and genome editing; and (6) citizen science or do-it-yourself biology (DIYbio). Since synthetic biology 'evolved from and shares many methodologies and tools of genetic engineering', the Scientific Committees suggests considering 'the assessment of risk guidance documents such as those issued by the GMO panel and/or the GMO unit of the European Food Safety Authority (EFSA) taking into account specific groups of organisms like microorganisms, plants and crops, or animals and guidance documents for environmental risk assessment of medicinal products' (ibid., p. 17). The Scientific Committees consider the approach developed for genetically modified organisms appropriate for the assessment of the contained use of synthetic biology derived organisms. 'For the identification of biological hazards and determination of the class of risk of the pathogen organism, classification lists of pathogenic microorganisms were established and are a useful tool for performing a risk assessment [...]. However, some pathogens and most GMOs are not classified into risk groups' (ibid., p. 20). The conclusion of the Scientific Committees is that '[w]ithin the scope of current GMO regulations, risk assessment is challenging, because of the lack of "comparators", e.g. the lack of comparator species, increasing the number of genetic modifications and engineered organisms' (ibid., p. 45). Since for risk assessment of genetically modified organisms it is necessary to identify the comparator, namely a similar organism produced without the help of genetic modification, this could not be possible for complex genetic systems engineered from genetic parts libraries where 'the emergent properties may present new challenges in predicting or testing for risks and in the identification of appropriate comparator organisms' (ibid., p. 27). Nevertheless, according to the Scientific Committees '[t]he comprehensive nature of the case-by-case risk assessment and mitigation procedures of the Directives [notably, Directive 2001/18/EC on the deliberate release of GMOS into the environment] is appropriate and adequate to manage the risks of SynBio activities and products associated with genetic parts libraries' (ibid.). In this sense, since synthetic biology 'evolved from and shares much of the methodologies and tools of genetic engineering', 'the risk assessment methodology of contained use activities and activities involving the deliberate release of' synthetic biology derived organisms 'are built on principles outlined in the Directives 2001/18/EC and 2009/41/EC and in Guidance notes published in Commission Decision 2000/608/EC[100]' (ibid., p. 45). Furthermore, the increase of information on nucleic acid proprieties, molecular products and data on the interaction of those molecules with other biochemical elements, as well as their accessibility, should enhance safety and risk assessment in the future, simultaneously reducing uncertainty. Finally, the improvement in the level of biosafety of synthetic biology applications could depend on the fact that several genetic circuit containment systems for single cell organisms have been developed and could be applied to organisms derived from synthetic biology in the future.

Conversely, the risk related to synthetic biology is not null. 'Although genetically engineered safeguard systems, e.g. engineered auxotrophic, induced lethality, gene flow prevention offer technical solutions to restrict engineered cells to laboratory or production settings, none function perfectly' (ibid., p. 42). Moreover, the 'lack of comparator species, increasing number of genetic modifications and engineered organisms' will probably soon challenge risk assessment

---

[100] European Commission (2000) *Commission Decision of 27 September 2000 Concerning the Guidance Notes for Risk Assessment Outlined in Annex III of Directive 90/219/EEC on the Contained Use of Genetically Modified Micro-Organisms* (notified under document number C(2000) 2736), *Official Journal of the European Union*, L258, 12.10.2000, pp. 43–48, http://eur-lex.europa.eu/legal-content/EN/TXT/PDF/?uri= CELEX:32000D0608&from= EN. Accessed 13 April 2015.

methodologies developed for genetically modified organisms (ibid., p. 45). But, there is more.

As said above, this framework of risk assessment procedures has some limits since some applications fall outside the scope of the provided operational definition. For example, protocells, which need to be integrated into living organisms and could be developed as autonomous cells only in the future, are excluded by the operational definition with the counterintuitive outcome that, since their long-term aim is to engineer a container or a chassis 'into which synthetic heritable could be introduced resulting in novel living, self-replicating organisms', (ibid., p. 31) they fall within the regulation of either nanotechnologies or chemicals (SCENIHR et al., 2014b, p. 9). In other words, they are (not alive) building blocks of 'new simple forms of living systems' that work 'from the bottom-up', although they are a clear product of synthetic biology (SCENIHR et al., 2015a, p. 31). In this instance, the risk should be nearly null, since both they cannot replicate, so as a dispersion is not possible, and the current state-of-the-art research is 'far from having commercial applications' (ibid.). Therefore, they fall under risk assessment methodologies provided for chemistry (e.g. REACH), nanotechnology or biology, although their risk assessment framework 'should draw on, but not necessarily be confined to, the methodology used for GMO risk assessment' with regard to allergenicity, pathogenicity, biological stability, etc. (ibid., p. 43). However, as recognized by the Scientific Committees, '[i]n the future, exposure to autonomous artificial cells that survive in the laboratory and in the environment might be possible' and '[r]isk mitigation measures must be put in place to prevent these scenarios' (ibid., p. 32). In this case we could wonder why we should not anticipate this future scenario since it is implicit in the aim of this line of research. Indeed, the conviction of the Scientific Committees is that '[i]f protocell research progresses towards autonomous, replicating chemical systems, which react dynamically to changes in their environment', they would fall within the risk assessment framework provided for GMOs and the 'hazardous properties of these cells should be assessed', by also considering also 'allergenicity, pathogenicity, biological stability, etc.' (ibid., p. 33). Herein, 'a case-by-case approach drawing upon a combination of existing frameworks including GMO and regulatory frameworks for chemicals and drugs' can be used (ibid.).

Xenobiological systems, namely biological systems engineered and produced with non-canonical biochemistries and/or alternative genetic codes (e.g. XNA instead of DNA or RNA) in order to produce economically interesting organisms with useful functions, also raise some concerns for the functioning of this model when we consider the risk assessment dimension. As recognized by the Scientific Committees, although the risk assessment framework provided for GMOs is deemed suitable in this case, since 'the GMO definition does not specify the biochemistry of biological material',[101] which could be thus partially synthetic, '[t]he existing pool of knowledge of risk characterization of GMOs cannot entirely be transferred to XB. For example, evolutionary fitness, ecological competitiveness, degree of horizontal gene flow, susceptibility to virus, diseases or predation, toxicity that are included in established risk assessment method-ologies for canonical biological systems, may not be compared to xenobiological systems' (ibid., p. 36). This implies that, firstly, 'the new variants have to be tested for their risk to human health and the environment and secondly, the xenobiological systems might be engineered to allow for improved biocontainment, i.e. the so-called genetic firewall which aims at avoiding exchange of genetic material through horizontal gene transfer or sexual reproduction between the xenobiological organisms and natural organisms' (ibid., p. 46). Thus, like a brand new field 'it is necessary to create and collect data sets and knowledge about the interaction between xenobiological and natural organisms for risk assessors to apply the established methodologies to xenobiological organisms' (ibid., p. 47).

In the case of genetic part libraries and methods, although existing risk assessment methodologies can be deemed appropriate, new tools are needed since 'it may be difficult to accurately assess the properties that emerge from interactions of many parts in more complicated systems' (ibid.). In this instance too, as recognized by the Scientific Committees, '[s]uch modifications would not neces-sarily be considered genetic modifications according to the current definitions of genetic modification in Directives 2001/18/EC and

---

[101] Therefore, they should be covered by Directives 2009/41/EC and 2001/18/EC.

2009/41/EC. This might create additional challenges from a risk assessment standpoint, in that organisms produced by these methods may contain more pervasive changes to the genomes of living organisms than traditional genetic modification techniques' (ibid., p. 48).

Also genetic firewall for synthetic biology derived organisms raises concerns. In fact, as said above, unfortunately nowadays 'available safety locks used in genetic engineering such as genetic safeguards (e.g. auxotrophy and kill switches) are not yet sufficiently reliable for SynBio' (ibid., p. 6). In fact, these genetic safeguards 'are not sufficiently reliable/robust (i.e. safe) for field release of engineered bacteria, because of mutation and positive selection pressure for mutants that may lead them to escape safeguards' (ibid., p. 7). To be clear about it, 'in the case of field releases of GMMs, there is high probability that engineered bacteria may escape various genetic safeguard systems due to mutation and positive selection pressure for mutants' (ibid., p. 49). For this reason, 'a clear strategy for analysis, development, testing, and prototyping applications based on new forms of biocontainment' are recommended (ibid., p. 7). Clearly, this fact should also have further undermined the operational definition provided by the Scientific Committees. In conclusion, '[t]he SC agrees that a "blue print" needs to start with a commitment to actively support the development of inherently safe applications. This commitment has so far been missing, but will be instrumental to keep SynBio in the realm of responsible innovation for the years to come' (ibid., p. 50).

Finally, although differently from genetic engineering, the development of synthetic biology will increase the possibility of it leading to a do-it-yourself biology (DIYbio), and this does not pose any new hazards for humans and the environment, the probability of unintentional harms might increase 'because DIYbio is more popular. [...Thus,] it is important that established safety practices among DIY biologists are maintained' and possibly verified 'by an independent biosafety entity' (ibid., pp. 46–47). Since 'good laboratory practices is a primary duty of the DIYbio community, according to their own Code of Ethics [...c]omplementary support by traditional institutional actors will help to achieve the highest training standards'. In this regard '[i]t is essential that the existing methodologies are applied even outside the traditional institutional settings' (ibid., p. 47). We can note that in this instance the creation of a database of all existing synthetic biology applications could be a useful means of exerting a form of control in this sector, also in the case of the so-called 'garage biology'.

In sum, '[n]ew challenges in predicting risks are expected due to emergent properties of SynBio developments and extensive genetically engineered systems including, (1) the integration of protocells into/with living organisms, (2) future developments of autonomous protocells, (3) the use of non-standard biochemical systems in living cells, (4) the increased speed of modifications by the new technologies for DNA synthesis and genome editing and (5) the rapidly evolving DIYbio citizen science community, which may increase the probability of unintentional harm' (ibid., p. 5).

While the existing framework of risk assessment methodologies established for common genetically modified organisms can also be used in the meantime for organisms derived from synthetic biology, 'there are specific cases in which new approaches may be necessary' due to the fast pace of development of the field (ibid., pp. 5–6). Indeed, in the view of the Scientific Committees notwithstanding '[t]he existing risk assessment methodologies, in particular for GMOs and chemicals, are applicable, however, several SynBio developments such as combining genetic parts and the emergence of new properties due to interactions (genetic parts libraries), combinations of chemical and biological assessments (protocells), interactions between xenobiological and natural organisms (xenobiology), acceleration of GM processes will require new methodological approaches' (ibid., p. 6). In particular, they address 'risks pertaining to (1) routes of exposure and adverse effects arising from the integration of protocells into living organisms and future developments of autonomous protocells, (2) new xenobiological variants and their risk on human health and the environment that should be engineered for improved biocontainment, (3) DNA synthesis and genome editing which enables by direct genome editing of zygotes genetic modifications in higher animals within a single generation, and (4) new multiplexed genetic modifications which increase the number of genetic modifications introduced in parallel by large-scale DNA synthesis and/or highly parallel genome editing will increase the distance between the resulting organism and any natural or previously modified organism' (ibid.).

This picture of the current framework of synthetic biology leads thus to the adaptation of existing risk assessment methodologies by: (a) supporting the characterization of the function of biological parts and the development of computational tools to predict emergent properties of synthetic biology organisms; (b) streamlining and standardizing the methods for submitting genetic modification data and genetic parts information to risk assessors; (c) encouraging the use of GMOs with a proven safety record as acceptable comparators for risk assessment; (d) aiming to ensure that risk assessment methods advance in parallel with synthetic biology advances; and (e) supporting the sharing of relevant information about specific parts, devices and systems to risk assessors (ibid., p. 29). Furthermore, research on genetic firewall for synthetic biology derived organisms also needs to be strengthened. However, one could also ask how in front of these risks a highly contested legislation such as this on GMOs, especially as regards the deliberate release into the environment, could be suitable for SynBio organisms that are engineered for performing new and unnatural functions (e.g. remediation), with the danger of altering the ecological balance. The risk here is to add a number of foreseeable risks to a highly fragile system by reducing spaces of public engagement for both fields.

The Scientific Committees also tackled specific issues on governance. Since 'the ability to engineer predictable outcomes of biological systems remains embryonic relative to most other fields of engineering' the approach recommended by the Scientific Committees with regard to risk governance at the EU level is to responsibly develop the sector with a 'continued evolution of governance mechanisms' under the *Responsible Research and Innovation* (RRI) framework (ibid., p. 14).[102] In other words, according to the Scientific Committees it should be applied a flexible, adaptive, participatory model which aims at involving all parties distributed at any level through both traditional and innovative means by fostering the spontaneous organization of

---

[102] On RRI see below.

responsibilities at an early stage. Thus, '[i]t is vital to recognise the importance of maintaining public legitimacy and support' and RRI, as recently endorsed by Community authorities at any level,[103] is the main route towards the development of flexible, but responsible, chassis for synthetic biology (ibid., p. 15). For this reason, scientific research needs to be accompanied by a public debate and 'potential applications should demonstrate clear social benefits' (ibid.). In this regard, not only must dystopic scenarios be neutralized through the involvement of civil society, but unrealistic hopes and overhyped utopic scenarios must also be avoided in order to reach more accurate and realistic information. The Scientific Committees recalled the suitable governance framework proposed by the International Risk Governance Council,[104] an independent organization, which provides a structure for how risks may be investigated, communicated and managed at the global level by taking into account not only scientific evidence but also risk perceptions, societal concerns and social values (ibid.).

Here, according to the Scientific Committees, key elements to be taken into account for governance are: actors in regulation, application areas, boundaries between synthetic biology and other emerging technologies (nanotechnologies, biotechnologies, information technologies, chemistry, etc.), the interrelation between different regulatory sectors (workers, environmental, medical protection, etc.) and the heterogeneity of this technological field. Furthermore, different levels (political, ethical, societal, professional, institutional and scientific) have to be considered. Not only public authorities need to be involved in developing risk governance structures, since room for developing self-governance initiatives is also to be preserved. As the success of synthetic biology will be determined by acceptance, which depends on: α) 'the scale of the risks or

---

[103] E.g. see European Union (2013) Regulation (EU) No. 1291/2013 of the European Parliament and the Council of 11/12/2013 *Establishing Horizon 2020 – the Framework Programme for Research and Innovation (2014-2020)*, *Official Journal of European Union*, L347/104, 20.12.2013, http://ec.europa.eu/research/participants/data/ref/h2020/legal_basis/fp/h2020-eu-establact_en.pdf. Accessed 11 March 2015.

[104] See http://www.irgc.org/risk-governance/irgc-risk-governance-framework/. Accessed 4 April 2015.

the risks in comparison with the perceived or actual benefits';
β) 'the potential/likelihood to control the risks and the trust in
this'; and γ) 'a widely accepted means of perceiving the risks,
e.g. benchmarking against familiar risks' (ibid., p. 16). To this
group probably we should also add the general perception of the
adequacy of the existing regulatory framework. According to this
view, only socioeconomic and ethical considerations can influence
societal acceptance. From the first aspect, those involved in this
sector should proactively be engaged in explaining both the pros
and cons of particular novel developments such as energy saving,
reduction of CO_2 emissions, as well as novel medicines (ibid.). In this
framework the early 'identification of developments' and the timely
explanation of 'their expected beneficial and potentially adverse
societal impacts in an understandable and transparent way' have
same importance. Thus: effective consultation mechanisms must be
established and maintained (ibid.). Moreover, as synthetic biology
is expanding and changing so rapidly, educational materials rapidly
loses their freshness. Accordingly, this 'means that education and
training should reflect this' rapid growth of knowledge (ibid.). In
this regard, a great effort needs to be done in this direction by both
training scientists and developing specific targeted educational programmes for schools, universities and 'workers involved with these
both manufacturing and disposal/recycling can safely managed or
prevent any risk' (ibid.).

From the ethical standpoint, according to the Scientific Committees four main considerations need to be addressed: (1) the blurring
of the distinction between life and non-life; (2) the interference with
nature; (3) the widening of the gap between have and have not
countries and sectors of society; and (4) misuse, leading to serious
threats to society (a biosecurity issue too). As emerging technology,
synthetic biology 'encourages the development of a transparent,
iterative process of risk governance, which includes risk assessment
and dialogue among stakeholders including civil society globally as
important elements' (ibid., p. 17). Instead, legal considerations on
the adequacy of the set of rules devised for old biotechnology are
not made since the update of existing methodologies is deemed as
sufficient in this context.

(iii) At the end of 2015 also the third opinion on synthetic biology was delivered (SCENIHR et al., 2015b).

It focused on research priorities and was also the object of a submission to a consultation before reaching the final stage. The opinion, written by the same Scientific Committees of the previous ones (SCCS, SCENIHR and SCHER), refers to present gaps in knowledge and risk assessment relevant to the environment in the near-term future (ibid., p. 8). Thus it does not refer to therapeutic applications of synthetic biology in clinical medicine such as immunotherapy or vaccines. According to Scientific Committees most of the medical applications of synthetic biology (viral vectors and/or genetically modified human cells), which fall out the SynBio definition, should undergo existing risk assessment methodologies currently used under GMO regulation (SCENIHR et al., 2015c, p. 7). In abstract, a 'precautionary approach in accordance with domestic legislation and other relevant international obligations is required to prevent the reduction or loss of biological diversity posed by organisms, components and products' (SCENIHR et al., 2015b, p. 12). The fact is that up to now synthetic biology is widely unregulated and there could be doubts that the GMO regulation can fit well in this case.

Scientific Committees identified both generic risk factors and specific risks to the environment. As regards the first aspect, Decision XI/11 of the Convention of Biological Diversity (CBD) requests to identify issues related to the conservation and sustainable use of biodiversity. Several applications aimed at reaching sustainable outcomes (biosensors of pollutants and bioremediation, next generation biofuels, new agricultural production able to reduce pesticides, de-extinction) can challenge biodiversity. In this context Scientific Committees thus underline risks 'related to accidental release, persistence of SynBio organisms intended for environmental release, such organisms becoming invasive or disruptive for food webs, transfer of genetic material from vertical gene flow or horizontal gene transfer' (ibid., p. 8). In this sense biodiversity can be increased by synthetic biology though artificially (ibid., p. 21). This replacement of biodiversity with an artificial one could raise concerns in civil society. As regards the latter aspect, specific risks are related to the five novel developments of synthetic biology previously addressed in opinion II: (1) genetic part libraries and methods; (2) minimal cells and designer chassis; (3) protocells and artificial cells; (4) xenobiology: (5) DNA synthesis and genome editing; and (6) citizen science (e.g. Do-It-Yourself Biology). The surge of new and uncharacterized biological functions, properties and products with the lack of any comparator, poses the need to develop new approaches to risk assessment. Major gaps identified are mostly two. First of all, the lack of knowledge and tools for predicting emergent properties of complex non-standard biological systems. Then the lack of tools for measurement of the structural differences between the original (natural) and the engineered organisms (ibid., p. 9).

Scientific Committees also provided a set of recommendations. As regards genetic parts Scientific Committees recommend to support research for (i) characterizing the interactions between modified and novel parts; (ii) developing computational tools to predict emergent new properties of SynBio organisms and their potential failures; and disseminating these tools; (iv) standardizing the methods for submitting genetic modification data and genetic parts information to risk assessors across EU member States; (v) developing guidelines for risk assessors on the evaluation of potential emergent properties of genetically engineered systems; (vi) studying the use of GMOs with a proven safety record as acceptable comparators (ibid., pp. 34–35). As regards minimal cells and designer chassis Scientific Committees recommend to develop (i) biosafety modules at the design stage; (ii) research on quantifying and qualifying the evolutionary change of phenotypes through time; a public repository of well-characterized engineered safe chassis and safety devices (SCENIHR et al., 2015b: 35). As regards protocells Scientific Committees recommend i) to implement knowledge to assess the implications, as well as the environmental and evolutionary consequences of interactions between non-living protocells and living organisms; (ii) to develop methods to assess the risk of allergenicity, pathogenicity and biological stability in the case protocells become life-like entities; (iii) to gain data about the ecological and evolutionary role of natural vesicles containing peptides, RNA and DNA (SCENIHR et al., 2015b: 36). As regards xenobiology Scientific Committees recommend that each individual chemical class of xeno-compounds should be characterized from the outset and tested comprehensively. They also recommend to establish a methodology to quantitatively and qualitatively characterize xenobiologic organisms as regards evolutionary fitness, ecological competitiveness, degree of horizontal gene flow, susceptibility to viruses, diseases and predation. Furthermore, it should be developed a clear and reliable metric to measure the escape frequency associated with different types of semantic containment (ibid., p. 37). As regards Do-It-Yourself Biology Scientific Committees recommend to increase the compliance of citizen scientists with harmonized European biosafety rules and codes of ethics (ibid., p. 38).

This great and sophisticate effort to reform current methodologies of risk assessment ended with some further recommendations such as to develop studies on impacts from accidental or intentional introduction of synthetic organisms into the environment with attention on the effects on habitats, food webs and biodiversity; research on the difference in physiology of natural and synthetic organisms, on vertical or horizontal gene flow, as well as on the consequences of de-extincted animals that could affect biodiversity. However, the question of the adequacy of the existing regulation remains unexpressed and largely presupposed.

Also the European Research Area was involved in the shaping of the rising governance of synthetic biology. Lately a consortium of research funding and policy organizations came together through the European Commission, funded the European Research Area Network in Synthetic Biology, with the aims of mapping 'existing activities in synthetic biology and, through consultations with leading researchers and other stakeholders, develop a strategic agenda to guide the emergence of the field within Europe' (ERASyn-Bio, 2014, p. 4). Several recommendations were provided by this organism (fostering investments in this field, building networks and multidisciplinary and transnational research communities, providing skilled, creative and interconnected workforces, securing openness of data), but, in particular, it tried to apply the Responsible Research and Innovation model to the sector of synthetic biology by implementing responsible and inclusive manners that involve scientists, community and industry 'to develop products and services that are needed by, and acceptable to, the public and other stakeholders' (ibid., p. 5). In this sense the push towards a flexible model able to responsibly engage the whole community involved in research into synthetic biology is clearly perceived at every level of the EU.

3.4.8 Let's make some provisional remarks. Although in fast evolution, the Community governance of synthetic biology seems to follow the path previously traced for other emerging technologies in Europe. It embraces a flexible, adaptive, experimental, as well as participatory, style, which can be framed within that of new governance. Here, the case-by-case method seems to be the polar star. While at the transnational and national level forms of self-governance are widespread within the community of market actors, at the EU level institutional actors took the initiative and tried to adapt the existing regulatory framework provided for genetically modified organisms. In this context self-arrangements by actors at stake are not excluded in principle, especially at the national level. This model joins together two approaches: one aimed at adapting existing governance structures and frameworks provided for GMOs, the other aimed at increasing the resilience of the system in order to produce rapid changes to governance where necessary. In this framework agencies and committees played a strategic role. Initially, the EGE drew the basis of the trajectory of the EU governance by using the GMO regulatory framework as a starting point, then a number of other EU bodies jointly acted in order to adapt it to the new field of synthetic biology and prepare the subsequent adjustments needed by the evolution of the sector. In this context the principle of substantial equivalence is central. However, according to a significant part of the scientific community there is a deep difference between an entity with just some enhanced features and a completely new entity since it can perform functions that do not exist in nature (such as destroying pollutants, producing biodrugs, bioweapons). The application of the substantial equivalence principle caused a shift of the focus on risk assessment procedures devised for genetically modified organisms. Thus, soon the process of updating the existing regulatory framework turned out into a process of revision of current methodologies of risk assessment by neglecting an evaluation of the adequacy of these rules to new challenges of synthetic biology. Accordingly, the existing GMO legislation is already applicable to the new field of synthetic biology provided that current risk assessment methodologies are updated. This can be deemed as problematic. For example, given the foreseeable non-homogeneity of risks raised by SynBio organisms (risks are different), as stated by the Scientific Committees, we expect that also the conditions for authorizing their deliberate release are different. In this regard, the mere reform of risk assessment methodologies does not seem sufficient. Moreover, also the different regime of labelling for marketed food and marketed feed that can be eat by livestock (without any information for consumers) does not seem adequate in the case of spread of SynBio products into the market. Here the lack of a centralized system of mapping market applications in the field of synthetic biology does not help. Also in this case the amendment of risk assessment procedures could be insufficient.

This situation could have also further consequences. If, as the Scientific Committees showed, these risk assessment procedures can be insufficient for several concerns in the near future, the limits of a regulation that is far to be accepted in Europe (that on GMOs) could reveal grave shortcomings in front of new SynBio organisms. As recognized, '[g]iven the European public's reaction to genetically modified food and crops, there is no guarantee that the problems that beset one area of biotechnology will not affect another' (Salter and Jones, 2002, 808). This could lead to the need of adopting abrupt shifts in regulation and, consequently, decrease the societal acceptance. In this context it is likely that some initiatives ascribable to the Classic Community Method of a hard law nature aimed at filling possible regulatory gaps could be provided in the future, triggering a rapid process of stiffening of governance within the EU.

Therefore, the European Union could likely to cope with three solutions one day: either (i) maintain the same regulatory framework provided for genetically modified organisms, and, then, stand by to abruptly change governance trajectory (e.g. the shift from soft to hard law); or (ii) start to draw a different regulatory framework under a different definition starting from the GMOs regulatory basis; or, finally, (iii) maintain the same regulatory framework under the same all-embracing definition, but start to draw some specific regulatory distinctions for synthetic biology. The risk of suddenly changing governance trajectory is implicit in any case-by-case approach, but we cannot exclude that the risks in the case of synthetic biology could be higher than in other technologies since, as expressly acknowledged, 'risk assessment criteria, methodology and risk management systems established for GMOs and pathogens provide a good basis for addressing potential risks' only for current and short-term developments of synthetic biology, namely until they are not diffuse into the market (SCENIHR, 2014, p. 19). In this sense, the proactive dimension of this model appears too fragile and it is unlikely to use the advantage given by the existing GMO regulation adequately.

3.5 The Rising Model of the *Responsible Research and Innovation*[105]

3.5.1 In a context where risks are distributed and responsibility is pulverized among a multitude of stakeholders, rules of innovation need to be rewritten. Today there is an increasing awareness that the success of innovation needs greater sharing of responsibility by all parties. In particular, in the field of the technoscientific progress.

First, this led to the spread of ELSI studies in order to strengthen the awareness of existing interconnections among disciplines put in question by emerging technologies. Since impacts of the technological development involve different fields, only the interdisciplinarity can cope with this new challenge.

This progressively paved the way to reshape governance processes in Europe.

The demand for the democratization of governance (Jasanoff, 2003) and the related demand for the democratization of ethics (Tallacchini, 2009) prepared the ground for the surge of the 'new governance' model in the field of emerging technologies and the provisory abandon of traditional tools of regulation (Kearnes and

---

[105]This paragraph retakes and slightly amends (Ruggiu, 2015a).

Rip, 2009; Widmer et al., 2010). In this framework, therefore, we can observe the spread of processes of responsibilization of all actors through the adoption of flexible tools aimed at fostering the participation to innovation (Dorbeck-Jung and Shelley-Egan, 2013). In this regard, soft forms of regulation, as well as spontaneous processes of self-regulation (i.e. self-governance), coexist in flexible frameworks which are fostered thanks to the usage of both old instruments, such as comitology, agency, networking, guidelines, ethical codes, and new ones, such as certification systems, social dialogue and consultations.

Within the European Union several examples of these flexible initiatives can be detected both in the field of nanotechnologies and in that of synthetic biology. These initiatives are thus aimed at: (i) implementing the process of responsibilization as natural support of research and innovation (Ruggiu, 2014b), (ii) at filling the gap of democratic legitimation of the unelected EU bodies, such as agencies (Trubeck et al., 2005, p. 16), and (iii) at preparing the passage to more traditional forms of regulation (Ruggiu, 2015b).

Now responsibility is the password for the advance in science and technology. And this outcome can be achieved only by anchoring governance on some key values at the centre of political communities. Today there is an increasing belief that the voluntary assumption of the responsibility is strictly connected to the ethical acceptability and the social desirability of research and innovation. Therefore, within the framework of the 'new governance' a novel paradigm of governance is rising in Europe.

This model, which is almost substituting 'new governance', is that of *Responsible Research and Innovation* (RRI).

'**Responsible Research and Innovation** (RRI) refers to the comprehensive approach of proceeding in research and innovation in ways that allow all stakeholders that are involved in the processes of research and innovation at an early stage (A) to obtain relevant knowledge on the consequences of the outcomes of their actions and on the range of options open to them and (B) to effectively evaluate both outcomes and options in terms of societal needs and moral values and (C) to use these considerations (under A and B) as functional requirements for design and development of new research, products and services' (van den Hoven et al., 2013, p. 3).

With RRI the incorporation of a number of strategic ethical values and societal inputs is deemed as the right way to transform the innovation process in a responsible action produced by the cooperation of all stakeholders. In this way innovation can feed the societal trust and stabilizes itself in a robust framework.

Within the research community there is the belief that only thanks to the support of key stakeholders for a responsible action of collective nature it is possible to successfully drive the innovation process and achieve outcomes that can be sharable within society. Trust is the non-eliminable ingredient of the recipe for the success of innovation. Today this belief is increasingly diffused within EU institutions. This is the reason why today the RRI paradigm already orients the EU practices in Europe, in particular in EU research framework programmes.

A clear example of this shift is Horizon 2020.

With Horizon 2020 research and innovation have 'been placed at the centre of the Europe 2020 strategy'.[106] Accordingly, Horizon 2020 seeks to develop this new model of governance within the European Union. According to EU authorities this model should boost for excellent science, a more competitive industry and a better society without compromising sustainability, ethical acceptability and a socially desirable framework.[107]

Notwithstanding this official endorsement, the RRI landscape appears to be quite multifaceted within the research community. This situation of epistemic disagreement can be also testified by the lack of a shared definition of RRI (Owen et al., 2013, p. 27). Notwithstanding this lack, or because of this, up to now at least two main tendencies characterize the RRI debate (Ruggiu, 2015a).

---

[106] European Union (2013) Regulation (EU) No. 1291/2013 of the European Parliament and the Council of 11 December 2013 *Establishing Horizon 2020 – the Framework Programme for Research and Innovation (2014–2020)*, of 20 December 2013, *Official Journal of European Union*, L347/104, http://ec.europa. eu/research/participants/data/ref/h2020/legal_basis/fp/h2020-eu-establact_ en.pdf. Accessed 13 July 2015.

[107] In this institutional framework we need to mention that both the Lund Declaration (2009) and the Council conclusions on the social dimension of the European Research Area (2010) underline the importance of integrating societal needs and ethical concerns into the research and development (van den Hoven et al., 2013, p. 3).

One, which I call *socio-empirical*, emphasizes the role of public engagement since governance 'would need to be based on the principle of inclusiveness, involving all actors at early stage' (Owen et al., 2012, p. 752). In this way this participatory model should produce a shift from science in society to a science *for* society, *with* society (ibid.).[108] Here the values on which to anchor European governance are created through democratic processes. Innovation is the product of a collective action, therefore the society has to decide how it must be, by addressing the societal values that need to be put at the centre of innovation. Values are therefore the end of a process of negotiation over innovation.

The other, which I call *normative*, aims at anchoring the process of decision making on some prefixed normative filters, such some shared goals expressed at the EU treaties level, in order to produce the ethical acceptability, sustainability and social desirability (van den Hoven et al., 2013, p. 23). In other words, to be responsible, namely ethically acceptable, the EU values must steer research and innovation. Values therefore are the starting point of processes of research and innovation.

These two approaches are only two different tendencies within the RRI framework (Ruggiu, 2015a). They are two abstract potentialities. These two versions of the same theoretical model express only two extreme possibilities among the entire range of available opportunities opened up by RRI. Therefore, they do not exclude mixes and contaminations. And, as a matter of fact, in the reality we can count more contaminations between these models than the manifestation of their pure version.

3.5.2 As said, so far no shared definition emerged in the research community (Owen et al., 2012, p. 27). Nevertheless, the von Schomberg's definition of RRI is often recalled in the academic debate on RRI (e.g. (Owen, 2014), p. 6).

According to von Schomberg it should be defined as 'a transparent, interactive process by which societal actors and innovators become mutually responsive to each other with a view to the

---

[108]This approach to RRI is also shared, among others, by Sutcliffe (2011), Blok, Lemmens (2014).

(ethical) acceptability, sustainability and societal desirability of the innovation process and its marketable products (in order to allow a proper embedding of scientific and technological advances in our society)' (von Schomberg, 2011b, p. 54).

It is worth noting that this definition was substantially followed by the Experts Group on the State of the Art in Europe on Responsible Research Innovation, which lastly maintained all its main features (van den Hoven et al., 2013). Here the ethical acceptability is the starting point of the model and on this the participation can be built according.

On the other side, Owen and his colleagues (2013) proposed a shorter (and broader) definition of responsible innovation, which seems to stress only the participatory tendency of the model that is deemed to be, alone, productive of shared ethical values.

According to Macgnathen, Owen and Stilgoe, in fact, 'Responsible innovation is a collective commitment of care for the future through responsive stewardship of science and innovation in the present' (Owen et al., 2013, p. 36; Silgoe et al., 2013, 3). As it emerges from this definition public engagement, here, is the only means to reach ethical acceptability in accordance with a socio-empirical approach.

These two definitions well express the above-mentioned approaches to RRI: the *normative* and the *socio-empirical*.

In general, the RRI model is characterized by four basic features.[109] (i) RRI is a process able to *involve* actors, mainly private ones, which are distributed in the global sphere (Stilgoe et al., 2013, p. 4; Owen et al., 2013, p. 38). (ii) It aims at *anticipating* regulatory choices by voluntary or spontaneous behaviour of stakeholders, mainly researchers, innovators and research funders (Barben et al., 2008). In this regard some talk of 'RRI by design' (Owen, 2014, p. 11). (iii) It produces a shift of the focus from a risk-assessment

---

[109]While there is no shared definition, there is no agreement too on the main features of RRI. See e.g. van den Hoven et al. (2013, pp. 57–58); Owen et al. (2013, p. 38); Silgoe et al. (2013, pp. 3–5). In 2012 the Directorate-General for Research and Innovation of the European Commission has addressed six dimensions in the RRI framework (engagement, gender equality, science education, open access, ethics, governance). See http://ec.europa.eu/research/science-society/document_library/pdf_06/responsible-research-and-innovation-leaflet_en.pdf. Accessed 7 August 2015.

to the assessment of innovation processes since the model is aimed at considering also the *loss of innovation opportunities*. In this sense, it is crucial the anticipation of the impact assessment at an earliest stage in order to make the trade-off of both negative and *positive* impacts of a given emerging technology before it is too late (von Schomberg, 2013, p. 55; Stilgoe et al., 2013, p. 3). (iv) Finally, it is to mention the socially oriented character since *ethical acceptability*, sustainability and societal desirability are irrenunciable ingredients of innovation (van den Hoven, et al., 2013, p. 58).

The integration of these components should reach a better level of both reflexivity, by asking researchers to think about their ethical, political and social assumptions, roles, as well as responsibilities; and responsiveness, by opening up the process to inputs stemming from stakeholders in order to change its direction when it does not meet societal needs and is ethically contested (Owen et al., 2013, p. 38).

3.5.3   However, the two definitions mentioned above are able to encompass two different modes that can legitimately interpret the RRI style. The first, of fully normative nature, which articulates processes of co-responsibilization of stakeholders around a set of normative filters in order to foster ethical acceptability (e.g. von Schomberg 2011a, 2013; van den Hoven et al., 2013). The second, of socio-empirical nature, which focuses on interaction processes among different stakeholders aimed at developing participatory forms of co-responsibility in a given field of innovation, namely inclusion (e.g. Owen et al., 2012; Owen et al., 2013; Owen, 2014; Sutcliffe, 2011).

It worth noting that, while the normative version talks of 'Responsible Research and Innovation', the socio-empirical approach prefers to use the term of 'Responsible Innovation'. In this regard, these two approaches would differ one another also nominally. The difference though is also conceptual.

According to the first, in fact, both the stage of innovation and the design of the research are the target of the shaping action of normative filters.

According to the second, innovation is responsible because research practices autonomously develop themselves giving rise to

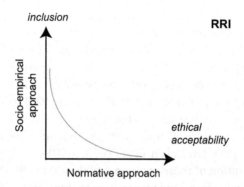

Figure 3.1 The normative approach tends to increase the ethical acceptability, while the socio-empirical approach tends to increase the level of inclusion. Within the RRI model the two approaches move in different directions.

responsible outcomes in the field of innovation. There is no need to shape practices from the external (top-down) because who take part to practices will transform them from the internal spontaneously, mutating their behaviour. And this will directly affect innovation (Groves, 2015).

The idea behind both approaches is that RRI can foster the surge of a responsible behaviour of a multitude of stakeholders by integrating the societal desirability of research and innovation and their ethical acceptability in flexible structures of governance. In this sense, they fully belong to the 'new governance' framework (Ruggiu, 2015a). In this conceptual framework participation plays in any event a crucial role. However, the relation between participation and ethical values is thought differently by these two approaches.

Notwithstanding both versions identify the same features and components of RRI (i.e. inclusiveness, anticipation, the focus on also positive impacts, the ethical acceptability), they differentiate each other for the different emphasis they put on some elements (i.e. normative anchor points, public engagement).

However, in this framework, by underlining exclusively the prescriptive nature of spontaneous responsible behaviour of stakeholders (i.e. they produce values), the socio-empirical approach tends thus to coincide with the current flexible, tentative and adaptive model of governance existing at EU level (Sutcliffe, 2011,

p. 5; Kearnes and Rip, 2009; Mandel, 2009). In this sense, it runs the risk of completely losing the novelty of the RRI model.

An example of this version can be deemed the consultation processes launched by the Commission in order to draft the code of conduct on nanotechnology research (Ruggiu, 2014b). In this context a number of sections of civil society including researchers, individuals, non-governmental organizations were involved through official consultations promoted at the EU level in order to identify principles of the code of conduct on nanoresearch.

Conversely, the normative approach means to create a responsible framework by anchoring policy choices on some principles of legal nature (van den Hoven et al., 2013, p. 23; Sutcliffe, 2011, p. 7), thus running the risk of stiffening governance processes.

Again, the case of the 2008 Commission code of conduct can be enlightening. In that occasion the final draft of the code on nanotechnology research was laid down on the basis of the EU objectives set out in the EU treaties that provided the basis for the subsequent consultations (Ruggiu, 2014b). In that case the institutional origin of 'normative anchor points' represented a difficulty to the communication process of values, which is strategic to rightly motivate stakeholders that will be requested to comply with rules stemming from those values. For example, consultation participants gave rise concerns on the vague reference to ethical principles and to responsibility for future generations, which was understood as legal liability and thus refused (European Commission, 2007b, p. 3). In this regard the socio-empirical version rightly points out that those values are better shared whether they are collectively addressed through the negotiation.

3.5.4 According to the socio-empirical version inclusion produces values, which have societal origin. Therefore, public engagement must be pursued to find values. The normative version, instead, thinks that pre-existing values produces inclusion. The ethical acceptability therefore is here the main value and this must be reached by identifying a set of ethical principles which are at the constitutional level.

Certainly, especially the socio-empirical version underlines the strategic role of public engagement of the parties at play (Sutcliffe,

2011, p. 3). Here, the starting point is the recognition of the limits existing in the idea of liability. The responsibility which is manifested through instruments of regulation is an important part of the responsible innovation 'but it has severe limitations' (Owen et al., 2013, p. 32). It expresses a *retrospective* logic unable to cope with challenges of the fast development of science and technology, which needs the ability of adapting choices and tools to the variability of scenarios. This capability is the resilience.

In this movable landscape neither actors, nor the society can wait for more data before commercializing a product. This fact creates a gap of responsibility since none can be deemed responsible for circumstances that cannot be foreseen and known at the stage of the action. Therefore, '[a]t the earlier stage of innovation we can have most of opportunities to shape and control innovation, with far fewer costs and vested interests; but it is precisely at these early stages that we have little or no evidence to make the case for control' (ibid., p. 34).

Against the precautionary principle, the risks of missed opportunities are the price of acting too early by regulating a field, which is rapidly growing up, but still unknown. In this framework it is better to rest on 'moral luck' since the moral responsibility is based on both what is known at the time of action and on what can be reasonably foreseen. The framework of responsible innovation must consider not only the products which would lead to a form of *retrospective responsibility*, but also the purposes of innovation which would pave the way to a form of *prospective responsibility* (ibid., p. 34). These purposes refer to what kind of future we want science and technology to face, to what challenges we want they cope with, and to what values we want they are anchored on (Owen, 2014, p. 3). In other words, this version of RRI 'asks how the targets for innovation can be identified in an ethical, inclusive, and equitable manner' (Owen et al., 2013, p. 35).

All societal actors, including researchers, policymakers and civil society, must be engaged in order to reach 'their joint participation in research and innovation, in accordance with the value of inclusiveness' which is expressly set out also in the Charter of fundamental rights of the European Union (Directorate-General for Research and Innovation of the European Commission, 2012, p. 1). It must be returned to society the possibility of deciding over its future. Therefore, autonomy of all stakeholder must be preserved. A sound framework for excellence in the research and innovation process entails that societal challenges are framed on the basis of widely representative social, economic and ethical concerns and common principles (ibid.). This means a public engagement of all parties on equal basis also as regards the gender representativeness (ibid.).

The equal participation is strategic for giving voice to all the represented interests and producing values that should be at the core of the process itself in a bottom-up manner. In this sense, the process (i.e. participation), which leads to the creation of values whereon the science and innovation can be anchored, is the sole value.[110] For this reason, the process must be non-partisan with regard to the values followed by all parties.

The emphasis of the socio-empirical approach on the structure of the process, led, for example, to use the stage gate architecture typical of management of innovation processes (Cooper, 1990).

Stage gate systems are frameworks where the process of decision-making is guided through the explicit provision of phases 'being subject to formal or informal approval at a decision "gate"' (Owen, 2014, p. 13). The stage gate architecture was used, for example, in the SPICE project (Stratospheric Particle Injection for Climate Engineering project), a project on geoengineering funded by UK Research Councils in 2010 which tried to develop a more democratic and legitimate framework for science and innovation by opening up 'to a broad anticipation, reflection and inclusive deliberation, with the aim of making policy more responsive' (Stilgoe et al., 2013, p. 10).[111]

This approach is not normative, 'in the sense that it has defined a process but not the values upon which this should rest' (Owen, 2014, p. 7). On the contrary, the normative dimension of values is the

---

[110]Thus, as far as the socio-empirical version underpins the value of the democratic participation, it can be deemed as prescriptive (like the normative one), but it does not move from some prefixed legal norms. It distinguishes from normative version since it eludes from the outset the normative dimension of rules and legal principles. In other words, it believes in the exclusive productivity of societal dimension, which tends to substitute that of (legal) norms.

[111]On the SPICE project see also Chapter 8 Part II on the right to healthy environment.

outcome of the process of deliberation. For this reason, values 'will differ according to the context in which the framework is applied and will be culturally-sensitive' (ibid., p. 7). In other words, values can differ according to the sociocultural and even technological context. Values differ not only according to the cultural context (for example, BRICS countries will focus on a certain set of values),[112] but they also differ according to which technology is concerned. Values in nanomedicine are different from those in SynBio applications in bioremediation and so forth.

In this framework, reflexivity and responsiveness are key features of this model since it leads all the concerned parties to reflect on purposes of science and innovation (on what innovation has to do and not to do); it maintains opened all options, increasingly includes new perspectives and permits to correct errors while they occur (resilience). 'This introduces the principle of deliberative democracy into the dimension of responsiveness' (Owen et al., 2013, p. 35).

The possibility that all parties contribute to the definition of values at the core of the process is thus crucial.

In context of uncertainty the negotiation, inclusive and democratic, is the main value at stake and it testifies to the commitment in setting the agenda of innovation (Owen et al., 2012, p. 4). This is the reason why this version expresses a form of a rationality that focuses only on how the process (of negotiation) is conducted since the process as such is the means through which principles driving regulatory tools are chosen (Heydelbrand, 2003, p. 238). Here the process grounds (societal) values that are at the basis of a given field of science and technology.

Another aspect of this version is the lack of normativity.[113] Here the normative dimension of prefixed rules and principles is eluded. There is no way to understand in abstract what to responsibly act

---

[112] This objection is made by Groves (2015, p. 327).

[113] Namely, it does not address any normative dimension of both legal and moral nature on which reality ought to conform. In other words, the difference between the two versions cannot be interpreted in terms of the prescriptive/descriptive polarity since both versions are clearly prescriptive and identify a model able to transform the status through, on the one hand, the process (of negotiation), through norms, on the other.

means, in as much as the prefixed values in a given context (e.g. EU goals) can be nothing more of a provisional starting point which can much differ from the final outcome (Owen, 2014, p. 7). Only the reality can teach us what responsible innovation means, meaning that rules and principles are the final outcome of the process of negotiation of actors at play. In these terms it can be deemed as socio-empirical. Then, there is the need to collect a multitude of cases, i.e. a number of *de facto* governance arrangements in place,[114] wherein this model has been realized in view of extrapolating its main features (e.g. (Sutcliffe, 2011, pp. 19–26)).[115] There is no normative dimension upstream, no principle or value is prefixed in this framework, except that of the mere negotiation, since the reality is the only normative dimension able to build the theoretical model of RRI. In this sense, the dimension of normativity can only implode and collapse into that of reality.

3.5.5 Instead, the normative approach focuses mainly on the role of these normative filters established, for example, in the EU law, meant as factors of steering European policies towards anticipatory, participatory and responsible outcomes (von Schomberg, 2010, 2011a, 2011b, 2013; van den Hoven et al., 2013). In this framework the EU objectives emerge as the inescapable starting point of any initiative of governance in the field of emerging technologies. These goals should work as 'normative anchor points' in the context of governance by allowing EU institutions to anticipate choices of policy and to make decisions through their right balancing.

The role of these filters appears almost central in the normative version of RRI since they tend to bind European governance on the achievement of some goals laid down by EU treaties. Thanks to the interaction of these constitutional goals the success of the chance of both anticipating policy choices and fostering the stakeholder engagement would be ensured. Therefore, not only 'normative

---

[114]See Stilgoe et al. (2013, p. 7).

[115]A concrete example of this version is the SPICE project on geoengineering (i.e. the deliberate manipulation of the earth's climate) funded by UK Research Councils which tried to develop a more democratic and legitimate framework for science and innovation by opening up 'to a broad anticipation, reflection and inclusive deliberation, with the aim of making policy more responsive' (Stilgoe et al., 2013, p. 10).

anchor points' shape EU research funding programmes, documents of policy, guidelines, codes of conduct, but they also work as a centre of gravity of processes of public engagement such as social dialogue and consultations. In other words, they are a source of structuring of the entire governance. In this sense, here (constitutional) values also found processes of participation.

These basic values are laid down in constitutions and, in Europe, they are set out at the heart of the EU law.

According to von Schomberg (2013, p. 57), for example, they are 'normative targets which we can find in EU Treaty on the EU'. As noted by some, this points out the mainly European origin of the RRI model.[116] And this could represent a limit in abstract outside from this context. 'Normative anchor points' are goals institutionalized at the EU level within EU treaties. In other words, they are legal norms, which are at the summit of the EU law and steer (should steer) the action of all EU authorities.

But what are these constitutional guides of governance in Europe?

By referring to Article 3 of the Treaty on the European Union[117] 'anchor points' are EU goals, namely: the promotion of technoscientific advance'; market competitiveness; sustainability (which includes also precautionary principle); the promotion of social justice, equality, solidarity and fundamental rights; the protection of human health and environment. In particular, the RRI framework provides that 'research and innovation must consider the principles on which the European Union is founded, i.e. the respect of human dignity, freedom, democracy, equality, the rule of law and the respect of human rights, including the rights of persons belonging to minorities' (Directorate-General for Research and Innovation of the European Commission, 2012, p. 1).

In this regard, the reference to fundamental rights among other common values at the core of the EU makes this model an interesting

---

[116]http://www.scidev.net/global/innovation/opinion/responsible-innovation-european.html. Accessed 15 December 2016.

[117]European Union (2010) *Treaty on the European Union* (consolidated versions 2010/C 83/01) *Official Journal of the European Union* 2010/C/83/01, 30 March 2010 (TEU post-Lisbon).

case of a *rights-based model of governance* (Ruggiu, 2013, p. 211; 2015a, p. 224; Arnaldi and Gorgoni, 2016).

Notwithstanding the normative version means to emphasize the integration of some EU targets within the EU action, it is not problem-free with regard to their role, function and scope, especially with regard to just fundamental rights.

In abstract the interaction of these normative goals should overtake a case-by-case approach, which is notably the current approach of EU (Stoke and Bowman, 2012), and lead to reach simultaneously anticipatory and participative outcomes. In fact, they anticipate the directions of the EU policy by pre-determining the space of possible and legitimate choices for policymakers and identify 'positive impacts' at the early stage, contributing in steering the allocation of research funding (von Schomberg 2011b, p. 53; 2013, p. 59). Their action in the distribution of funding through the discernment of ethically sound research projects from those that are outside the space of legitimate choices of policy, makes 'anchor points' a powerful anticipation factor in EU governance. In this way, the compliance with fundamental rights of the EU Charter should determine the ethical acceptability of research and, once those projects are commercialized, of innovation (von Schomberg, 2011b, p. 50).

Their contribution to the social dialogue and participation is due to the rationale, which legitimates their presence in democratic societies, in particular in the EU law. The justification of the recourse to 'normative anchor points' is rooted on the peculiar context of the public discourse in modern societies (von Schomberg, 2010). Since 'the consequences of technological innovation are usually the result of collective action or effects of social systems', rather than resulting from the actions of individuals, there is the need of an ethics of co-responsibility (Ibid., p. 61). Therefore, given both the current state of scientific uncertainty and the current lack of consensus even in the scientific community, as well as within the society, the collective responsibility can be based only on fundamental constitutional principles such as fundamental rights. In pluralistic societies divided in a multitude of views and opinions, fundamental rights are the only common ground rooted on consensus, democratic, able to overcome disagreement in matter of science.

If in the case of conflict, the disagreement can be overcome thanks to the recourse to constitutional rights, this should happen even when the disagreement affects innovation. In this sense these principles, understood as procedural norms of the public deliberation according to the Habermasian theoretical framework (Habermas, 1992), found the public discourse also in technoscientific field.

In this sense, if any conflict can be solved with the reference to these principles, even public engagement has to be based necessarily on them. In this theoretical framework 'dimensions of responsibility [...] are value- and not rules-based', meaning that they are anchored on some EU fundamental goals (Owen et al., 2012, p. 756). Even though these normative filters 'are in themselves results of public and policy deliberation and enable consensual decision making at the policy level [...] they need to be consciously applied and be subject of public monitoring' (von Schomberg, 2011b, p. 48). In this sense, 'normative anchor points' need to be accordingly concretized in EU instruments of policy, as well as governance.

In this framework, the right balance among 'normative anchor points' acquires though a strategic role since the final outcome of EU governance depends exclusively on it.

3.5.6 A particular interpretation of the 'new governance' turn sees in the usage of flexible forms of government the rise of a new model of rationality counterpoised to the old one understood as founded on a goal rationality. This new logic is now based on a *process rationality* (Heydelbrand, 2003). In this sense, while the socio-empirical approach mainly reflects a process rationality, the normative one should follow a *goal rationality*, namely a type of rationality that focuses mainly the objectives (the values) which should steer any process so that the process is fair in as much as the principles at its basis are fair (ibid., p. 236). The transformation of the global framework led to 'the eclipse of regulation and the decentralization of state and economy' and the rise of a 'new mode of governance based on a logic of informal negotiated processes within social and socio-legal networks' (ibid., p. 326).

The current age is governed by this rising paradigm. This shift led to substitute the solutions of conflicts through rules that are an application of principles set out at the summit of our constitution with solutions based on processes of negotiation of all parties at stake. In this latter case what counts is how the process is devised and enacted. 'Colloquially, this is often interpreted as getting the right people at the table, and one will get substance' (ibid., p. 328). In other words, while the final outcome depends in the first case on goals (namely values) laid down at the beginning of the process, in this latter case it depends on the forming process of the deliberation on what values should govern at the end (Ruggiu, 2015a, p. 220). Only the democratic nature of the process of negotiation can ensure that a fair solution is reached in a societal conflict, whatever it is. On the contrary, the goal rationality believes that only constitutional principles can ensure the right working of democracy especially when the conflict affects crucial questions even in the field of science and technology.

This confirms the actuality of the debate, which counterpoised advocates of democracy (Waldron, 1999) and advocates of constitutional rights (Dworkin, 1996). Are good outputs founded on decisions based on democratic processes or on right principles?

Although the role of 'normative anchor points' within the normative version appears quite clear and would permit to solve problems concerning policy choices at the institutional level, it is not problem-free.

These normative targets for innovation 'embed tensions, complex dilemmas, as well as areas of contestation and outright conflict' (Owen et al., 2013, p. 37). As recognized by Weber (1922, p. 332ff.), values are intrinsically conflictual. Therefore, in regime of moral pluralism the risk is to pave the way to a state of permanent conflict about values, leading to the paradox that the more they are institutional, the more they are contested in society. For this reason, only the openness of the democratic debate in the field of science and innovation can embed new perspectives in front of an ever evolving context, granting thus the capacity of the system to revise its decisions and trajectories (Sutcliffe, 2011, p. 10; Holbrook and Briggle, 2014, p. 54).

According to this argument only the parties at stake are legitimized to choose values under which to develop their activity

(Holbrook and Briggle, 2014, p. 62). The top-down method of values setting can only stiffen the debate and lead to weak decisions.

Against the opinion that these values are already legitimized by democratic processes, this argument points out the increasing lack of legitimation which exists in modern societies (ibid., p. 60). This legitimation crisis now affects also founding values, for example fundamental rights.

However, one could note that the values addressed by the normative version are very general and refer to the current interests at stake within the civil society, though expressed in a bureaucratic manner (Ruggiu, 2015a, p. 224). For example, enterprises and governments have the interest in the more competitiveness of the market. Research institutions and funding organizations, as well as policymakers, have the interest in the advance of science and technology. Civil society organizations and the citizens in general can have the interest in the increment of occupation, the growth of sustainable activities of firms, as well as in the protection of health, safety of products, and, in general terms, in individual rights. It is hard to imagine different goals at stake, although we can imagine that the technoscientific progress affects them in different ways. For example, the rights affected by Electronic Health Record Systems (the right to the protection of personal data, privacy, the patient's right to consent for electronic health exchange) are presumably different from those affected by enhancing technologies (right to bodily integrity, self-determination, human dignity). Moreover, the same right can be affected differently depending to the technological context. This should also imply a different impact assessment, even in the same field (e.g. ICTs) depending the application at stake (Internet, healthcare).

As said, another problem is the moral disagreement existing within modern societies (Owen et al., 2013). Even if we may agree on these goals, in concrete situations we may disagree on their meanings and applications. As expressed by only Article 3 (TEU post-Lisbon), they can be quite indeterminate and semantically ambiguous so as they need to refer to other more specific norms of the EU treaties. For example, the goal of the market competitiveness can be linked to free circulation of goods and services such as Article

26,2 (Internal market),[118] Article 28ff. (Free movement of goods)[119] and Article 56 (Free movement of services)[120] of the Treaty on the functioning of the European Union.[121]

We need also to bear in mind that some ambiguities can exist among 'normative anchor points' themselves.

For example, the 'anchor point' of the 'quality of life, high level of protection of human health and environment' can cover both the State's interest of the Public health (e.g. Article 168 TFEU),[122] namely a public interest, and the right to health, namely a fundamental right of the individual (Art. 31 of the EU Charter – Fair and just working conditions[123]; Article 35 of the EU Charter – Health care[124]).

Moreover, the same article can lead to conflicting uses and applications. For example, Article 35 of the EU Charter deals with the same interest both as an individual right and as a legitimate EU aim of public nature.[125] The problem is that public interests on health and safety entail a given trade-off by public authorities and can conflict with individual rights, as well as with the level of

---

[118]'The internal market shall comprise an area without internal frontiers in which the free movement of goods, persons, services and capital is ensured in accordance with the provisions of the Treaties'.

[119]'The Union shall comprise a customs union which shall cover all trade in goods and which shall involve the prohibition between Member States of customs duties on imports and exports and of all charges having equivalent effect, and the adoption of a common customs tariff in their relations with third countries'.

[120]'Within the framework of the provisions set out below, restrictions on freedom to provide services within the Union shall be prohibited in respect of nationals of Member States who are established in a Member State other than that of the person for whom the services are intended'.

[121]European Union (2008) *Treaty on the functioning of the European Union* (consolidated versions 2010/C 83/01) *Official Journal of the European Union* 2008/C/115/08, 9 May 2008 (TFEU).

[122]'A high level of human health protection shall be ensured in the definition and implementation of all Union policies and activities'.

[123]'Every worker has the right to working conditions which respect his or her health, safety and dignity'.

[124]'Everyone has the right of access to preventive health care and the right to benefit from medical treatment under the conditions established by national laws and practices'.

[125]'A high level of human health protection shall be ensured in the definition and implementation of all the Union's policies and activities'.

protection required by supreme courts in the application of national constitutions and human rights treaties (Ruggiu, 2015a, p. 229).

In this sense, the public interest in the 'high level of protection of health' should not be confused with the individual right to health as such. These ambiguities, content indeterminacy and semantic unclearness can affect the understanding of stakeholders, even when the values at stake can be deemed as sharable in abstract. For example, semantic concerns on EU goals clearly emerged in the instance of the code of conduct for responsible nanoscience and nanotechnology research[126] (EC CoC), which represents an interesting case of 'normative anchor points' in action (von Schomberg, 2010; Sutcliffe, 2011, p. 22; Ruggiu, 2014b).[127]

3.5.7 RRI leads to the overcoming of the liability paradigm. According to the normative version, since the benefits of technology are (eventually) demonstrated by the market success, the market finally decides what counts as an 'improvement' in current societies (von Schomberg, 2013, p. 54). In this context unpredictable and positive impacts run the risk of being solely justified in economic terms. While, thanks to the retrospective paradigm of legal liability there is anyway a responsibility for negative impacts after the launching of products into market, '[t]here is no equivalent for a formal evaluation of the benefits' (von Schomberg, 2013, p. 55). In this framework a retrospective approach based on paradigms of accountability, liability and legal responsibility tends to prevail but the value of opportunities is neglected. Risk regulation is an important framework for the growth and development of innovation. However, the ambition of RRI is to anticipate the

---

[126] European Commission (2008) Recommendation *on a Code of Conduct for Responsible Nanosciences and Nanotechnologies Research* C(2008) 424 final, available at http://ec.europa.eu/research/science-society/document_library/pdf_06/nanocode-apr09_en.pdf. Accessed 29 September 2017.

[127] For example, translation problems of the word 'accountability' were apparent during the NanoCode survey. Indeed, 'the French and the German translations of the "accountability" principle as "responsibility" earned mistrust as they were interpreted with a connotation of *implying legal liabilities* as well as suggesting that scientists are held responsible for what is done with their work by decision outside their control or by other actors in the future' (Meili et al., 2011a, p. 6 – italics mine).

assessment of positive impacts of research and innovation by creating a responsibility framework for all actors.

To evaluate positive impacts there is the need of public engagement.

Forms of co-responsibility of stakeholders can be built only if all parties are involved from the outset in search of 'right impacts' by turning the retrospective standpoint into a prospective and proactive one. In this sense, the public investment in research and innovation cannot be justified only in macro-economic terms any longer. But since we cannot appeal to the Aristotelian ideal of good life, we need to resort to another basis. This normative dimension, which allows us to distinguish right impacts from unintended negative consequences, is expressed, as we know, by the values embedded in the Treaty on European Union referring to, in its turn, fundamental rights of the Nice Charter[128] (von Schomberg, 2013, p. 58).

Elements of this new perspective can be found in FP7 and other research funding programs, such as Horizon 2020 (van den Hoven et al., 2013, p. 21). As Horizon 2020 shows, European policy is increasingly legitimized in terms of public values and these values are currently expressed in the Charter of fundamental rights of the European Union. In fact, they draw the normative framework needed to define the impacts as legitimate.

According to the Shared Value Creation Theory there is an alternative way to create value into the market between for-profit and non-profit (Kramer, 2011).

'Society's needs are huge – health, better housing, improved nutrition, help for aging, greater financial security, less environmental damage' (ibid., p. 7). Surprisingly they represent the greatest unmet needs in the global economy. In advanced economies there is an increasing demand for products and services that meet societal needs which can support an alternative way of market. The development of enterprise forms that address these needs represent thus an alternative way of increasing productivity and expanding

---

[128] European Union (2000) *Charter of Fundamental Rights of the European Union*, adopted in Nice on 7 December 2000 (came into force on 1 December 2009). After the entry into force of the Lisbon Treaty (1 December 2009) the Nice Charter is now legally binding according to Art. I-6 §2 Lisbon Treaty.

their markets through responsible innovation. It joins together the self-interested behaviour to create economic value with the creation of societal value (ibid., p. 17). By fostering innovation, improving production techniques and building supportive industry clusters at company locations in order to increase firms' efficiency, yields, product quality and sustainability, the market can simultaneously solve social problems and creates profit (ibid., p. 5).

The premise is that both economic and social progress must be addressed by using value principles, better connecting the companies' success with the societal improvement (ibid., p. 7). A feature of the shared value is that it focuses 'on the right kind of profits – profits that create societal benefits rather than diminish them' (ibid., p. 17).

This conceptual framework can be deemed the ground which leads the innovation to meet 'Grand Challenges' addressed by the Lund Declaration (2009): namely 'global warming, tightening supplies of energy, water and food, aging societies, public health, pandemics, and security' (ibid., p. 1). Market and innovation can respond to societal needs by pursuing success and profit and addressing these fundamental goals that are at the summit of the EU law. In this way the 'Grand Challenges' can be seen as 'normative ends of responsible innovation' (Stilgoe et al., 2013, p. 10), i.e. a manifestation of those 'normative anchor points' expressed in the Treaty on the European Union.

In this sense, 'Grand Challenges' represent 'an alternative justification for investing in research and innovation' from the standpoint of societal needs (von Schomberg, 2013, p. 59). Although they tend to maximize the impacts of science and innovation on society, by concentrating funding in some specific areas, 'Grand Challenges' express that normative dimension where European governance needs to be anchored. But, differently from 'anchor points' they speak the language of the society by referring to needs and ambitions, instead of that of (ethical) values and goals.

In sustainable science there is the need to 'define criteria for R&D processes that are more problem-oriented and transdisciplinary, [to] take into account social needs and therefore [to] contribute to the solution of Grand Challenges' (van den Hoven et al., 2013, p. 20). This implies the increasingly integration of studies on ethical, legal, and societal aspects of emerging technologies in the public debate at the earlier stage in order to widen the scope of issues at play (Kearnes and Rip, 2009, pp. 9, 12ff.). The integration of social and ethical aspects in the research and innovation process can foster the quality of research, the development of more successful products and improve the market competitiveness (van den Hoven, et al., 2013, p. 22).

This integrative effort would tend thus to shift the focus of the discussion from the consideration of mere risks to the evaluation of opportunities of science and innovation embedding views of futures into the debate on responsible innovation. 'Grand Challenges' focus on some specific opportunities (supply of energy, water and food, etc.). Therefore, according to the normative approach also opportunities of development need to be prefixed.

As noted, while the normative version tends to propose a fixed anchorage of the European governance (values and 'Grand Challenges'), the anchorage should be movable and variable for the socio-empirical one (Owen, 2014).

According to the socio-empirical version, dynamic processes of society need to be free in order to express needs that are emerging within the societal body. In this sense there is no 'Grand Challenge' upstream since only the democratic participation of all parties can addresses the most urgent social needs, by allowing to maintain the dialogue open on current trajectories of science and innovation and to correct possible errors of direction (Owen et al., 2013, p. 37).

Innovation must be a collective effort of *co-design* by users whose needs have to be taken into account from the outset by shaping direction of research, as well as of development of products that will be commercialized in the future (Groves 2015, p. 328). Only users are in the best position for evaluation which needs the consumers have. This is the deep meaning of quest for the democratization of innovation of socio-empirical approaches.

As rightly pointed out, to fix some upstream goals would only mean to stiffen research and innovation and loose opportunities of improving the societal conditions (ibid., p. 329; Ruggiu, 2016, p. 115). To focus on just some opportunities of innovation would mean to lose all other opportunities of innovation at stake.

In this context the socio-empirical version casts the doubt that the best economic growth in science and innovation needs the convergence of funding in some given areas. It would be a deprivation of the process of science and technology.

Instead of focusing on 'Grand Challenges', the course of innovation needs to be articulated in terms of visions and expectations that would take place in public and democratic fora (Owen et al., 2013, p. 30). Visions are the result of the listening of needs of users. Only thanks to visions needs of recipients of innovation can be highlighted and can shape its trajectories.

This could be a way to make the society as creator of innovation caring about its products by bearing the responsibility of its destiny (Grinbaum and Groves, 2013, p. 132). It is a process of re-appropriation of innovation by the society. In this sense, visions and expectations are the privileged location for considerations of responsibility as long as they express a proactive attitude in the development of science and technology (Simakova and Coenen, 2013, p. 252). However, by anchoring research on public engagement exactly like 'Grand Challenges' do, here the risk is to tell science *what* to must do. In other words, the collection of inputs from the society runs the risk of taking the place of science and technology in their tasks, while it could only be a good opportunity for new studies and new products (Ruggiu, 2016, p. 115).

3.5.8 In this framework we cannot but notice that the socio-empirical and the normative approaches tend to address also two different modes of anticipation of the innovation impacts.

Here a clarification is needed by distinguishing three modes of our relation with the future.

First of all, we need to distinguish anticipation from forecasts and foresight (Poli, 2015).

Forecast is data-based. It has a predictive scope and calculable nature. It tends 'to adopt either a very short – as with econometric models – or a very long – as with climate change models – temporal window' (Poli, 2015, p. 90). It is often quantitative and develops under assumptions of continuity. This model of calculating the future works since 'its structure remains essentially the same or the laws governing it remain the same' (ibid.).

Instead, foresight is not predictive and it aims at addressing a number of possible futures. These futures take form of scenarios. Foresight has mainly qualitative nature by focusing on discontinuity. In other words, it is the field of possible scenarios, which can have an explorative aim (they develop from the present to the future) or a normative aim (they shift from the future to the present).

Finally, anticipation is different from the data elaboration from main trends (forecast), as well as from the exploration of possible scenarios (foresight). Anticipation aims at identifying systemic models able to change human behaviours in order to better cope with futures. Anticipation refers to processes and structures able to regulate future negative consequences still uncertain. This is the field of governance and policy-making and it is linked to the resilience of the system to increasing societal insecurities (Poli, 2014, p. 17). 'A system behaving in an anticipatory way – an anticipatory system – takes its decisions in the present according to "anticipations" about something that may eventually happen in the future' (Poli, 2015, p. 97).

New governance models therefore address specific anticipation models. In this sense, the two RRI versions also address two different anticipation models of governance.

According to the socio-empirical version the main problem of innovation is due to the fact that only a part of risks is known.

These risks are correctly treated by risk assessment tools since they are most of all foreseeable. In this sense they are a typical case of forecast. However not all risks of innovation are foreseeable. Some risks cannot be foreseen at the stage of research and tend to be detected several years after the commercialization of products. This create not only a state of mass experimentation since '[s]ociety becomes the laboratory' of innovation (Felt and Wynnne, 2007, p. 26), but also reveals a recursive move of the innovation process, called 'reflexive uncertainty' (Groves, 2015, p. 322).

Risks, especially those unknown, can always come back. To treat only known risks is not sufficient. Therefore, to make innovation really fair, there is the need a criterium for sharing those unknown risks before they occur. In this sense, society must be involved in a reflection on the purposes of science and innovation in order to clarify what risks can be reasonably uphold by the whole society

(Stilgoe et al., 2013, p. 2). This process of common thinking is 'aimed at increasing resilience while revealing new opportunities for innovation and shaping of agendas for socially-robust risk research' (ibid., p. 3).

This does not mean that risk assessment is completely outdated.

Traditional tools are still confirmed within this anticipation model, but since innovation produces also unknown risks that will be discovered much later, there is the need to novel tools to face this unforeseen baggage of the uncertainty (Groves, 2013, p. 133).

In this framework risk assessment shifts from a mode wherein it is dealt with in a technocratic manner according to a traditional linear cause-effect model, to a mode which requires a socio-empirical approach when new futures are being formed such as in the case of foresight. This new model of risk assessment is called by some the Analytical-Deliberative model (Renn, 2016; Rosa et al., 2013).

Here risk assessment is melt with theories of visions and tends to be absorbed in a broader framework in which expectations and societal practices converge. The craftwork of creating scenarios is the main tool now. This skill is a craft-like knowledge ($\tau \chi \nu \eta$, téchne) which must transcend individual boundaries. Society therefore must be involved in this collective work.

Shared imaginaries are built by imaginative and non-representational practices that underline the influence of images and texts on societal vision of the future (Groves, 2013, p. 186). This could also be deemed a venue for democratizing the thinking about future (Miller and Bennet, 2008). This process of selection and strengthening of visions would allow the identification of right impacts, meant as those impacts that affected parties feel as more desirable according to their individual views. This process of sharing of visions is therefore the final result of the anticipatory attitude of the system in front of the challenges of emerging technologies. This should also absorb risk assessment processes, since in a context where risks of innovation are mainly unknown, it is primary the question of choosing which risks the society wants to uphold (Groves, 2015, p. 324). In this framework risk assessment does not disappears technically speaking, but maintains a residual role mainly confined on known risks. Therefore, this model of

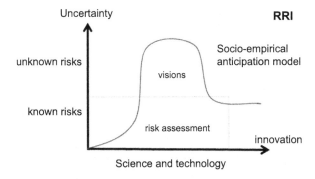

Figure 3.2 As innovation moves forward, known risks are subject to traditional risk assessment while unknown risks are treated via public engagement. As knowledge increases and the unknown diminishes, also the space of risk assessment increases.

anticipation ultimately rests on the capacity of the system to build a shared framework of imaginaries of the future, namely foresight.

If, according to this perspective, the production of values rests on the democratic engagement of stakeholders against any prefixed value, the position of stakeholders appears fragile. What stakeholder is legitimate to take part to the process? Which modes of public engagement can lead to responsible outcomes? Participation does not mean necessarily more democracy nor good outcomes (Smismans, 2008). This logic correlation does not exist. For example, the participation of some parts of civil society can lead to a given outcome, which could have been different if some other parts of civil society would have taken part.[129] In this regard the composition of the sample of stakeholders appears decisive in view of the final outcome. Besides, also the modes, wherein the involvement of the parts at stake occurs, appear crucial. For example, in the Italian case of Stamina,[130] the great involvement of citizens through media

---

[129] For example, animal rights organizations took part to the first consultation aimed at drafting the EC CoC in 2007, while neither trade union, nor any consumers' or patients' organization were involved (European Commission, 2007).

[130] Stamina is a protocol for the extraction, manipulation and re-infusion of stem cells in patients with diverse diseases ranging from Parkinson's disease, Alzheimer's and muscle-wasting disorders. Davide Vannoni, the inventor of the Stamina protocol, has been sentenced in 2015. On this see Abbott (2013).

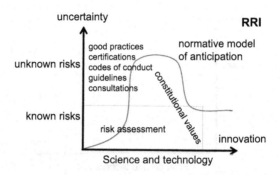

Figure 3.3 As innovation moves forward, known and unknown risks are subject to the same constitutional values, which are also embedded in the risk assessment tools and forms of soft law, aimed at fostering public engagement.

distorted the final outcome, giving legitimacy to an unproven and unscientific process (Abbott, 2013, p. 418). Notwithstanding the great involvement of some parts of civil society (e.g. media, patients' families, journalists), the outcome was completely distorted and considerably diverges from the paradigm of RRI. Therefore, this could be deemed as an emblematic case of 'irresponsible innovation' despite the participation.

According to the normative version traditional, instead, tools of risk assessment need to be reshaped on the basis of the prefixed values of the system. Research funding programmes, for example, clearly have this aim. In this model of anticipation known and unknown risks produced by innovation are handled according to the homogeneous criteria that shape not only means of risk assessment, but also any other governance tool, even public engagement. Certification systems, guidelines, codes of conduct, consultation, social dialogue and good practices need to be articulated around of a clear set of goods that are protected at the summit. This produces coherence and a more coordinated action of governance. In this context 'Grand Challenges' represent a clear example of optimizing processes and structures of governance towards a common goal.

However, in the RRI framework, even in its normative version, the process of balancing of institutional goals (fundamental rights and other EU objectives) does not provide any particular guarantee for the protection of individual rights. Especially, in a situation of scientific uncertainty there is no reason of penalizing an interest of any nature (e.g. the internal market, competitiveness, or even public health or sustainability) for avoiding a potential or unknown risk. There is no reason to consider one responsible for consequences that cannot be foreseen at the time of action (Owen et al., 2013, p. 32). And without data, which are not available per definition in the field of emerging technologies (Ruggiu, 2013c), and are fundamental for establishing any causal relation of harm, no precaution could be justified, even in the name of rights. As seen, the framework of anticipation drawn by the RRI normative version is strongly conditioned by the process of balancing among anchor points. If the process of balancing among anchor points means finding a compromise between opposite interests, this compromise risks being unsatisfactory every time individual rights are at play (Ruggiu, 2015a).

3.6 Rights-Based Models of Governance of Emerging Technologies

3.6.1 RRI is the final stage of a process of evolution which has involved the European governance of emerging technologies. This process started with the rise of biotechnology as promising field in the world and continued with nanotechnologies first, and synthetic biology then.

With biotechnology we find the seeds of the current case-by-case approach and ethics gained a crucial role in the development of governance of emerging technologies in the European context.

This premise was consolidated with nanotechnologies when the new governance paradigm arose. In this framework ELSI studies were joined by a mature governance strategy aimed at using flexible tools of governing such as EP resolutions, EC communications, agencies, networking, comitology, social dialogue and consultations for preparing the subsequent consolidation in hard law instruments (revision of existing regulations, new regulations and directives according to the classic Community Method or CCM). In this context the new governance paradigm was used by the EU as a lung for allowing the amendment of the existing regulation and producing a new sectorial regulation of traditional style.

In the same way the rising governance of synthetic biology is founded on the work of EU committees and agencies by triggering a process of adaptation of the existing GMO legislation with an attempt which is unique in the world. It is a formidable effort to use the existing structures and (probably) to delay unavoidable modifications when information is complete and there is a better awareness on what rules are most appropriate. In this context the case-by-case approach reached its summit by creating a robust framework able to adapt the present regulation according to the foreseeable challenges of a rising field.

The RRI paradigm is surging internally to this model by putting the question of the responsibility of collective behaviour involving research and innovation. Only by transforming behaviour of who takes part to the innovation process is possible to anticipate intended and unintended consequences of the technoscientific advance and make the society consciously holding risks of its choices.

However, the case-by-case approach exposes this model to unforeseen shifts and to contradict its initial premises. This internal weakness is also increased by the issue of individual rights.

The lack of protection of individual rights from the outset can lead to abrupt changes of direction in governance (Ruggiu, 2015a). In this sense, resilience is both one of the main features of the RRI model and the origin of its weakness. Paradoxically this issue is affecting not only the current case-by-case approach existing within the EU but also the RRI model in its two versions, namely a model which aims at grounding itself on two features that were neglected in the case of biotechnology: participation and anticipation.

Both socio-empirical and normative approaches to RRI are not immune vis-à-vis concerns over the protection of individual rights. In this sense, the RRI model runs the risk of considering cases that do not give priority to the protection of individual rights as an example of responsible research and innovation.

Can they be deemed good since they are merely participative and anticipatory?

3.6.2 Von Schomberg (2013) provided several examples of failures of innovation. Some of them are due to the neglect of the normative dimension of 'anchor points' in particular individual rights. In these instances, the prior and prompt consideration of the legal and ethical framework would have avoided a considerable economic loss. Instead, by promptly integrating the issue of the impact on individual rights from the stage of the design of a given project, negative outcomes can be avoided. It emerges therefore the need of a 'RRI by design' (Owen, 2014, p. 11) that focuses on the loss affecting individuals indiscriminately.

It must be underlined here the existence of a linkage between the lack of implementation of individual rights and the economic loss of a given choice in the field of innovation (Ruggiu, 2015a). It emerges in several cases, therefore, as the lack of protection of individual rights can have disastrous consequences on governance.

For instance, in 2011 the abandon of the project on an electronic patient record system (EPRS) by the Dutch government occurred when the Senate voted down due to the fact that the project still manifested several unresolved privacy issues (von Schomberg, 2013, p. 61). In this case the failure was caused by a late integration of the legal analysis on privacy within a project fully technology-driven at the earliest stage. A number of issues, such as those about the ownership of the data or about the responsibility for mistakes, were neglected and then they were dealt with too late. This fact led to a foreseeable economic and financial failure (up to €300 million).

Another example of system failure due to the neglect of legal issues on fundamental rights comes from stem cell research (Ruggiu, 2013a). In 2011 the patentability of isolated and purified neural progenitor cells for the treatment of neural defects (stem cells) was severely limited by a decision of the European Court of Justice[131] which excluded from patentability some uses of human embryos (with industrial, commercial purposes, or scientific research) and

---

[131]See the judgement of the Court of Justice (Grand Chamber), *Oliver Brüstle v. Greenpeace eV* (Case C-34/10) 18 October 2011 *European Court Reports*, 2011, p. I-09821.

those techniques of production of stem cells the mere production of which implied the prior destruction of human embryos. In this instance, where human dignity was at stake, the courts' activity relating to fundamental rights severely affected the governance direction in stem cell reserach. Innovation was inevitably affected by this shift giving rise to an economic loss.

The partial failure of GMOs in Europe can be explained in terms of lack of participation (von Schomberg, 2013, p. 61), but also in terms of rights (Ruggiu, 2015a, p. 230). In a recent case the Court of appeal of Lyon found the US Monsanto legally responsible for poisoning a French farmer who inhaled Lasso, a powerful pesticide (a weedkiller), which was used to accompany the Monsanto corn cultivation. Subsequent medical tests found a high amount of the substance contained in the pesticide in the farmer's body. Although the product was banned in France only in 2007, before it was withdrawn from sale in Belgium and the UK in 1992, in Canada in 1985 by Monsanto itself.[132] Nevertheless, in the end, Monsanto was found guilty for having failed to give warning of the presence of the hazardous chemical chlorobenzene on the Lasso label.

Also the European system of human rights can be relevant for the success of a given technology: for example, assisted reproductive technology. In 2012 the Strasbourg Court declared illegal the prohibition of preimplantation genetic diagnosis for those couples suffering from a genetic transmissible diseases such as cystic fibrosis especially when the only alternative for having a healthy child unaffected by the parents' genetic disease would have been to naturally select healthy embryos and use the abortion for those unhealthy.[133] This clearly means that although the trade-off of fundamental rights made by national or EU authorities can be correct according to existing rules, nevertheless it can find a subsequent obstacle in the work of the Strasbourg Court on human rights.

Clearly in these cases individual rights can be a cause of systemic failure of governance arrangements (Ruggiu, 2015a). This can

---

[132] http://www.bbc.com/news/world-europe-17024494. Accessed 16 December 2016.
[133] *Costa and Pavan v. Italia* (Appl. no. 54270/10) judgement of 28 August 2012.

happen both when constitutional rights are at stake (the Dutch case of the EPRS, the Italian case on preimplantation genetic diagnosis) and when EU fundamental rights are at play (the German case on stem cell research).

In all these cases the trade-off done by public authorities was not a sufficient guarantee for the protection of individual rights, as well as able to avoid, therefore, unexpected judicial complaints.

This is true also where the EU law is at stake, namely when we consider a framework which is generally deemed as one of the most demanding with regard to the protection of rights of the European citizens.

3.6.3 This outcome leads us to consider a not secondary element under the EU law: the method of balancing as the rule of application of fundamental rights when they compete with other public interests such ad market competition, technoscientific advance, public health, sustainability, etc.

Differently from what happens with regard to human rights under the European Convention on Human Rights (ECHR),[134] in fact, under the Community law there is no rule of prevalence for EU fundamental rights in front of the other EU or national goals (Robles Morchón, 2001, p. 263). Both elements can be balanced one with another.

This means that under certain circumstances the protection of fundamental rights can be put behind by the promotion of given public interests which are established as EU objectives in its treaties. For example, in the case of synthetic biology the prevalence of the goal of sustainability might lead to use SynBio organisms for the environmental remediation, but this solution could be not suitable with regard to the protection of individual rights since up to now there is no guarantee that we have the control of those artificial organisms once released into the environment. This environmental friendly solution could breach, in fact, rights of

---

[134]Council of Europe (1997) *Convention for the Protection of Human Rights and Dignity of the Human Being with regard to the Application of Biology and Medicine* (Convention on Human Rights and Biomedicine or the Oviedo Convention) (CETS n. 164), adopted in Oviedo on 4 April 1997 (came into force on 1 December 1999).

fishermen, consumers (freedom of enterprise), or population in the neighbours (right to health, right to a healthy environment).

This situation could happen whenever a EU interest different from rights is at stake.

As the European Court of Justice stated, indeed, 'restrictions may be imposed on the exercise of those rights (i.e. fundamental rights), in particular in the context of a common organization of a market, provided that those restrictions in fact correspond to objectives of general interest pursued by the Community'.[135]

As seen, this framework is also confirmed by RRI, even in its normative version, where the responsible action of stakeholders should be the outcome of the balancing of all 'normative anchor points', thus including public interests such as EU goals.

If balancing 'anchor points' means to find a compromise among opposite interests, this compromise cannot be satisfactory whenever individual rights are at play: individual rights can be totally or partially breached. In other words, also within the RRI framework sometimes EU fundamental rights might be, even only partially, sacrificed in view of boosting some relevant public interests (internal market, competitiveness, public health, sustainability, etc.), whereas the acceptable outcome should be to reach a given interpretation of a public interest which also realizes individual rights.

This also means that, after the development of a given trajectory of governance in science and technology, national or supranational courts might alter the initially taken course of governance. That is, what exactly happened in the cases addressed by von Schomberg (2013) and me (2015a). This outcome cannot but transform fundamental rights in a 'negative externality' of European governance by making to rise a retrospective standpoint at the heart of the RRI framework (Ruggiu, 2015a). In other words, under this modulation of the RRI framework the proactive dimension runs the risk of imploding and the anticipatory dimension collapses.

---

[135] See the judgement of the Court of Justice (Third Chamber), *Waachauf v. Bundesamt für Ernährung und Forstwirtschaft* (Case 5/88) 13 July 1993 *European Court Reports 1989-02609*, §18.

According to its anticipation models, in fact, all tools need to be reshaped by 'normative anchor points': risk assessment procedures, advises, reports, tools of social dialogue, consultations, etc. But without the prior consideration of the dimension of rights this set of tools appears to be intrinsically weak. If the risk of adverse courts' decisions exists with regard to governance, at least in principle, we need to acknowledge that this risk is increased whenever legal issues on fundamental rights are more or less unconsciously neglected (Ruggiu, 2015a).

This result is paradoxical, especially, if we consider RRI according to the normative approach. A possible solution can come from a legal consideration concerning rights.

This result is common also to the socio-empirical version.

In this context 'anchor points' are the outcome of public engagement. Values are the final point of the stakeholder participation. Therefore, there is no guarantee that the values chosen via the process of negotiation coincide with those at the basis of the system, in particular with fundamental rights. Here principles of democracy prevail over constitutional principles and a process of sharing visions of the future is deemed essential to cope with uncertainty, especially that of unknown nature. In this context the anticipation model tends to overlap risk assessment procedures with tools of foresight aimed at incorporating the scenarios logic into governance structures. Risk assessment tends to dissolve into visions and becomes available again when the progress stabilizes itself and risks are known again. This also means that this model of governance does not cultivate any specific measure concerning rights and has the same shortcoming of the normative approach.

A possible solution to this impasse comes from human rights.

3.6.4 Contrary to the common opinion, we need to distinguish human rights from fundamental rights and this structural distinction can be for the RRI framework an opportunity of strengthening its foundations.

While human rights are those rights affirmed at the international law level that belong to the person as such, fundamental rights are those rights acknowledged at the EU or national level that belong only to the citizen (Pariotti, 2013, pp. 4–5).

This conclusion is relevant with regard to, for example, the field of clinical experimentation where limits concerning only citizens could be set out under the EU or national law (at least in abstract), while those limits cannot be laid down under the human rights law, which protects all individuals (not only citizens).[136] Furthermore, if human and fundamental rights normally coincide as content, they may diverge sometimes. This is the case, for example, of the right to keep and bear arms set forth by the Second Amendment of the US Constitution that is a fundamental right but not a human right. Finally, a last difference is, as said, the possibility of balancing EU fundamental rights with public interests of either the State or EU, while within the ECHR framework this is not possible (with some provided exceptions). In other words, human rights cannot be balanced with State's public interests except some exceptional cases (Geer, 2000).

In the ECHR ambit the application of rights is often made under the doctrine of the margin of appreciation (Donati and Milazzo, 2003; Benvenisti, 1998; Geer, 2000) which basically differ from EU balancing.

This doctrine was developed especially by the Strasbourg Court as basis for justifying the State's discretion in the protection of individual rights. It is also recalled by the Court of Justice of the European Union (Anrò, 2009, pp. 14–18) and by other supranational courts.

Due to its jurisprudential origin, the doctrine of margin of appreciation was variously criticized by several scholars (Benvenisti, 1998; Kratochvìl, 2011; Geer, 2000) and now it has been embedded in an Additional Protocol to the ECHR which has not entered into force yet.[137]

First of all, the margin of appreciation must not be confused with cases of suspension of the ECHR application when the life of the nation is threatened (e.g. civil war, terrorism) regulated by

---

[136]See Art. 5 of the Oviedo Convention which protects third countries' nationals that are subjected to a clinical experimentation.

[137]Council of Europe (2013) *Additional Protocol No. 15 amending the Convention on the Protection of Human Rights and Fundamental Freedoms* (CETS 213), adopted in Strasbourg 24 June 2013, which has not entered into force yet, http://www. echr.coe.int/documents/protocol_15_eng.pdf. Accessed 20 December 2016.

Article 15 ECHR. In this instance derogatory measures can be taken by the State only in case of public emergency threatening the life of nation provided that Articles 2 (right to life), 3 (prohibition of torture, inhuman and degrading treatment), 4 §1 (prohibition of slavery), the principle of no punishment without law (or in Latin *nulla poena sine lege*, Article 7 ECHR) cannot be derogated. For this reason, these rights are called absolute rights (Donati and Milazzo, 2003).

The doctrine of margin of appreciation, according to which a right protected by the ECHR can be derogated by a State's discretionary measure, was affirmed only with regard to some specific ambits. These are: respect for private and family life (Art. 8 ECHR), freedom of thought (Art. 9 ECHR), freedom of expression (Art. 10 ECHR), freedom of association (Art. 11 ECHR), freedom of movement (Art. 2,3 P4), propriety (Art. 1 P1) (Geer, 2000, 24–26). Moreover, the margin of appreciation is acknowledged in ethically controversial cases (such as the status of embryo, medically assisted procreation, surrogacy, end-life decisions, assisted suicide), namely whenever there is no single position among member States of the Council of Europe (Ruggiu, 2018).

In other words, it is not a general principle within the ECHR framework. Nor the use of the State's discretion is unfettered since it is subjected under strict scrutiny of the Strasbourg Court which applies some criteria of jurisprudential origin. This means that the State's discretion must respect those criteria to be considered correctly used.

There is no balancing between human rights and public interest, but when there are the conditions provided by the ECtHR, those rights can be limited by a State's measure.

Contrary to what one could think, the established margin of appreciation is not, therefore, completely free (the State's decision is not unfettered) and it does not entail a lack of control by the Strasbourg Court. The ECtHR normally scrutinizes these matters closely,[138] assessing (in the case of Article 8 ECHR for example) whether the State's interference with ECHR rights was made in accordance with the law (the principle of legality), pursuing

---

[138] *Costa and Pavan v. Italia* (Appl. 54270/10) judgement of 28 August 2012, par. 68.

a legitimate aim (provided for by the ECHR), necessary in a democratic society, and through the adoption of a measure that is proportionate to the pursued legitimate aim (the 'proportionality' test).[139] In this sense, the margin of discretion does not alleviate the need for the State to prove they have respected the rights and freedoms set forth by the ECHR.

These differences represent a limit for the EU law, but they can also be an opportunity of development within of the framework of the European governance because they can enhance standards of protection of EU fundamental rights in Europe.

While there is a concurring plurality of tools of the protection of individual rights in Europe, where the system set forth by the Council of Europe coexists with that led down by the European Union, the two main supranational courts in Europe, namely the European Court of Human Rights (ECtHR) with regard to the Council of Europe, and the Court of Justice of European Union (CJEU), have developed a spontaneous mechanism of coordination (Spielmann, 1999; Bultrini, 2008). Given the current process of accession of EU to the ECHR triggered by the Lisbon Treaty, this spontaneous self-coordination between their jurisprudence appears of paramount importance. This mechanism of self-coordination, in fact, can further strengthen the level of protection of fundamental rights within the EU by extending the level of protection of human rights to the EU law. This process not only aims at avoiding conflicting decisions and inconsistencies between their jurisprudences. This can have a further consequence by subtracting EU fundamental rights from the balancing with other competing objectives of EU.

This strengthening process can be extended by involving the work of respective ethical advisory boards, namely the Committee

---

[139]E.g. *Lambert and others v. France* (Appl. 46043/14) judgement of 5 June 2015 *Reports of Judgements and Decisions*, 2015. While the State enjoys a broader discretion in the environmental matter, where a nation's strategic choices are at stake (e.g. nuclear, transport, industry), this margin is quite strict and generally subject to a deeper scrutiny of the Court in the immigration and health matters, where individuals are directly affected. See e.g. *Costa and Pavan v. Italia* (Appl. 54270/10) judgement of 28 August 2012, par. 79; *Menneson v. France* (Appl. 65192/11) judgement of 26 June 2014 *Reports of Judgements and Decisions*, 2014, par. 77; *Open Door Counselling Ltd and Dublin Well Woman v. Ireland* (Appl. 14234/88, 14235/88) judgement of 29 October 1992 *Series A*, No. 246-A, par. 68.

on Bioethics with regard to the Council of Europe (DH-BIO)[140] and the EGE with regard to EU, in order to coordinately cope with challenges of technoscientific advance and increase the efficiency of technological governance.

A weak coordination among ethical advisory bodies in Europe already exists. This involve both national ethical advisory boards and the DH-BIO and the EGE. The implementation of this process, however, might avoid normative short circuits to which the lack of protection of individual rights at the governance level can lead. Given the strategic role of advisory bodies in policy-making, as well as in building governance processes, this anticipate risks for rights at the phase of design. This further integration of issues concerning human rights within the ethical, legal and social analysis of the impact of emerging technologies can thus strengthen the anticipative attitude of governance at the European level, by fostering that key feature of the RRI framework: anticipation.

In this way individual rights would shift from a mere negative externality to a proactive element of governance.

We need to acknowledge that in the model of RRI, as expressed by both its versions, the action of 'anchor points' does not necessarily produce the outcome of the protection of individual rights. This led to raise the question of rights, of human rights, within the debate on RRI.

The key question is therefore to transform human rights in a proactive tool of technological governance.

3.6.5 There is an increasing attention being paid to the role of European human rights law in the field of governance of technoscientific progress (Ruggiu, 2013a, 2013b, 2015a; Koops et al., 2013; Arnaldi et al., 2016; Arnaldi and Gorgoni, 2016; Leenes et al., 2017). This increasing interest takes as its starting point the idea that human rights can play a positive role in governance frameworks, for example in RRI. According to this scholarship the 'regulatory challenges of [emerging technologies] can thus be situated in a framework of common overarching principles that constitute the European sphere of rights and freedoms. [...] These

---

[140]http://www.coe.int/en/web/bioethics/dh-bio. Accessed 3 December 2017.

are embedded in, inter alia, the European Convention on Human Rights and the EU Charter of Fundamental Rights' (Leenes et al., 2017, p. 30).

These rights are also established in national constitutions, as well as in EU treaties (Arnaldi and Gorgoni, 2016, p. 16).

However, as seen, constitutional and EU fundamental rights are in concurrence with other public interests such as market competitiveness and scientific research which can be legitimately prioritized in the process of balancing (Robles Morchòn, 2001). This can cause the sacrifice of individual rights which can give rise to forms of system failure of innovation (von Schomberg, 2013, p. 61; Ruggiu, 2015a).

This leads to the counterintuitive outcomes that principles at the heart of our constitutional orders can be sacrificed sometimes for sake of the technoscientific advance. Against this outcome, human rights, whose protection States are obliged, are established at the international law level, especially in Europe with the ECHR. Human rights can have the paramount function of strengthening the protection of individual rights in the balancing with other public interests at the national and EU level, avoiding cases of system failure of innovation and the abrupt shift of governance arrangements. Thanks to them and their practice some principles applicable to the technoscientific progress can be promptly identified and proactively used for consistently steering governance of emerging technologies.

The main feature of the abstract norms from which human rights derive is that, since these norms are indeterminate, their content 'is not established once for all in the law-making process, but must be shaped, also in a bottom-up manner and by several relevant actors during the application stage, like judges but also private actors promoting tools of self-regulation' (Arnaldi et al., 2016, p. 28). This means that because of this original indeterminacy, 'the judicial stance contributes to the definition of the content of rights' (Arnaldi et al., 2016, p. 29). Therefore, it becomes crucial to undertake a preliminarily analysis of the practice of human rights where we can understand how they can be violated.

In order to understand the meaning of a human right therefore it is crucial to think of its violations. In this sense, the study of the framework of decisions of supranational courts, such as the ECtHR, is crucial in order to understand what the impact of technoscience might be on human rights, namely what violations it might entail.

This is necessary for hypothesizing scenarios made up of possible judicial complaints, which involve the technoscientific advance in a given sector.

This *regulatory foresight* (Blind, 2008) is a type of short-term vision, which is based on present research in its testing phase, and on current developments in the market. Therefore, the study of the judgements of the ECtHR represents the first non-eliminable step for identifying rules and legally binding principles which States must adhere to, principles that can also be extended to the field of research and innovation. This further (legal) analysis of the human rights framework is needed 'to address unexpected side effects that regulatory interventions aimed at safeguarding certain rights', as well as 'to stay alert to the need for updating, expanding or changing the framework in light of changes in society and value systems that are brought about through the mutual shaping process of technologies, social processes, and normative outlooks' (Leenes et al., 2017, p. 33).

The 'rights-based framework as outlined may provide a fruitful backdrop for assessing and regulating [emerging] technologies' (Leenes et al., 2017, p. 33).

There are few, yet enlightening instances of the use of this approach to tackle the challenges of technoscientific progress.

As known, this is the method followed in the reports of, for example, the DH-BIO, where it systematizes the ensemble of decisions by the ECtHR which may be pertinent in certain fields such as health,[141] bioethics in general covering reproductive rights, medically assisted procreation, consent to medical treatment, retention of personal data and bio-samples and end-of-life decisions,[142] internet and data protection.[143]

---

[141] Committee on Bioethics of the Council of Europe (DH-BIO) (2008) *The Court's Case-Law Concerning the Health in General*, 18 July 2008.

[142] Committee on Bioethics of the Council of Europe (DH-BIO) (2009) *Bioethics and the Case-Law of the Court*, 14 October 2009.

[143] http://www.echr.coe.int/Documents/Research_report_internet_ENG.pdf. Accessed 3 December 2017.

There are also periodical reports issued by the secretariat of the Strasbourg Court on its jurisprudence, which cover several fields including health in general,[144] end of life,[145] the environment,[146] persons with disabilities,[147] detention conditions of prisoners,[148] personal data protection,[149] etc., which can be extremely useful when we need to assess a given technological flied in the light of human rights.

This regulatory framework is also used as model in the practice in human rights in which some non-governmental organizations such as Amnesty International, Greenpeace, Doctors without borders, Human Rights Watch are committed. On these principles is built the paradigm of Corporate Social Responsibility which develops in parallel and not in contrast with the that of legal liability of the ECtHR. The NGOs' action represents, we might say, the 'invisible hand' of the retrospective logic expressed by liability.

A precious example, of how this practice can influence the consideration of the rights of those invisible workers who make possible the realization of our high-tech products, is the 2017 Amnesty International report titled 'Time to recharge'.[150]

This report 'ranks industry giants including Apple, Samsung Electronics, Dell, Microsoft, BMW, Renault and Tesla on how much they have improved their cobalt sourcing practices since January 2016'. It assesses company practices according to a set of criteria that reflect international standards, namely the protection of human rights, including the requirement that companies carry out what are known as 'due diligence' checks on their supply chain and the

---

[144] http://www.echr.coe.int/Documents/FS_Health_ENG.pdf. Accessed 3 December 2017.

[145] http://www.echr.coe.int/Documents/FS_Euthanasia_ENG.pdf. Accessed 3 December 2017.

[146] http://www.echr.coe.int/Documents/FS_Environment_ENG.pdf. Accessed 3 December 2017.

[147] http://www.echr.coe.int/Documents/FS_Disabled_ENG.pdf. Accessed 3 December 2017.

[148] http://www.echr.coe.int/Documents/FS_Detention_conditions_ENG.pdf. Accessed 3 December 2017.

[149] http://www.echr.coe.int/Documents/FS_Data_ENG.pdf. Accessed 3 December 2017.

[150] https://www.amnesty.it/scarica-report-time-to-recharge-corporate-action-and-inaction-to-tackle-abuses-the-cobalt-supply-chain/. Accessed 3 December 2017.

requirement that they are transparent about the associated human rights risks.

This practice assessment works before and beyond the paradigm of liability is enacted according to human rights standards. In fact, the category of stakeholders here identified work out of Western countries where enforcement mechanisms, such as that of the ECtHR, cannot function but they can however influence the whole chain of innovation. Notwithstanding, the human rights law works as a clear guidance of innovation practices.

This tool is devised in order to identify a set of clear actions which can constitute a human rights violation and the needed correctives (namely behaviour) which can contribute to implement their protection. This clear framework is of paramount importance for those enterprises which act abroad in countries where human rights law has no grasp, since they did not sign or ratify international treaties on human rights or they do not comply with their provisions (Pariotti and Ruggiu, 2012).

The usefulness of this approach is that it permits to consider in advance a possible breach of individual rights, avoiding innovation failures, for example due to judicial complaints, at any level (national, EU, international) or due to broadcast media campaign of sensitization which can affect the reputation of companies engaged in innovation.

In this framework at the same time inclusion (of those who are often excluded from participatory processes) is also embodied, as is ethical acceptability.

'These strategies try to ensure that a responsible and antici-patory attitude will reduce risks of harms to rights and values, and therewith provide guidance to the actors involved' (Leenes et al., 2017, p. 30). This consideration, in fact, is crucial not only when entrepreneurs have to take decisions on their business, but also when the policymaker takes strategic decisions in the field of technoscientific progress, when the regulator has to devise a governance framework for a new emerging technology or, finally when, judges have to decide a hard case involving the technoscientific progress.

To have in advance the possible area of impact of a given technology on individual rights permits to reduce the risk of future

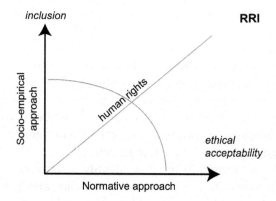

Figure 3.4 The balance between the two approaches within in the RRI model is gained through the implementation of human rights. Here inclusion and ethical acceptability meet in the implementation of human rights practices.

litigations and of abrupt shifts of a particular research field, as well as market.

This preliminary work on the ECtHR jurisprudence, therefore, allows to pass from a conception of rights as mere negative externality of technoscientific developments to a proactive conception of rights since we can derive from the case law which limits rights pose (e.g. the ban of human cloning, genetic germline enhancement, the creation of human–animal chimeras)[151] but also what pathways they address (e.g. human enhancement of, in particular, persons with disability,[152] data protection and information technologies.[153])

In this regard, there is the need of a brokering work able to link this preliminary analysis to the challenges stemming from a given technoscientific field in order to find limits and possible opportunities according to a rights-based approach. Rights, in fact, also have a positive role in governance ('a general orienting function') since they 'help to identify priorities and therefore justify rules that favour one application, responding to values and needs deemed fundamental, over others' 'by pointing to innovations [...]

---

[151] On this see Chapter 4 of Part II on human dignity.
[152] On this see Chapter 7 of Part II on the right to bodily integrity.
[153] On this see Chapter 9 of Part II on privacy.

that should be fostered through regulation' (Leenes et al., 2017, p. 31). In this way, this 'constitutional framework could point towards developments [. . .] that would better fulfil fundamental values and ensure the implementation of rights, so as to impress socially beneficial connotations onto the scientific endeavour' (Leenes et al., 2017, p. 31).

3.6.5 Almost all models of governance can be articulated in a rights-based fashion. There is not a privileged model of governance which protects human rights. In particular, both the normative and the socio-empirical approaches can be articulated in sense rights-oriented.

In the RRI framework, in its normative version, the key feature of anticipation of the normative approach is tightly connected to the work of 'normative anchor points' and how they interact. In this framework fundamental rights play a role far to be secondary.

In order to be really anticipatory however EU fundamental rights should be able to shape governance choices at an early stage by prioritizing not only those policies which are most respectful of individual rights (i.e. of workers, consumers, patients, the environment) and leading towards the adoption of some basic precautionary measures in a state of scientific uncertainty (e.g. labelling in the case of nanofood[154] and cosmetics).[155] Fundamental rights need to be taken into account also proactively, by shaping current policies in terms of rights, with a better focus on those aspects that can be framed through the language of rights. This can be reached thanks to the human rights law.

---

[154] European Union (2011) Regulation (EC) No. 1169/2011 of the European Parliament and of the Council of 25 October 2011 *on the provision of food information to consumers, amending Regulations (EC) No 1924/2006 and (EC) No 1925/2006 of the European Parliament and of the Council, and repealing Commission Directive 87/250/EEC, Council Directive 90/496/EEC, Commission Directive 1999/10/EC, Directive 2000/13/EC of the European Parliament and of the Council, Commission Directives 2002/67/EC and 2008/5/EC and Commission Regulation (EC) No 608/2004, OJ L304/18.* In this case the entry into force of the provision which introduces the mandatory labelling for foodstuffs containing nanomaterials will be only by December 2014 (Art. 18.3).

[155] European Union (2009) Regulation (EC) No. 1223/2009 of the European Parliament and of the Council of 30 November 2009 *on cosmetics products, OJ L349/59.*

In other words, the human rights approach can foster that interpretation of EU objectives that is more respectful of fundamental rights.

For example, in the case of human enhancing technologies (HETs) many provisions of the ECHR framework lead to push those applications that vulnerable people such as persons with disability (Ruggiu, 2018). Today the field of disability is already an ambit where HETs are often developed before being experimented on healthy people. The Council of Europe framework tends to foster this current trend. But to do this, there is the need of studies that previously analyse the framework of provisions stemming both from other soft law instruments of the Council of Europe (such as the Oviedo Convention) and from the set of decisions of the Strasbourg Court that can be pertinent in a particular technoscientific field. This assessment, prior to the action of public authorities and simultaneous to research, can strengthen therefore the anticipation model with the RRI framework.

As regards socio-empirical approaches, anticipation rests on tools of foresight such as visions. In this framework risk assessment tends to be melt way with the analysis of possible scenarios. But, since they are not rights-oriented, the final identification of values via public engagement might diverge from those set out by the system at the institutional level. In other words, societal values might differ from fundamental rights.

The reason of this unsatisfactory outcome is that the process of participation, even though structured according to a stage gate architecture, mainly depends on the extent, the nature and organization of participants. As broad and participative can be the process of negotiation, nothing pushes participants to argument in favor of rights (especially of other persons not involved in the process). Nor the participation of non-governmental organizations ensures this outcome alone, since each one defend the interests for which was established. For example, the participation of environmental organizations does not ensure that consumers' or patients' rights can be represented in the final deliberation. Not only aspects related to the quantity of participation processes are important, also those concerning the quality of participants.

Moreover, the protection of human rights in front of technoscientific advances needs resources, knowledge, time and organizational structure that only some independent entities of international nature can have. There is both the need of knowledge on how emerging technologies impact regulation and how human and fundamental rights work in a given context, such the European one.

These features can be owned only by organizations of independent nature from both the State and the market that are established only at the international level such as within the Council of Europe and are engaged in the protection of human rights. Only these independent organizations can ensure the consideration of those interests which are not addressed either because some parts are excluded from participatory tools from the beginning (e.g. anonymous workers, especially abroad) or because those who participate did not raise them (civil society, lay public). This does not preclude the participation of other NGOs. There is no aut between the participation of NGOs and the engagement of organizations that protect human rights at the international level.

The participation of these entities therefore could be a way to transform also socio-empirical RRI arrangements in sense rights-oriented rearticulating their structure.

The impact assessment on individual rights, in fact, can be easily inserted in a *stage gate architecture*, namely a framework where the whole process of decision making is guided through the prevision of phases 'being subject to formal or informal approval at a decision "gate"' (Owen, 2014, p. 13). If we include in any stage and phase the analysis of rights as condition for passing to the subsequent stage, also informal processes of negotiation can be rights-based. This is what 'RRI by design' could mean from the standpoint of rights.

Both the normative and the socio-empirical approach can be transformed in sense rights-oriented.

There is no impediment to this outcome in both cases of the normative and socio-empirical approach to RRI. No one of these frameworks shows obstacles to this process of systemic integration and of strengthening of the dimension of anticipation according to human rights.

3.6.6 At the end, this process of better integration between the two systems of the EU and Council of Europe reaches a significant outcome: the transformation of the current governance arrangement following the case-by-case logic in a *rights-based model of governance* (Ruggiu, 2013a; 2013b; 2015a).

This approach rooted on the human rights law discovers a consistency principle, according to which policy decisions, as well as regulation, need to be coherent with the principles of justice of liberal-democratic societies that are, thanks to the national and international law; legally binding in most of modern countries (Dworkin, 1986, p. 212). According to this perspective, all State's levels, legislative, administrative and judiciary need to be coherent with fundamental constitutional principles. These principles in liberal-democratic countries are mainly expressed by human rights and the Rule of Law. And it would be paradoxical that those principles that we claim in all ambits (from political liberties, to immigration, to terrorism, etc.), could not inform also the field of emerging technologies (e.g. Francioni, 2007, p. 4). This consistency principle cannot but be applied also in the EU framework to which the main European States belong. Within this pattern, thanks to the legal obligation of the respect for human rights, governance reaches a better consistency with the goals that are at its basis and avoids normative short circuits caused by the lack of protection of individual rights (Ruggiu, 2015a). This makes governance to reach its system stability.

This conceptual framework is particularly meaningful especially in Europe where all EU countries are bound to comply with norms of the ECHR, even in the field of emerging technologies. All the more so the EU treaties provided for an accession process of EU to ECHR.

In this model thanks to the reference to human rights by both law-making and decision-making processes, the dimension of anticipation of governance becomes consolidated, gaining a long-term duration and a possibility of coherent evolution in the future. Therefore, a rights-based model of governance represents a pattern available for almost any governance model in EU, where the principle of integrity is applied at each level (political, regulatory, judiciary), namely a system where policy decisions, rules and their administrative application are in accordance with the (legal) principles of human rights (Dworkin, 1986, p. 160; Ruggiu, 2013b; 2015a).

References

Abbott, A. (2103). Stem-Cell Ruling Relies Researchers, *Nature*, **495**, pp. 418–419.

Aitken, R.A, Bassan, A., Friedrichs, S., Hankin, S.M., Hansen, S.F., Holmqvist, J., Peters, S.A.K., Poland, C.A., and Tran, C.L. (2011). *Specific Advice on Exposure Assessment and Hazards/Risk Characterisation for Nano-materials under REACH (RIP-oN 3) – Final Project Report, RCN/ RIP-oN 3/FPR/1/FINAL*, 07 July 2011, European Commission, http://ec. europa.eu/environment/chemicals/nanotech/pdf/report_ripon3.pdf. Accessed 1 March 2015.

Anrò, I. (2009). Il margine di apprezzamento nella giurisprudenza della Corte di giustizia dell'Unione europea e della Corte europea dei Diritti dell'Uomo", in: A. Oddedino, E. Ruozzi, A. Viterbo, F. Costamagna, L. Mola, and L. Poli (eds.) *La funzione giurisdizionale nell'ordinamento internazionale e nell'ordinamento comunitario*, Napoli: Edizioni Scientifiche Italiane, pp. 7–28.

Arkin, A., Arnold, F., Berry, D., Boldt, J., Church, G., Ellington, A., Endy, D., Fussenegger, M., Gold, E.R., Greenwood, J., Lee, S.Y., Lim, W., Minshull, J., Murray, T.H., Poste, G., Prather, K.L.J., El-Samad, H., Smolke, C., and Weiss, R. (2009). 'What's in a Name?', *Nature Biotechnology*, **27**(12), pp. 1071–1073.

Arnaldi, S., and Gorgoni, G. (2016). Turning the tide or surfing the wave? Responsible Research and Innovation, fundamental rights and neoliberal virtues, *Life Sciences, Society and Policy*, **12**(6), pp. 1–19.

Arnaldi, S., Gorgoni, G., and Pariotti, E. (2016). RRI as Governance Paradigm: What's New? in: Lindner, R., Khulmann, S., Randles, S., Bedsted, B., Gorgoni, G., Griessler, E., Loconto, A., and Mejlgaard, N. (eds.), *Navigating Towards Shared Responsibility*, Karlsruhe: Fraunhofer Institute for Systems and Innovation Research ISI, pp. 23–29.

Arrigo, E. (2006). Code of Conduct and Corporate Governance, *Symphonya. Emerging Issues in Management*, **1**, pp. 93–109, http://symphonya. unimib.it/article/view/2006.1.07arrigo/8776. Accessed 18 February 2015. Accessed 29 September 2017.

Barben, D., Fischer, E., Selin, C., and Guston, D.H. (2008). Anticipatory Governance of Nanotechnology: Foresight, Engagement, and Integration, in: Hackett, E., Lynch, M., and Wajcman, J. (eds.), *The Handbook of Science and Technology Studies* (3rd ed.), Cambridge, MA: MIT Press, pp. 979–1000.

Barr, A., and Wilson, R. (2014). Google's Newest Search: Cancer Cells, *The Wall Street Journal*, Oct. 29, 2014, http://www.wsj.com/articles/google-designing-nanoparticles-to-patrol-human-body-for-disease-1414515602. Accessed 6 March 2015.

Beck, U. (1986). *Risikogesellschaft: Auf dem Weg in eine andere Moderne*, Frankfurt am Main: Suhrkamp; It. trans. (2000). *La società del rischio. Verso una seconda modernità*, Roma: Carocci.

Bennet, G., Gilman, N., Stavrianakis, A., and Rabinow, P. (2009). From Synthetic Biology to Biohacking: Are We Prepared?, *Nature Biotechnology*, **27**(12), pp. 1109–1111.

Benvenisti, E. (1998). Margin of Appreciation, Consensus, and Universal Standards, *Journal of International Law and Politics*, **31**(4): 843–854.

Berg, P. (2008). Meetings that Changed the World: Asilomar 1975: DNA Modification Secured, *Nature*, **455**, pp. 290–291.

Berger, M. (2013). An Overview of the European Union's Nanotechnology Project, *Nanowerk*, http://www.nanowerk.com/spotlight/spotid=31109.php. Accessed on 7 March 2015.

Blind, K. (2008). Regulatory foresight: Methodologies and selected applications, *Technological Forecasting & Social Change*, **75**, pp. 496–516.

Blok, V., and Lemmens, P. (2015). The Emerging Concept of Responsible Innovation: Three Reasons Why It Is Questionable and Calls for a Radical Transformation of the Concept of Innovation, in: Koops, E.J., Oosterlaken, I., Romijn, H.A., Swierstra, T.E., and Van den Hoven, J. (eds.), *Responsible Innovation 2: Concepts, Approaches and Applications*, Cham: Springer, pp. 19–35.

Bowman, D.M., van Calster, G., and Friedrichs, S. (2010). Nanomaterials and Regulation on Cosmetics, *Nature Nanotechnology*, **5**(2), p. 92.

Brownsword, R. (2008). *Rights, Regulation and the Technological Revolution*, Oxford: Oxford University Press.

Carro-Ripalda, S. and Macnaghten, P. (2014). *GMFuturos Policy Brief. A New Approach to Governing GM Crops? Global Lessons for UK and EU*, Institute of Hazard, Risk and Resilience, Durham University https://www.dur.ac.uk/resources/ihrr/GMFuturosPolicyBriefFinal.pdf. Accessed 9 June 2015.

Chessa, A., Morescalchi, A., Pamolli, F., Penner, O., Petersen, A.M., and Riccaboni, M. (2013). Is Europe Evolving Toward an Integrated Research Area? *Science*, **339**, pp. 650–651.

Coglianese, C., and Mendelson, E. (2010). Meta-regulation and Self-regulation, in: Baldwin, R., Cave, M., and Lodge, M. (eds.) *The Oxford Handbook on Regulation*, Oxford: Oxford University Press, pp. 146–168.

Cooper, R.G. (1990). Stage Gate Systems: A New Tool for Managing New Products, *Business Horizons*, **33**(3), pp. 44–54.

Cosmetics Europe (2011). *Colipa Guidelines on Cosmetic Products. Compliance with Regulation 1222/2009 on Cosmetics products*, https://www.cosmeticseurope.eu/publications-cosmetics-europe-association/guidelines.html?view= item&id=84. Accessed 27 March 2015.

Cosmetics Europe (2012). Roles and Responsibilities Along the Supply Chain: A Practical Guide. Compliance with Regulation 1222/2009 on Cosmetics Products, https://www.cosmeticseurope.eu/publications-cosmetics-europe-association/guidelines.html?view=item&id=89. Accessed 27 March 2015.

Christoforou, T. (2008). Genetically Modified Organisms in European Union Law, in N. De Sadeleer (ed.), *Implementing the Precautionary Principle: Approaches from the Nordic Countries, EU and USA*, London: Earthscan, pp. 197–228.

de Búrca, G. and Scott, J. (2006). *Introduction: New Governance, Law and Constitutionalism*, in G. De Burca, and J. Scott (eds.), *Law and New Governance in the EU and the US*, Oxford, Hart Publishing, pp. 1–12.

De Saille, S. (2015). Innovating Innovation policy: the emergence of 'Responsible Research and Innovation, *Journal of Responsible Innovation*, **2**(2): 152–168.

de Salvia, M. (1997). Ambiente e Convenzione europea dei diritti dell'uomo, *Rivista internazionale dei diritti dell'uomo*, **10**, pp. 246–257.

Deibel, E. (2014). Open Source and Synthetic Biology: Openness & the Translation of DNA into Informatic Formats and *Vice Versa*, *Notizie di Politeia*, **30**(113), pp. 126–146.

Dohertya, T.S., Glenc, A.S., Nimmod, D.G., Ritchiea, E.G., and Dickmane, C.R. (2016). Invasive Predators and Global Biodiversity Loss, *Proceedings of the National Academy of Sciences*, http://www.pnas.org/content/early/2016/09/13/1602480113. Accessed 28 September 2016.

Donati, F. and Milazzo, P. (2003). La dottrina del margine di apprezzamento nella giurisprudenza della Corte europea dei diritti dell'uomo, in: P. Falzea, A. Spadaro, and L. Ventura (eds.) *La Corte costituzionale e le Corti d'Europa. Atti del seminario svoltosi a Capannello (CZ) il 31 maggio-1 giugno 2002*, Torino: Giappichelli, pp. 65–117.

Dorbeck-Jung, B. and Shelley-Egan, C. (2013). Meta-regulation and Nanotechnologies: The Challenge of Responsibilisation within the European Commission's Code of Conduct for Responsible Nanosciences and Nanotechnologies Research, *Nanoethics*, **7**(1), pp. 55–68.

Dworkin, R. (1986). *The law empire*, Cambridge (Mass.): The Belkamp Press of Harvard University Press; It. trad. (1989). *L'impero del diritto*, Milano: Il Saggiatore.

Dworkin, R. (1996). *Freedom's law. The Moral Reading of the American Constitution*, Oxford, Harvard University Press.

European Academies Science Advisory Council (EASAC) (2010). *Realising European Potential in Synthetic Biology: Scientific Opportunities and Good Governance*, Halle (Germany): European Academies Science Advisory Council, https://www.cbd.int/doc/emerging-issues/emergingissues-2013-10-EASAC-SyntheticBiology-en.pdf. Accessed 28 March 2015.

European Commission (2000). Commission Decision of 27 September 2000 *Concerning the Guidance Notes for Risk Assessment Outlined in Annex III of Directive 90/219/EEC on the Contained Use of Genetically Modified Micro-organisms* (notified under document number C(2000) 2736), *Official Journal of the European Union*, L258, 12.10.2000, pp. 43–48, http://eur-lex.europa.eu/legal-content/EN/TXT/PDF/?uri=CELEX: 32000D0608&from= EN. Accessed 13 April 2015.

European Commission (2004). Commission Communication of 12/05/2004 *Towards a European Strategy for Nanotechnology*, COM(2004) 338 final. Luxembourg: Commission of the European Communities, http:// ec.europa.eu/research/industrial_technologies/pdf/nanotechnology_ communication_en.pdf. Accessed 23 February 2015.

European Commission (2005). Commission Communication of 07/06/2005 *Nanosciences and Nanotechnologies. An Action Plan 2005–2009*, COM(2005) 423 final. Brussels: European Commission, http://ec. europa.eu/research/industrial_technologies/pdf/nano_action_plan_ en.pdf. Accessed 23 February 2015.

European Commission (2007a). Commission Communication of 19/07/2007 *'Towards a Code of Conduct for Responsible Nanosciences and Nanotechnologies Research and Nanotechnologies Research' Consultation Paper*. Brussels: European Commission, http://ec.europa.

eu/research/consultations/pdf/nano-consultation_en.pdf. Accessed 24 February 2015.

European Commission (2007b). *'Code of Conduct for Responsible Nanosciences and Nanotechnologies Research'—Detailed Analysis of Results from the Consultation.* Brussels: European Commission, http://ec.europa.eu/research/science-society/document_library/pdf_06/consultation-nano-sinapse-feedback_en.pdf. Accessed 24 February 2015.

European Commission (2008a). Commission Recommendation of 07/02/2008 *on a Code of Conduct for Responsible Nanosciences and Nanotechnologies Research*, C(2008) 424 final. Brussels: European Commission. http://ec.europa.eu/research/participants/data/ref/fp7/89918/nanocode-recommendation_en.pdf. Accessed 23 February 2015.

European Commission (2008b). Commission Communication of 17/06/2008 *on Regulatory Aspects of Nanomaterials*, COM(2008) 366 final, Brussels: European Commission, http://eur-lex.europa.eu/LexUriServ/LexUriServ.do?uri=COM:2008:0366:FIN:en:PDF. Accessed 1 March 2015.

European Commission (2010). *Recommendation on a Code of Conduct for Responsible Nanosciences and Nanotechnologies Research: 1st Revision. Analysis of Results from the Public Consultation*, http://ec.europa.eu/research/consultations/nano-code/results_en.pdf. Accessed 24 February 2015.

European Commission (2011). Commission Recommendation of the European Commission of the 18 October 2011 *on the definition of nanomaterial*, 2011/696/EU, 2011, *Official Journal of the European Union*, L275/38, http://eur-lex.europa.eu/legal-content/EN/TXT/PDF/?uri=CELEX:32011H0696&from= EN. Accessed 29 February 2015.

European Commission (2012a). Commission Staff Working Paper *on Types and Uses of Nanomaterials, Including Safety Aspects*, SWD(2012) 288 final, 3.10.2012, Brussels: European Commission, http://eur-lex.europa.eu/LexUriServ/LexUriServ.do?uri=SWD:2012:0288:FIN:EN:PDF. Accessed 1 March 2015.

European Commission (2012b). Commission Communication of 3/10/2012 *Second Regulatory Review on Nanomaterial*, COM (2012) 572 final. Brussels: European Commission, http://eur-lex.europa.eu/legal-content/EN/TXT/PDF/?uri=CELEX:52012DC0572&from=EN. Accessed 20 February 2015.

European Commission (2013). *Nanotechnology: The Invisible Giant Taking Europe's Future Challenges*, Luxembourg: Publications Office of the European Union, http://bookshop.europa.eu/en/nanotechnology-pbKI3013325/downloads/KI-30-13-325-EN-C/KI3013325ENC_002 .pdf;pgid=y8dIS7GUWMdSROEAlMEUUsWb0000Gho_fC38;sid=Fn1fd 9slL9Zfeor1JQnB0LkAtL4jJ5ZGCA8=?FileName=KI3013325ENC_002. pdf&SKU=KI3013325ENC_PDF&CatalogueNumber=KI-30-13-325-EN-C. Accessed 7 March 2015.

The European Consumer Organisation (BEUC) (2014). *EFSA's Policy on Independence and Scientific Decision-Making Processes*, BEUC Statement 25th EFSA Stakeholder Platform meeting, Brussels 25 June 2014, http://www.beuc.org/publications/beuc-x-2014-044_ipa_efsa_policy_ on_independence_and_scientific_decision-making_processes.pdf. Accessed 24 June 2015.

EFSA Scientific Committee (2011). Scientific Opinion on Guidance on the Risk Assessment of the Application of Nanoscience and Nanotechnologies in the Food and Feed Chain, *EFSA Journal 2011*, **9**(5), 2140 (36 pp.).

ERASynBio (2014). *Next Steps for European Synthetic Biology: A Strategic Vision for ERASynBio*, https://www.erasynbio.eu/lw_resource/ datapool/_items/item_58/erasynbiostrategicvision.pdf. Accessed 20 March 2015.

European Food Safety Authority (EFSA) (2010). Guidance on the Environmental Risk Assessment of Genetically Modified Plants, *EFSA Journal 2011*, **8**(11), 1879, pp. 1–111. http://www.efsa.europa.eu/it/scdocs /doc/1879.pdf. Accessed 29 March 2015.

European Food Safety Authority (EFSA) (2011). Guidance for Risk Assessment of Food and Feed from Genetically Modified Plants, *EFSA Journal 2011*, 9(5): 2150, pp. 1–37. http://www.efsa.europa.eu/en/search/ doc/2150.pdf. Accessed 29 March 2015.

European Food Safety Authority (EFSA) Panel on Genetically Modified Organisms (GMOs) (2012). Scientific Opinion Addressing the Safety Assessment of Plants Developed through Cisgenesis and Intragenesis, *EFSA Journal*, **10**(2), p. 2561 (1–33).

European Food Safety Authority (EFSA) (2015). Annual report of the EFSA Scientific Network of Risk Assessment of Nanotechnologies in Food and Feed for 2014, *EFSA supporting publication 2014*: EN-762, pp. 1–11.

European Group of Advisers on Ethical Implications of Biotechnology to the European Commission (GAEIB) (1993). *Opinion on Ethical Aspects Arising from the Commission Proposal for a Council Directive on Legal*

Protection for Biotechnological Inventions – Opinion N. 3 – 30 September 1993.

European Group of Advisers on Ethical Implications of Biotechnology to the European Commission (GAEIB) (1995). *Ethical Aspects of the Labelling of Foods Derived from Modern Biotechnology – Opinion N. 5 – 5 May 1995.*

European Group of Advisers on Ethical Implications of Biotechnology to the European Commission (GAEIB) (1996). *Ethical Aspects of Genetic Modification of Animals– Opinion N. 7 – 21 may 1996* http://ec.europa. eu/environment/chemicals/lab_animals/pdf/genetic_modification .pdf. Accessed 31 October 2016.

European Group on Ethics in Science and New Technologies (EGE) (2007). *Opinion on the Ethical Aspects of Nanomedicine – Opinion N. 21 – 17 January 2007*, Luxemburg: Publications Office of the European Union, http://ec.europa.eu/archives/bepa/european-group-ethics/docs/ publications/opinion_21_nano_en.pdf. Accessed 22 December 2016.

European Group on Ethics in Science and New Technologies (EGE) (2008). *Ethics of Modern Developments in Agricultural Technologies – Opinion N. 24 – 17 December 2008*, Luxemburg: Publications Office of the European Communities 2009.

European Group on Ethics in Science and New Technologies (EGE) (2009). *Ethics of Synthetic Biology – Opinion N. 25 – 17 November 2009*, Luxemburg: Publications Office of the European Union, http://ec.europa. eu/archives/bepa/european-group-ethics/docs/opinion25_en.pdf. Accessed 22 December 2016.

European Parliament (2006). Parliament Resolution of 28/09/2006 *Nanosciences and Nanotechnologies. An Action Plan 2005–2009* (2006/2004 (INI)). Strasbourg: European Parliament, http://www. europarl.europa.eu/sides/getDoc.do?pubRef=-//EP//TEXT%2BTA% 2BP6-TA-2006-0392%2B0%2BDOC%2BXML%2BV0//EN. Accessed 23 February 2015.

European Parliament (2009). Parliament Resolution of 24/04/2009 *on Regulatory Aspects of Nanomaterials* (2008/2208(INI)), Brussels: European Parliament, http://www.europarl.europa.eu/sides/getDoc. do?pubRef= -//EP// NONSGML+ TA+ P6- TA- 2009 -0328+ 0+ DOC+ PDF+ V0//EN. Accessed 29 February 2015.

European Parliament (2009). Regulation (EC) No. 1223/2009 of the European Parliament and of the Council of 30 November 2009 *on cosmetics products, Official Journal of the European Union*, L349/59.

European Parliament (2012). Resolution of the European Parliament of 19 January *2012 on the Council's position at first reading with a view to the adoption of a regulation of the European Parliament and of the Council, concerning the making available on the market and use of biocidal products* (05032/2/2011 – C7-0251/2011-2009/0076 (COD)), P7_TA-PROV(2012)0010.

European Union (2006). Decision (EC) No. 1982/2006 of the European Parliament and of the Council of 18/12/2006 *Concerning the Seventh Framework Programme of the European Community for Research, Technological Development and Demonstration Activities (2007-2013)*, *Official Journal of the European Union*, L412/1, 30.12.2006. http://cordis.europa.eu/documents/documentlibrary/90798681EN6.pdf. Accessed 11 March 2015.

European Union (2008). Regulation (EC) No. 1272/2008 of the European Parliament and of the Council of 16 December 2008 *on Classification, Labelling and Packaging (CLP) of Hazardous Substances and Mixtures,* amending and repealing Directives 67/548/EEC and 1999/45/EC, and amending Regulation (EC) No. 1907/2006 *Official Journal of the European Union*, L353/1, 31.12.2008.

European Union (2011). Regulation (EC) No. 1169/2011 of the European Parliament and of the Council of 25 October 2011 *on the Provision of Food Information to Consumers*, amending Regulations (EC) No. 1924/2006 and (EC) No. 1925/2006 of the European Parliament and of the Council, and repealing Commission Directive 87/250/EEC, Council Directive 90/496/EEC, Commission Directive 1999/10/EC, Directive 2000/13/EC of the European Parliament and of the Council, Commission Directives 2002/67/EC and 2008/5/EC and Commission Regulation (EC) No. 608/2004, *Official Journal of the European Union*, L304/18.

European Union (2011). Directive 2011/65/EU of the European Parliament and of the Council of 08/06/2011 *on the Restriction of the Use of Certain Hazardous Substances in Electrical and Electronic Equipment*, *Official Journal of the European Union*, L174/88, http://eur-lex.europa.eu/legal-content/EN/TXT/PDF/?uri=CELEX:32011L0065& from=EN. Accessed 28 February 2015.

European Union (2012a). Regulation (EU) No. 528/2012 of the European Parliament and the Council of 22/05/2012 *Concerning the Making Available on the Market and Use of Biocidal Products*, *Official Journal of the European Union*, L167/1, 27/06/2012, pp. 1–123, http://eur-lex.europa.eu/legal-content/EN/TXT/PDF/?uri=CELEX:32012R0528 &from= EN. Accessed 20 February 2015.

European Union (2012b). Directive 2012/19/EU of the European Parliament and the Council of 04/07/2012 *on Waste Electrical and Electronic Equipment (WEEE), Official Journal of the European Union*, L197/38, 24/07/2012, http://eur-lex.europa.eu/legal-content/EN/TXT/PDF/?uri=CELEX:32012L0019&from= EN. Accessed 14 March 2015.

European Union (2013). Regulation (EU) No. 1291/2013 of the European Parliament and the Council of 11/12/2013 *Establishing Horizon 2020 – the Framework Programme for Research and Innovation (2014-2020)*, *Official Journal of European Union*, L347/104, 20.12.2013, http://ec.europa.eu/research/participants/data/ref/h2020/legal_basis/fp/h2020-eu-establact_en.pdf. Accessed 11 March 2015.

European Union (2014). Regulation (EU) No. 334/2014 Amending Regulation (EU) No. 528/2012 of the European Parliament and the Council of 22/05/2012 *Concerning the Making Available on the Market and Use of Biocidal Products, Official Journal of the European Union*, L103, 05/04/2014, pp. 22–32.

Falkner, R. (2009). The Troubled Birth of the 'Biotech Century': Global Corporate Power and Its Limits, in: Clapp, J. and Fuchs, D. (eds.) *Corporate Power in Global Agrifood Governance*, Cambridge, MA: MIT Press, pp. 225–252.

Felt, U., and Wynnne, B. (2007). *Taking European Society Seriously. Report of the Expert Group on Science and Governance to the Science, Economy and Society Directorate, Directorate-General for Research, European Commission*, Luxembourg: Office for Official Publications of the European Communities.

Ferrarese, M.R. (2010). *La governance tra politica e diritto*, Bologna: Il Mulino.

Fontana, G. (2010). Genetically Modified Micro-organisms: The EU Regulatory Framework and the New Directive 2009/41/EC on the Contained Use, *Chemical Engineering Transactions*, **20**, pp. 1–6.

Francioni, F. (2007). Genetic Resources, Biotechnology and Human Rights, in: Francioni, F. (ed.) *Biotechnologies and International Human Rights*, Oxford and Portland, Oregon: Hart Publishing, pp. 3–32.

Greer, S. (2000). *The Margin of Appreciation: Interpretation and Discretion under the European Convention on Human Rights*, Strasbourg: Council of Europe Publishing.

Gibson, D.G., Glass, J.I., Lartigue C., Noskov, V.N., Chuang, R.Y., Mikkel, A.A., Benders, G.A., Montague, M.G., Ma, L., Moodie, M.M., Merryman, C.,

Vashee, S., Krishnakumarm, R., Assad-Garcia, N., Andrews-PfannKoch, C., Denisova, E.A., Young, L., Qi, Z.Q., Segall-Shapiro, Th.H., Calvey, Ch.H., Parmar Prashanth, P., Hutchinson III, C.A., Smith Hamilton, O., and Venter, C. (2010). Creation of Bacterial Cell Controlled by Chemically Synthesized Genome, *Science Express*, 20 May 2010, pp. 1–12.

Grande, E. (2001). The Erosion of State Capacity and the European Innovation Policy Dilemma. A Comparison of German and EU Information Technology Policies, *Research Policy*, **30**, pp. 905–921.

Grinbaum, A., and Groves, Ch. (2013). What Is 'Responsible' about Responsible Innovation? Understanding the Ethical Issues, in: Owen, R., Heintz, M. and Bessant, J. *Responsible Innovation: Managing the Responsible Emergence of Science and Innovation in Society*, London: John Wiley, pp. 119–142.

Grobe, A., Kreimberger, N., and Funda, O. (2011). *NanoCode WP2 Synthesis Report on Stakeholder Consultations*, March 2011.

Groves, C. (2013). Horizons of Care: From Future Imaginaries to Responsible Research and Innovation, in: K. Konrad, Ch. Coenen, A.B. Dijkstra, C. Milburn, and H. van Lente Shaping (eds.) *Emerging Technologies. Governance, Innovation, Discourse*, Berlin: IOS Press/AKA-Verlag, pp. 185–202.

Groves, Ch. (2015). Logic of Choice and Logic of Care? Uncertainty, Technological Mediation and Responsible Innovation, *Nanoethics*, **9**(3): 321–333.

Guerra, G., Muratorio, A., and Ruggiu, D. (2014). Introduction, *Notizie di Politeia*, **30**(113), pp. 3–7.

Habermas, J. (1992). Faktizität und Geltung. Beitrage zur Diskurstheorie des Rechts und des demokratischen Rechtsstaat, Suhrkamp Verlag, Frankfurt am Main; It. trad. (1996). Fatti e norme. Contributi a una teoria discorsiva del diritto e della democrazia, Milano: Guerini e Associati.

Hafner, A., Lovric, J., Lakos, G.P., and Pepic, I. (2014). Nanotherapeutics in the EU: An Overview of the Current State and Future Directions, *International Journal of Nanomedicine*, **9**, pp. 1005–1023.

Heydebrande, W. (2003). Process Rationality as Legal Governance: A Comparative Perspective, *International Sociology*, **18**(2), pp. 325–349.

Hankin, S.M., Peters, S.A.K., Poland, C.A., Foss Hansen, S., Holmqvist, J., Ross, B.L., Varet, J., and Aitken, R.J. (2011a). *Specific Advice on Fulfilling Information Requirements and Exposure Assessment for Nanomaterials under REACH (RIPoN 2) – Final Project Report, RCN/ RIP-oN 2/FPR/1/FINAL, 1 July 2011,* European Commission, http://ec.europa.eu/environment/chemicals/nanotech/pdf/report_ ripon2.pdf. Accessed 29 September 2017.

Holbrook, J.B., and Briggle, A. (2014). Knowledge Kills Action: Why Principles Should Play a Limited Role in Policy-Making, *Journal of Research Innovation,* **1**(1), pp. 51–66.

International Association of Synthetic Biology (IASB) (2009). *The IASB Code of Conduct for Best Practices in Gene Synthesis,* Cambridge (MA.), 3 November 2009, http://www.ia-sb.eu/tasks/sites/synthetic-biology/assets/File/pdf/iasb_code_of_conduct_final.pdf. Accessed 18 March 2015.

J. Craig Venter Institute (JCVI) (2014). *Synthetic Biology and the U.S. Biotechnology Regulatory System: Challenged and Options,* http:// www.jcvi.org/cms/fileadmin/site/research/projects/synthetic-biology-and-the-us-regulatory-system/full-report.pdf. Accessed 28 March 2015.

Journeay, W.S., and Goldman, R.H. (2014). Occupational Handling of Nickel Nanoparticles: A Case Report, *American Journal of Industrial Medicine,* **57**(9), pp. 1073–1076.

Kagan, V.E., Konduru, N.V., Feng, W., Allen, B.L., Conroy, J., Volkov, Y., Vlasova, I.I., Belikova, N.A., Yanamala, N., Kapralov, A., Tyurina, Y.Y., Shi, J., Kisin, E.R., Murray, A.R., Franks, J., Stolz, D., Gou, P., Klein-Seetharaman, J., Fadeel, B., Star, A., and Shvedova, A.A. (2010). Carbon Nanotubes Degraded by Neutrophil Myeloperoxidase Induce Less Pulmonary Inflammation, *Nature Nanotechnology,* **5**, pp. 354–359.

Kearnes, M.B., and Rip, A. (2009). The Emerging Governance Landscape of Nanotechnology, in: Gammel, S., Losch, S., and Nordmann, A. (eds.) *Jenseits Von Regulierung: Zum Politischen Umgang Mit Der Nanotechnologie,* Berlin: Akademische Verlagsgesellschaft, pp. 97–121.

Kjølberg, K.L., and Strand, R. (2011). Conversations About Responsible Nanoresearch, *Nanoethics,* **5**(1), pp. 99–113.

Kooiman, J., and Van Vliet, M. (1993). Governance and Public Management, in: Eliassen, K., and Kooiman, J. (eds.), *Manging Public Organizations* (2nd ed.), London: Sage.

Kramer, M.R. (2011). Creating Shared Value: How to Reinvent Capitalism and Unleash a Wave of Innovation and Growth, *Harvard Business Review,* January–February, pp. 1–17.

Kratochvìl, J. (2011). The Inflation of the Margin of Appreciation Doctrine of the European Court of Human Rights, *Netherlands Quarterly of Human Rights*, **29**(3): 324–357.

Krupp, F., and Holliday, C. (2005). Let's Get Nanotech Right, *Wall Street Journal*, Tuesday, June 14, 2005, Management Supplement, p. B2.

Kuhlbus, Th. A.J., Asbach, Ch., Fissan, H., Göhler D., and Stintz, M. (2011). Nanoparticle Exposure at Nanotechnology Workplaces: A Review, *Particle and Fiber Toxicology*, **8**(22), pp. 1–18. http://www.particleandfibretoxicology.com/content/pdf/1743-8977-8-22.pdf. Accessed 12 March 2015.

Kurath, M., Nentwich, M., Fleischer, T., and Eisenberger, I. (2014). Cultures and Strategies in the Regulation of Nanotechnology in Germany, Austria, Switzerland and the European Union, *Nanoethics*, **8**(2), pp. 121–140.

Leone, L. (2014). *Hard e soft law*? L'Europa e le scelte normative sulle tecnologie emergenti, *Notizie di Politeia*, **30**(113), pp. 147–168.

Lusser, M., Parisi, C., Plan, D., and Rodrìguez-Cerezo, E. (2011). *New Plant Breeding Techniques. State-of-the-Art and Prospects for Technological Development*. JRC Technical Report, Luxemburg: Publications Office of the European Union, http://ipts.jrc.ec.europa.eu/publications/pub.cfm?id= 4100. Accessed 29 September 2017.

Lyall, C., and Tait, J. (2005). Shifting Policy Debates and the Implications for Governance, in Lyall, C., and Tait, J. (eds.) *New Modes of Governance. Developing an Integrated Policy Approach to Science, Technology, Risk and the Environment*, Adelshot: Ashgate, pp. 1–17.

Mandel, G.N. (2009). Regulating Emerging Technologies. Legal Studies Research Paper Series, Research Paper N. 2009–18, 4-8-2009, *Law, Innovation & Technology*, **1**, p. 75, https://papers.ssrn.com/sol3/papers.cfm?abstract_id= 1355674. Accessed 20 February 2015.

Mandel, G.N. (2013). Emerging Technology Governance, in: Marchant, G.E., Abbott, K.W., and Allenby, B. (eds.) *Innovative Governance Models for Emerging Technologies*, Cheltenham (UK), Northampton, MA. (USA): Edward Elgar, pp. 44–62.

Mantovani, A., Porcari, A., and Azzolini, A. (2010). *NanoCode WP1 Synthesis Report on code of Conduct, Voluntary Measures and Practices Towards a Responsible Development of N&N*, September 2010.

Marchant, G.E., Sylvester, D.J., and Abbott, K.W. (2008). Risk Management Principles for Nanotechnology, *Nanoethics*, **2**(1), pp. 43–60.

Marrani, D. (2013). Nanotechnologies and Food in European Law, *Nanoethics*, **7**(3), pp. 177–188.

MASIS (Monitoring Activities of Science in Society in Europe Experts Group) (2009). Challenging Futures of Science in *Society, Emerging Trends and Cutting-Edge Issues*, Luxembourg: European Commission, Directorate-General for Research.

Meili, C., Widmer, M., Schwarzkopf, S., Mantovani, E., and Porcari, A. (2011). *NanoCode MasterPlan: Issues and Options on the Path Forward with the European Commission Code of Conduct on Responsible N&N Research*, November 2011.

Mehta, M.D. (2004). From Biotechnology to Nanotechnology: What Can We Learn from Earlier Technologies? *Bulletin of Science and Technology & Society*, **24**(1), pp. 34–39.

Miller, C.A., and Bennet, I. (2008). Thinking Longer Term about Technology: Is There Value in Science Fiction-Inspired Approaches to Constructing Futures? *Science and Public Policy*, **35**, pp. 597–606.

Money, H.A. (ed.) (2005). *Invasive Alien Species: A New Synthesis*, Washington: Island Press.

Murphy, T. (ed.) (2009). *New technologies and Human Rights*, Oxford: Oxford University Press.

Mwilu, S.K., El Badawy, A.M., Bradham, K., Nelson, C., Thomas, D., Scheckel, K.G., Tolaymat, T., Ma, L., and Rogers, K.R. (2013). Changes in Silver Nanoparticles Exposed to Human Synthetic Stomach Fluid: Effects of Particle Size and Surface Chemistry, *Science of the Total Environment*, **447**, pp. 90–98.

New Techniques Working Group (NTWG) (2012). *Final Report*, Luxemburg: Publications Office of the European Union, ftp://ftp.jrc.es/pub/EURdoc/JRC63971.pdf. Accessed 2 April 2015.

Owen, R. (2014). *Responsible Research and Innovation: Options for Research and Innovation Policy in the EU*, European Research and Innovation Area Board (ERIAB), Foreword Visions on the European Research Area (VERA), http://ec.europa.eu/research/innovation-union/pdf/expert-groups/Responsible_Research_and_Innovation.pdf. Accessed 7 March 2015.

Owen, R., Macnaghten, Phil., and Stilgoe, J. (2012). Responsible Research and Innovation: from Science in Society to Science for Society, with Society, *Science and Public Policy*, **39**, pp. 751–760.

Owen, R., Stilgoe, J., Macnaghten, Phil., Gorman, M., Fisher, E., and Guston, D. (2013). A Framework for Responsible Innovation, in: Owen, R., Bessant,

J. and Heintz, M. (eds.) *Responsible Innovation*, London: John Wiley & Sons Ltd, pp. 27–50.

Pariotti, E. (2013). *I diritti umani: concetto, teoria, evoluzione*, Padova: CEDAM.

Pariotti, E., and Ruggiu, D. (2012). Governing Nanotechnologies in Europe: Human Rights, Soft Law, and Corporate Social Responsibility, in: Van Lente, H., Coenen, C., Konrad, K., Krabbenborg, L., Milburn, C., Seifert, F., Thoreau, F., and Zülsdorf, T. (eds.), *Little by Little. Expansions of Nanoscience and Emerging Technologies*, Heidelberg: IOS press/AKA, pp. 157–168.

Parker, C. (2007). Meta-Regulation: Legal Accountability for Corporate Social Responsibility, in: McBarnet, D., Voiculescu, A., and Campbell, T. (eds.), *The New Corporate Accountability: Corporate Social Responsibility and the Law*, Cambridge: Cambridge University Press, pp. 207–241.

Pauwels, K., Willemarck, N., Breyer, D., and Hermann, Ph. (2012). *Synthetic Biology: Latest Developments, Biosafety Considerations and Regulatory Challenges*, Brussels: Biosafety and Biotechnology Unit, http://www.biosafety.be/PDF/120911_Doc_Synbio_SBB_FINAL.pdf. Accessed 30 March 2015.

Plan, D., and Van den Eede, G. (2010). *The EU Legislation on GMOs. An Overview*, Luxembourg: Publications Office of the European Union.

Poland, C.A., Duffin, R., Kinloch, I., Mynard, A., Wallace, W.A.C., Seaton, A., Stone, V., Brown, S., Mac Nee, W., and Donaldson, K. (2008). Carbon Nanotubes Introduced into the Abdominal Cavity of Mice Show Asbestos-Like Pathogenicity in a Pilot Study, *Nature Nanotechnology*, **17**, pp. 423–428.

Poli, R. (2014). Anticipation: What About Turning the Human and Social Sciences Upside Down? *Futures*, **64**, pp. 15–18.

Poli, R. (2015). Social Foresight, *On the Horizon*, **23**(2), pp. 85–99.

Presidential Commission for the Study of Bioethical Issues (PCSBI) (2010). *New Directions: The Ethics of Synthetic Biology and Emerging Technologies*, Washington D.C.: Presidential Commission for the Study of Bioethical Issues, http://bioethics.gov/sites/default/files/PCSBI-Synthetic-Biology-Report-12.16.10_0.pdf. Accessed 20 March 2015.

Renn, O. (2016). Inclusive Resilience: A New Approach to Risk Governance, in: IRGC (ed.) *Resource Guide on Resilience*, Lausanne: EPFL International Risk Governance Center. v29-07-2016.

Rip, A. (2002). *Coevolution of Science, Technology and Society. An Expert Review for the Bundesministrium Bildung und Forshung's Förderinitiative Politik, Wissenschaft und Gesellschaft* (Science Policy Studies), http://www.google.it/url?sa= t&rct= j&q= &esrc= s&frm= 1&source= web&cd= 2&ved= 0CEEQFjAB&url= http%3A%2F%2 Fciteseerx.ist.psu.edu%2Fviewdoc%2Fdownload%3Fdoi%3D10.1.1. 201.6112%26rep%3Drep1%26type%3Dpdf&ei=g7vKUfj0NoWpOr G5gbgP&usg=AFQjCNH9NKjk8E5pIoFqJsMVSPofWw-TLw&sig2= GjdoftjztHB-vqW160HMEA&bvm= bv.48340889,d.ZWU. Accessed 20 February 2015.

Robles Morchón, G. (2001). La protezione dei diritti fondamentali nell'Unione Europea, *Ars interpretandi. Annuario di ermeneutica giuridica*, **6**, pp. 249–269.

Roco, M. (2006). Progress in Governance of Converging Technologies Integrated from the Nanoscale, *Annals of the New York Academy of Science*, **1093**, pp. 1–2.

Rhodes, R. (1996). The New Governance: Governing without Government, *Political Studies*, **44**, pp. 652–667.

Rosa, E.A., Renn, O., and McCright, A.M. (2013). *The Risk Society Revisited. Social Theory and Governance*, Philadelphia: Temple University Press.

Ruggiu, D. (2012a). Synthetic Biology and Human Rights in the European Context: Health and Environment in Comparison of the EU and the Council of Europe Regulatory Frameworks, *Biotechnology Law Report*, **31**(4), pp. 337–355.

Ruggiu, D. (2012b). *Diritti e temporalità. I diritti umani nell'era delle tecnologie emergenti*, Bologna: Il Mulino.

Ruggiu, D. (2013a). A Rights-Based Model of Governance: The Case of Human Enhancement and the Role of Ethics in Europe, in: Konrad, K., Coenen, C., Dijkstra, A., Milburn, C., and van Lente, H. (eds.), *Shaping Emerging Technologies: Governance, Innovation, Discourse*, Berlin: IOS Press/AKA-Verlag, pp. 103–115.

Ruggiu, D. (2013b). Temporal Perspectives of the Nanotechnological Challenge to Regulation. How Human Rights Can Contribute to the Present and Future of Nanotechnologies, *Nanoethics*, **7**(3), pp. 201–215.

Ruggiu, D. (2013c). Dominating Non-Knowledge. Rights, Governance and Uncertain Times, *Cosmopolis: Robotics and Public Issues*, **9**(2), http://www.cosmopolisonline.it/articolo.php?numero=IX22013&id =19. Accessed 23 October 2017.

Ruggiu, D. (2014a). I diritti umani alla sfida della biologia sintetica tra Unione europea e Consiglio d'Europa: il caso delle applicazioni sull'essere umano, *Notizie di Politeia*, **30**(113), pp. 9–28.

Ruggiu, D. (2014b). Responsibilisation Phenomena: The EC Code of Conduct for Responsible Nanosciences and Nanotechnologies Research, *European Journal of Law and Technology*, 5(3), pp. 1–16, http://ejlt.org/article/view/338. Accessed 29 September 2017.

Ruggiu, D. (2015a). Anchoring European Governance: Two versions of Responsible Research and Innovation and EU Fundamental Rights as 'Normative Anchor Points'. *Nanoethics*, **9**(3), pp. 217–235.

Ruggiu, D. (2015b). The Consolidation Process of the EU Regulatory Framework on Nanotechnologies: Within and Beyond the EU Case-By-Case Approach, *European Journal of Law and Technology*, **6**(3), pp. 1–35.

Ruggiu, D. (2016). A Reply to Groves, *Nanoethics*, **10**(1), pp. 111–116.

Ruggiu, D. (2018). Implementing a Responsible, Research and Innovation Framework in Human Enhancement According to Human Rights: The Right to Bodily Integrity and the Rise of 'Enhanced Societies', *Journal of Law, Information & Technology*, **10**(1), pp. 1–40.

Salter, B., and Jones, M. (2002). Human Genetic Technologies, European Governance and the Politics of Bioethics, *Nature*, **3**, pp. 808–814.

Schiemann, J., and Hartung, F. (2014). EU Perspectives on New Plant-Breeding Techniques, in: A. Eagelsham, and R.W.F. Hardy (eds.), *NABC Report 26: New DNA-Editing Approaches: Methods, Applications and Policy for Agriculture*, Ithaca, New York: North American Agricultural Biotechnology Council, pp. 201–210, http://nabc.cals.cornell.edu/Publications/Reports/nabc_26/NABC%20Report%2026.pdf. Accessed 2 July 2015.

Scientific Committee on Consumer Safety (SCCS) (2012). *Guidance on the Safety Assessment on Nanomaterials in Cosmetics*, Brussels: European Commission.

Scientific Committee on Emerging and Newly Identified Health Risks (SCENIHR) (2009). *Risk Assessment of Products of Nanotechnologies – Opinion of 19 January 2009 –* Brussels: European Commission.

Scientific Committee on Emerging and Newly Identified Health Risks (SCENIHR), Scientific Committee on Health and Environmental Risks (SCHER), Scientific Committee on Consumer Safety (SCCS) (2014a). *Opinion on Synthetic Biology I: Definition – Opinion of 25 September 2014*, Luxembourg: European Commission.

Scientific Committee on Emerging and Newly Identified Health Risks (SCENIHR), Scientific Committee on Consumer Safety (SCCS), Scientific Committee on Health and Environmental Risks (SCHER) (2014b). *Results of the Public Consultation on SCENIHR's Preliminary Opinion on the Synthetic Biology I – Definition*, http://ec.europa.eu/health/scientific_committees/emerging/docs/followup_cons_synbio_en.pdf. Accessed 23 October 2017.

Scientific Committee on Emerging and Newly Identified Health Risks (SCENIHR), Scientific Committee on Health and Environmental Risks (SCHER), Scientific Committee on Consumer Safety (SCCS) (2015a). *Opinion on Synthetic Biology II: Risk Assessment Methodologies and Safety Aspects – Opinion of 4 May 2015*, Luxembourg: European Commission.

Scientific Committee on Emerging and Newly Identified Health Risks (SCENIHR), Scientific Committee on Health and Environmental Risks (SCHER), Scientific Committee on Consumer Safety (SCCS) (2015b). *Opinion on Synthetic Biology III: Risks to the Environment and Biodiversity Related to Synthetic Biology and Research Priorities in the Field of Synthetic Biology - Opinion of 27 November, 3 and 4 December 2015*, Luxembourg: European Commission.

Scientific Committee on Emerging and Newly Identified Health Risks (SCENIHR), Scientific Committee on Health and Environmental Risks (SCHER), Scientific Committee on Consumer Safety (SCCS) (2015c). *Results of the public consultation on SCENIHR's preliminary Opinion on the Synthetic Biology – Research Priorities*, http://ec.europa.eu/health/scientific_committees/consultations/public_consultations/scenihr_consultation_28_en.htm. Accessed 23 February 2016.

Scott, J., and Trubek, D.M. (2002). Mind the Gap: Law and New Approaches to Governance in the European Union, *European Law Journal*, **8**(1), pp. 1–18.

Secretariat of the Convention on Biological Diversity (2015). *Synthetic biology*, Montreal, Technical Series No. 82, 118 pages, https://www.cbd.int/doc/publications/cbd-ts-82-en.pdf. Accessed 7 December 2016.

Séralini, Gilles-Eric, Clair, E., Mesnage, R., Gress, S., Defarge, N., Malatesta, M., Hennequin, D., and Spiroux de Vendômois, J. (2014). Republished Study: Long Term Toxicity of a Roundup Herbicide and a Roundup-tolerant Genetically Modified Maize, *Environmental Sciences Europe*, **26**(14), pp. 1–17, http://www.enveurope.com/content/pdf/s12302-014-0014-5.pdf. Accessed 9 June 2015.

Shvedova, A.A., Kisin, E.R., Mercer, R., Murray, A.R., Johnson, V.J., Potapovich, A.I., Tyurina, Y.Y., Gorelik, O., Arepalli, S., Schwegler-Berry, D., Hubbs, A.F., Antonini, J., Evans, D.E., Ku, B.K., Ramsey, D., Maynard, A., Kagan, V.E., Castranova, V., and Baron, P. (2005). Unusual Inflammatory and Fibrogenic Pulmonary Responses to Single-Walled Carbon Nanotubes in Mice, *American Journal of Physiology Lung Cellular Molecular Physiology*, **289**(5), pp. L698–L708.

Simakova, E., and Coenen, Ch. (2013). Visions, Hype, and Expectations: A Place for Responsibility, in R. Owen, J. Bessant, and M. Heintz (eds.) *Responsible innovation*, London: John Wiley & Sons, pp. 241–266.

Smismans, S. (2008). New Modes of Governance and the Participatory Myth, *West European Politics*, **31**(5), pp. 874–895.

Song, Y., Li, X., and Du, X. (2009). Exposure to Nanoparticles is Related to Pleural Effusion, Pulmonary Fibrosis and Granuloma, *European Respiratory Journal*, **34**(3), pp. 1–31.

Sørensen, E., and Triantafillou, P. (2009). Introduction, in: Sørensen, E., and Triantafillou, P. (eds.), *The Politics of Self-Governance*, Farnham: Ashgate, pp. 1–22.

Stilgoe, J., Owen, R., and Macnaghten, P. (2013). Developing a Framework for Responsible Innovation, *Research Policy*, **42**(9), pp. 1568–1580, http://www.sciencedirect.com/science/article/pii/S0048733313000930. Accessed 15 July 2015.

Stoke, E., and Bowman, D.M. (2012). Looking back to the Future of Regulating New Technologies: The Case of Nanotechnology and Synthetic Biology, *European Journal of Risk Regulation*, **2**, pp. 235–241.

Stoke, G. (1998). Governance as Theory: Five Propositions, *International Social Science Journal*, **155**, pp. 17–28.

Sutcliffe, H. (2011). *A Report on Responsible Research Innovation for European Commission*, Matter, http://ec.europa.eu/research/science-society/document_library/pdf_06/rri-report-hilary-sutcliffe_en.pdf. Accessed 15 July 2015.

Synthetic Biology 2.0 (2006). *Declaration of the Second International Meeting on Synthetic Biology*, Berkeley: California (USA), http://dspace.mit.edu/handle/1721.1/32982. Accessed 17 March 2015.

Tallacchini, M. (2009). Governing by Values. EU Ethics: Soft Tool, Hard Effects, *Minerva*, **47**, pp. 281–306.

Throne-Host, H., and Rip, A. (2011). Complexity of Labelling of Nanoproducts on the Consumers Market, *European Journal of Law and Technology*, **2**(3).

Throne-Host, H., and Strandbakken, P. (2009). 'Nobody Told Me I Was a Nano-consumer': How Nanotechnologies Might Challenge the Notion of Consumer Rights, *Journal of Consumer Policy*, **32**(4), pp. 393–402.

Tiberghien, Y. (2006). *The Battle for the Global Governance of Genetically Modified Organisms: The Roles of the European Union, Japan, Korea, and China in a Comparative Context*, Centre d'études et de recherches internationals, Les Etudes du CERI, 124, avril 2006, pp. 1–49, http://www.sciencespo.fr/ceri/sites/sciencespo.fr.ceri/files/etude124.pdf. Accessed 3 June 2015.

Trubek, D.M., Cottrell, M.P., and Nance, M. (2005). 'Soft Law', 'Hard Law', and European Integration: Toward a Theory of Hybridity, *University of Wisconsin Legal Studies Research Paper No. 1002*.

Turker, J.B., and Zilinskas, R.A. (2006). The Promise and Perlis of Synthetic Biology, *The New Atlantis: A Journal of Technology & Society*, 25, http://www.thenewatlantis.com/docLib/TNA12-TuckerZilinskas.pdf. Accessed 19 March 2015.

van Calster, G. (2006). Regulating Nanotechnology in the European Union, *European Environmental Law Review*, August/September, pp. 238–247.

Van den Belt, H. (2009). Playing God in Frankenstein's Footsteps: Synthetic Biology and the Meaning of Life, *Nanoethics*, **3**, pp. 257–268.

van den Hoven, J., Jacob, K., Nielsen, L., Roure, F., Laima, R., and Stilgoe, J. (eds.) (2013). *Options for Strengthening Responsible Research and Innovation. Report of Experts Group on the State of the Art in Europe on Responsible Research Innovation*, Luxemburg: European Commission, http://ec.europa.eu/research/science-society/document_library/pdf_06/options-for-strengthening_en.pdf. Accessed 15 July 2015.

Venter, G. (2013). *Life at the Speed of Light: from the Double Helix to the Dawn of the Digital Life*, New York: Viking; it. trans. (2014). *Il disegno della vita. Dalla mappa del genoma alla biologia digitale: il mio viaggio nel futuro*, Milano: Rizzoli.

von Schonberg, R. (2010). Organising Public Responsibility: On Precaution, Code of Conduct and Understanding Public Debate, in: Fiedeler, U., Coenen, C., Davies, S.R., and Ferrari, A. (eds.) *Understanding Nanotechnology: Philosophy, Policy and Publics*, Amsterdam: Ios Press, pp. 61–70.

von Schomberg, R. (2011a). Introduction, in: R. von Schomberg (ed.), *Towards Responsible Research and Innovation in the Information* and Communication Technologies and Security Technologies Fields, Luxembourg: Publications Office of the European Union, pp. 7–16

von Schonberg, R. (2011b). Prospects for Technology Assessment in a Framework of Responsible Research and Innovation, in: Dusseldorp, M., and Beecroft, R. (eds.), *Technikfolgen abschätzen lehren: Bildungspotenziale transdisziplinärer Methoden*, Wiesbaden: Vs Verlag, pp. 39–61.

von Schonberg, R. (2013). A Vision of Responsible Innovation., in: Owen, R., Heintz, M., and Bessant, J. (eds.), *Responsible Innovation: Managing the Responsible Emergence of Science and Innovation in Society*, London: John Wiley.

Vos, E. (2016). EU Agencies and Independence, in: Rettleng, D. (ed.), *Independence and Legitimacy in the Institutional System of the European Union Agencies and Independence*, Oxford: Oxford University Press, pp. 206–229.

Waldron, J. (1999). *Law and disagreement*, Oxford, Oxford University Press.

Weber, M. (1922). *Gesammelte Aufsätze zur Wissenschaftslehre*, Tübingen: Mohr; It. Trad. (1958). *Il metodo delle scienze storico-sociali*, Torino: Einaudi.

Widmer, M., Meili, C., Mantovani, E., and Porcari, A. (2010). *The FramingNano Governance Platform: A New Integrated Approach to the Responsible Development of Nanotechnologies*, February 2010.

Wynne, B., and Felt, U. (eds.) (2007). *Taking European Knowledge Society Seriously: Report of the Expert Group on Science and Governance to the Science, Economy and Society Directorate, Directorate-General for Research*, Brussels, European Commission, http://ec.europa.eu/research/scincesociety/document_library/pdf_06/european-knowledge-society_en.pdf. Accessed 4 June 2015.

Zhang, X., Sun, H., Zhang, Z., NiuQian, C.Y., and Crittenden, J.C. (2007). Enhanced Bioaccumulation of Cadmium in Carp in the Presence of Titanium Dioxide Nanoparticles, *Chemosphere*, **67**, pp. 160–166.

PART II

Introduction

In order to transform human rights in a governance tool, it is crucial to overcome two shortcomings that affect them: the indeterminacy and vagueness of the norms from which they stem. In this sense, if a subject (such as a State) has to answer the question 'how should I behave in order to respect human rights?' the norms on human rights cannot be helpful in the form as they are shaped.

The main obstacle to consider human rights a governance tool is their inescapable indeterminacy due to the fact that human rights stem from abstract norms whose content is vague and merely allusive. Since these norms are indeterminate, the main feature of the abstract norms from which human rights derive is that their content, 'is not established once for all in the law-making process, but must be shaped, also in a bottom-up manner and by several relevant actors during the application stage, like judges but also private actors promoting tools of self-regulation' (Arnaldi et al., 2016, p. 28). This means that because of this original indeterminacy, 'the judicial stance contributes to the definition of the content of rights' (Arnaldi et al., 2016, p. 29), it becomes crucial to undertake a preliminarily analysis of the framework of decisions of supranational courts in order to understand what the impact of technoscientific developments might be on human rights. In other words, we need to consider their past violations and hypothesize scenarios made up of possible judicial complaints, which involve them in relation to technoscientific advance. Developing this *regulatory foresight* is necessary in order understand what impact the technoscientific advance can have on the law in the future (Blind, 2008). However, in this context, it has to be made considered in the light of individual rights. It is a type of short-term vision, which is based on present research in its testing phase, and on current developments in the market (Ruggiu, 2015, p. 229).

In this regard, the current framework of decisions of the Strasbourg Court can be a fruitful field where we can find principles and rules that might be applied to analogous cases about technoscientific developments and individual rights. The study of the judgements of the ECtHR represents the first non-eliminable step for identifying rules and legally binding principles which States must adhere to, principles that can also be extended to the field of research and innovation. This further (legal) analysis of the human rights framework is needed 'to address unexpected side effects that regulatory interventions aimed at safeguarding certain rights', as well as 'to stay alert to the need for updating, expanding or changing the framework in light of changes in society and value systems that are brought about through the mutual shaping process of technologies, social processes, and normative outlooks' (Leenes et al., 2017, p. 33).

The 'rights-based framework as outlined may provide a fruitful backdrop for assessing and regulating [emerging] technologies' (Leenes et al., 2017, p. 33). There are few but enlightening instances of the use this approach to tackle challenges of the technoscientific progress. This is the method followed by, for example, the Committee on Bioethics of the Council of Europe (DH-BIO) in its reports where it systematizes the ensemble of decisions by the ECtHR which may be pertinent in certain technological fields such as, in particular, health,[1] bioethics in general covering reproductive rights, medically assisted procreation, consent to medical treatment, retention of personal data and bio-samples, end-life decisions.[2] There are also periodical reports issued by the secretariat of the Strasbourg Court on its jurisprudence, which cover several fields such health in general,[3] end life,[4] the environment,[5] persons with

---

[1] Committee on Bioethics of the Council of Europe (DH-BIO) (2008) *The Court's Case-Law Concerning the Health in General*, 18 July 2008.

[2] Committee on Bioethics of the Council of Europe (DH-BIO) (2009) *Bioethics and the Case-Law of the Court*, 14 October 2009.

[3] http://www.echr.coe.int/Documents/FS_Health_ENG.pdf. Accessed 29 September 2017.

[4] http://www.echr.coe.int/Documents/FS_Euthanasia_ENG.pdf. Accessed 29 September 2017.

[5] http://www.echr.coe.int/Documents/FS_Environment_ENG.pdf. Accessed 29 September 2017.

disabilities,[6] detention conditions of prisoners,[7] etc., that can result in extremely useful when on the basis of human rights we need to assess applications of nanotechnology, the rise of synthetic biology, freedom of choice and human enhancement.

This approach, which particularly focuses on the ECtHR jurisprudence, begins to be considered also at the EU level, for example with regard to the field of the data protection law. In this context it is worthwhile mentioning the work of the ARTICLE 29 Data Protection Working Party, which explicitly took into account the ECHR framework, and the decisions of ECtHR as well as the joint work of the European Union Agency for Fundamental Rights (FRA) and the Council of Europe together with the Registry of the Strasbourg Court on European data protection law, which specifically considered a wide number of ECtHR decisions as pertinent to this matter (Boillat and Kjaerum, 2014).

Even EGE has begun to adopt this approach. Recently, in one of its opinions, that on the ethical implications of new health technologies and citizens' participation, it has expressly followed this method of analysis directly using the framework of the Strasbourg Court decisions as reference point of its legal analysis, in order to determine regulatory limits and provisions in a given ambit. With regard to the use of mobile health (mHealth), digital health (ehealth) and other health technologies EGE highlighted the human rights borders related to the freedom to participate in discussions regarding matters of public health (EGE, 2015, pp. 45–47). The normative legitimation of processes of participation in science and technology was thus found in the ECHR framework and its associated jurisprudence. In this last regard, however, this work can be deemed only partially successful since it only focused on the case law related to freedom of scientific research without considering, though, that related to the right to health and the right to data protection, namely two crucial aspects in the field of new health technologies. In my view, the focus on only one right,

---

[6]http://www.echr.coe.int/Documents/FS_Disabled_ENG.pdf. Accessed 29 September 2017.
[7]http://www.echr.coe.int/Documents/FS_Detention_conditions_ENG.pdf. Accessed 29 September 2017.

neglecting the other aspects involved, could have altered the final outcome of the opinion by giving the readers a distorted image. Furthermore, EGE could have better connected the implications of this regulatory analysis on the EU legislation in this field taking prescriptive conclusions from the human rights law. However, this opinion can be deemed as a positive shift in the EGE's approach. In fact, also on the basis of these reports, it was adopted the current EU legislation on data protection, having a strong focus on all dimensions of the right to privacy: General Data Protection Regulation[8] and the Directive 2016/680 on the protection of natural persons with regard to the processing of personal data.[9]

The usefulness of this approach is that it permits to consider in advance a possible breach of individual rights which can be taken into account at the supranational level. This consideration is crucial when the policymaker takes strategic decisions in the field of technoscientific progress and when the regulator has to devise a governance framework for a new emerging technology. To have in advance the possible area of impact of a given technology on individual rights permits to reduce the risk of future litigations and of abrupt system failures of a particular research field, as well as market (Ruggiu, 2015, p. 230). This preliminary work on the ECtHR jurisprudence allow to pass from a conception of rights as mere negative externality of technoscientific development to a proactive conception of rights vis-à-vis the progress since we can derive from the case law which limits rights pose (e.g. the ban of human cloning, genetic enhancement, the creation of human–animal chimeras) but also what pathways they address (human enhancement of, in particular, persons with disability). In this regard there is the need of a brokering work able to link this preliminary analysis to the challenges stemming from a given technoscientific field in order to find limits and possible opportunities according to a rights-based approach. Rights, in fact, also have a positive role

---

[8]European Union (2016) Regulation EU 2016/679 *on the protection of natural persons with regard to the processing of personal data and on the free movement of such data, and repealing Directive 95/46/EC, Official Journal of the European Union,* 4.5.2016, L 119/1.

[9]http://eur-lex.europa.eu/legal-content/EN/TXT/PDF/?uri= CELEX: 32016L0680&from= EN1. Accessed 29 September 2017.

in governance, which we can call it 'a general orienting function', since they 'help to identify priorities and therefore justify rules that favour one application, responding to values and needs deemed fundamental, over others' 'by pointing to innovations [...] that should be fostered through regulation' (Leenes et al., 2017, p. 31). In this way, this 'constitutional framework could point towards developments [...] that would better fulfil fundamental values and ensure the implementation of rights, so as to impress socially beneficial connotations onto the scientific endeavour' (Leenes et al., 2017, p. 31).

In this Second Part of the book, therefore, I mean to analyse the jurisprudence of the Strasbourg Court which can be relevant for a specific technological sector. In order to systemizing the framework of decisions of the ECtHR it is crucial to launch an *act of mapping* of its jurisprudence articulated on the basis of rights at the centre of the ECHR, as well as on the basis of their possible violations by a given technology. Since its jurisprudence develops according to the principle of stare decisis,[10] it is useful to consider decisions of the Strasbourg Court that might be relevant for given sector, in order to identify possible binding principles in its case law (called *rationes decidendi*), that could be applied to future cases on emerging technologies. This would enable us not only to recognize potential constraints for technological and scientific advance in this field, and to provide a basis for ensuring protection of the rights of individuals, but also to explore possible avenues of development in the technoscientific field which are more in accordance with human rights. This would avoid, for example, future adverse court decisions on individual rights, as well as strengthening governance processes (Ruggiu, 2013, p. 2010; 2015, p. 230). In this way I think scholars can have a concrete example of the work which I consider necessary to efficaciously tackle problems put by the technoscientific progress and develop a governance framework that uses rights for concretely giving a response to the rise of a new emerging technology. In this

---

[10]*Stare decisis* or the *doctrine of binding precedent* is the legal principle (*ratio decidendi*) according to which courts are obliged to respect the precedent established by prior decisions. 'A judge deciding *a later, similar case* (one with similar fact to the precedent) *must* apply the same principles established by the courts.' See Riley (2008, p. 212).

regard, it is to say that, as rightly noted by some, ethical questions are context-laden (Owen, 2014, p. 7). They depend on the particular cultural environment wherein they rise and, most of all, on the specific technological field which is implied. This means that the analysis of a given set of court decisions can give a certain outcome with regard to nanotechnology and another with regard human enhancement. There is a space of variability within same sample subjected to the regulatory foresight articulated on the basis of rights. Therefore, although it focuses the same set of decisions, the study has to be repeated every time on the light of the new questions raised by the specific technological context concerned.

In this Part of book, the study of the case law will be articulated on rights: human dignity, right to health, autonomy and the consent, right to bodily integrity, right to a healthy environment, privacy, freedom of scientific research. First, we need to take into account that the ECHR articles do correspond to a single right, but at the same time each human right can be covered by a plurality of articles of the Convention. Therefore, it is not useful to develop our analysis article by article, but right by right, considering thus a multitude of provisions for the same individual right concerned. For example, the right to health is covered by Article 2 of the ECHR, which protects the right to life, Article 3, prohibition of inhuman and degrading treatment, and Article 8 (respect for private life). Secondly, the impact on each right varies depending on the specific technological sector which is concerned. Therefore, these rights will be analysed on light of a specific challenge in the technoscientific field. For example, dignity with regard to genetic enhancement, health with regard to nanotechnologies, bodily integrity with regard to human enhancement, the environment with regard to geoengineering, etc.

The case law is not relevant per se, but on the light of the fact that the entire framework of the Council of Europe are filtered through the decisions of the Strasbourg Court. Therefore, the Court not only uses the norms of the ECHR which are applicable to the case concerned, but also those instruments of the Council of Europe without legally binding force such as resolutions of the Parliamentary Assembly of the Council of Europe, guidelines of the Committee on Bioethics, as well as the Oviedo Convention and its Protocols, which often deal with issues related to the scientific progress, and can therefore help the Court in solving hard cases. From this case law, therefore rules and legally binding principles emerge as a reference point for future cases raised by emerging technologies.

References

Arnaldi, S., Gorgoni, G., and Pariotti, E. (2016). RRI as a Governance Paradigm: What is New? in: Lindner, R., Kuhlmann, S., Randles, S., Bedsted, B., Gorgoni, G., Griessler, E., Loconto, A., and Mejlgaard, N. (eds.), *Navigating towards Shared Responsibility in Research and Innovation: Approach, Process and Results of the Res-AGorA Project*, Fraunhofer ISI: Karlsruhe, pp. 23–29.

Blind, K. (2008). Regulatory Foresight: Methodologies and Selected Applications, *Technological Forecasting & Social Change*, **75**, pp. 496–516.

Boillat, P., and Kjaerum, M. (2014). *Handbook on European Data Protection Law*, Luxembourg: Publications Office of the European Union.

The European Group on Ethics in Science and New Technologies (EGE) (2016). *Opinion No. 29 on The Ethical Implications of New Health Technologies and Citizen Participation*, Brussels, 13 October 2015. Luxembourg: Publications Office of the European Union.

Leenes, R., Palmerini, E., Koops, B.-J., Bertolini, A., Salvini, P., and Lucivero, F. (2017). Regulatory Challenges of Robotics: Some Guidelines for Addressing Legal and Ethical Issues, *Law, Innovation & Technology*, **9**(1), pp. 1–44.

Owen, R. (2014). *Responsible Research and Innovation: options for research and innovation policy in the EU. European Research and Innovation Area Board (ERIAB), Foreword Visions on the European Research Area (VERA)*, Available at: https://ec.europa.eu/research/innovation-union/pdf/expert-groups/Responsible_Research_and_Innovation.pdf. Accessed 28 September 2017.

Riley, A. (2008). *Legal English and the Common Law*, Padova: CEDAM.

Ruggiu, D. (2013). Temporal Perspectives of the Nanotechnological Challenge to Regulation. How Human Rights Can Contribute to the Present and Future of Nanotechnologies, *Nanoethics*, **7**(3), pp. 201–215.

Ruggiu, D. (2015). Anchoring European Governance: Two versions of Responsible Research and Innovation and EU Fundamental Rights as 'Normative Anchor Points', *Nanoethics*, **9**(3), pp. 217–235.

Chapter 4

Human Dignity[1]

4.1 Introduction

4.1.1 The concept of human dignity, although it is one of the most central in bioethics, is particularly hard to be defined due to the multitude of different applications and interpretations to which it can pave the way. Basically, by reflecting the Kantian categorical imperative in modern times it prohibits using another person merely as a means to an end (Jacobson, 2007, p. 293).

The UNESCO Universal Declaration on Bioethics and Human Rights[2] recalls this idea in several parts, in particular with its Article 3[3] and it links it to the idea of equality (Art. 10)[4] and

---

[1]This chapter partially retakes and deepens the analysis of Ruggiu (2018).

[2]UNESCO (2005) *Universal Declaration on Bioethics and Human rights*, adopted by acclamation by the 33rd session of the General Conference of the UNESCO in 19 October 2005.

[3]Article 3 – Human dignity and human rights 1. Human dignity, human rights and fundamental freedoms are to be fully respected. 2. The interests and welfare of the individual should have priority over the sole interest of science or society.

[4]Article 10 – Equality, justice and equity The fundamental equality of all human beings in dignity and rights is to be respected so that they are treated justly and equitably.

Human Rights and Emerging Technologies: Analysis and Perspectives in Europe
Daniele Ruggiu
Copyright © 2018 Pan Stanford Publishing Pte. Ltd.
ISBN 978-981-4774-93-2 (Hardcover), 978-0-429-49059-0 (eBook)
www.panstanford.com

non-discrimination (Art. 11).[5] It has also been linked to human genome by the UNESCO Universal Declaration on Human Genome and Human Rights.[6]

Dignity is not only at the basis of the international human rights law, it is also at the very core of the Council of Europe framework, with the Oviedo Convention.[7] The Convention mainly aims at protecting human dignity from the 'misuse of biology and medicine'.[8] In this regard, Article 1 of the Oviedo Convention explicitly protects the dignity of the individual. It is worthwhile noting that the idea of dignity was affirmed in relation to challenges of biomedicine for the first time. The ECHR, for example, does not mention it, even in its preamble (Beyleveld and Brownsword, 2001, p. 12).

Instead, dignity is recalled three times in the preamble of the Oviedo Convention: when the Convention stresses the need for respecting the human being both as individual and as member of human species; when it addresses the dangers of the misuse of biology and medicine for human being; and when it requests efficacious measures to safeguard rights and the dignity of human beings. The Convention places respect for human dignity first in the list of principles that should govern the biomedical field. It solemnly starts by recalling that States 'Parties [. . .] shall protect the dignity and identity of all human beings and guarantee everyone, without discrimination, respect for their integrity and other rights and fundamental freedoms with regard to the application of biology and medicine' (Art. 1).

As said dignity is a divisive concept on which we can address two opposite conceptions (Beyleveld and Brownsword, 2001). While one

---

[5]Article 11 – Non-discrimination and non-stigmatization No individual or group should be discriminated against or stigmatized on any grounds, in violation of human dignity, human rights and fundamental freedoms.

[6]UNESCO (1997) *Universal Declaration on the Human Genome and Human rights*, adopted by the UNESCO General Conference in 11 November 1997 and subsequently endorsed by the United Nations General Assembly in 1998.

[7]Council of Europe (1997) *Convention for the Protection of Human rights and Dignity of the Human Being with regard to the Application of Biology and Medicine* (Convention on Human rights and Biomedicine or the Oviedo Convention) (CETS n. 164), adopted in Oviedo on 4 April 1997 (entered into force on 1 December 1999).

[8]Preamble of the Oviedo Convention.

develops around 'an empowerment conception of human dignity' that is geared for responsible rights-bearing agents, the other 'is a conception of human dignity as constraint' particularly geared for the protection of vulnerable individuals (Brownsword, 2009, p. 129). In this regard, the former conception, which is a liberal understanding of human dignity, appears to be implicit in several provisions of the ECHR such as the provisions relating to the right to life (Art. 2), prohibition of inhuman and degrading treatments (Art. 3), and the non-discrimination principle (Art. 14). The latter, instead, which is a more conservative understanding of human dignity, is recalled by many key provisions of the Oviedo Convention, such as those relating to the use of human embryos as research tools (Art. 18) or the recognition of proprietary rights over removed body parts and tissues (Arts. 19 and 21) or, finally, the use of technologies for the purpose of genetic enhancement (Arts. 13 and 14). In other words, in the empowerment conception the system of human rights is aimed at preserving the dignity of the person through the protection of individual rights. Therefore, the rights-bearing agent contributes to the protection of dignity (their own dignity) via the judicial action of claiming of their individual rights. Conversely, the conception of dignity as constraint aims at defending the dignity of each human being, especially the most vulnerable ones, even against the individual's will if it is necessary. This can happen in the case of people lacking of capacity because of an illness or disability such as a mental health problem when the state of vulnerability of a person legitimizes a third party (a relative, a legal representative, private associations, public institutions) to act on their behalf. In this instance other subjects such as relatives, public institutions, private associations intervene for protecting the dignity of a 'person' who cannot act by himself either because s/he is incapable (children, disable persons) or yet non-existent (unborn, future generations). In theory, this conception of dignity can lead to a paternalistic conception of human being and potentially it may conflict with the protection of individual rights proper of liberal countries. These two approaches with regard to human dignity were explicitly mentioned by the Strasbourg Court jurisprudence.[9]

---

[9] See, *Vo v. France* (App. 53924/00) judgement of 8 July 2004, par. 40.

These opposite conceptions of human dignity are variously present within both the ECHR and the Oviedo Convention. In other words, there is not a clear-cut polarization between the ECHR, meant as closer to an empowerment conception of dignity, and the Oviedo Convention, meant as closer to a conception as constraint. The seeds of both conceptions can be found in both documents, although in the Oviedo Convention the conception of dignity as constraint might appear prevalent. For example, we can find many references to the empowerment conception of dignity in several parts of the Oviedo Convention (those on the consent according to Art. 19ff., for example) and vice versa, namely references to the constraint conception can be found in the ECHR (the positive obligation to protect human life according to Art. 2, for example). This is not necessarily a shortcoming, since it can multiply the possibilities of solution of ethically controversial cases in a field such as that of genetic enhancement where the modification may affect both the same individual who enhances themselves and another person not yet existent (descendant). In plain English, within the Council of Europe framework where the empowerment conception of dignity has no solution through the ECHR (for example with regard to genetic modification of human being), one can refer to the conception as constraint through the Oviedo Convention. This seems to be the approach of the jurisprudence of the Strasbourg Court.

Whereas the EU Charter of Fundamental Rights[10] recognizes human dignity as an autonomous right which can be claimed by the rights-bearing agent (e.g. a private association such as Greenpeace), within the ECHR it is not recognized as a right. Namely, dignity is not an individual right within the ECHR neither thanks to the act of the Strasbourg Court. Therefore, the interests concerned are protected by other Convention rights (according to an empowerment conception of human dignity). For example, they fall within the broad scope of Article 8 ECHR (respect for private and family life), Article 3 ECHR (inhuman and degrading treatment), and (more rarely) Article 14 ECHR (non-discrimination).

---

[10] European Union (2000) *Charter of Fundamental Rights of the European Union*, adopted in Nice on 7 December 2000 (came into force on 1 December 2009). After the entry into force of the Lisbon Treaty (1 December 2009) the Nice Charter is now legally binding according to Art. I-6 §2 Lisbon Treaty.

4.2 The Challenge of Human Genetic Modification

4.2.1 Human dignity can be challenged in several ways: at the workplace (violation of privacy, slavery, etc.), in the field of control of immigration (deportation of persons which consists in a degrading treatment), in the ambit of technoscientific development. Human genetic modification is a field where human dignity is greatly at stake. In the future, thanks to new techniques such as CRISPR-Cas9 (Clustered Regularly Interspaced Short Palindromic Repeats) or synthetic biology, the spread of techniques aimed at the genetic modification of the animal world could lead to their application on human beings in order to make them more resistant to certain diseases, to extend their life expectancy, or to improve their mental performance. Several biotechnology applications are already being tested on animals and are near to be commercialized in order to have pets with specific genetic features or genetically modified animal products for the large consumption (Frankenfish, such as super-salmon[11]). This trend raises concerns on the extension of those same techniques to human beings. In particular, CRISPR-Cas9, which is simpler and more efficient than other techniques, might accelerate the race to change the DNA in the nuclei of reproductive cells that transmit information from one generation to the next. 'It means that we can imagine a day when human chromosomes may be modified in the sperm and egg to assure that one or another aspect of a child's inheritance is designed to order' (Pollack, 2015, p. 871).

Recently a Chinese team has become the first to inject a person with cells that contain genes edited using the revolutionary CRISPR–Cas9 technique for treating an aggressive lung cancer (Cyranoski, 2016). In 2017, a U.S. microbiologist, Brian Hanley, has just started testing an unregulated gene therapy with himself. 'The gene Hanley added to his muscle cells would make his body produce more of a potent hormone—potentially increasing his strength, stamina,

---

[11]Recently AquaBounty Technologies in Boston, Massachusetts, has patented a GM-salmon which can grow faster and become bigger than other species of salmon. According to FDA this novel transgenic fish is proved to be safe and can be commercialized in the U.S.

and life span' (Regalado, 2017). However, it is also possible the experimentation of these same techniques on human embryos raising enormous questions from the ethical standpoint. In fact, in a non-far future this will allow not only the early treatment of some incurable genetic diseases, but it will also pave the way to a world of tailored children. This worried scenario became true when a Chinese team has recently genetically modified human embryos for experimental purposes for the first time in the world (Puping et al., 2015). This stirred up a hot debate over human embryo editing in the researchers' community raising the issue of the establishment of some clear-cut limits to scientific research (Cyranoski and Reardon, 2015). Recently an U.S. team of Oregon Health and Science University would have perfected the gene-editing technique CRISPR on human embryos (Ma et al., 2017), 'demonstrating that it is possible to safely and efficiently correct defective genes that cause inherited diseases' (Connor, 2017). A further evolution of CRISPER techniques of gene editing is 'base editing' by which scientists can correct anomalies of a single DNA basis. DNA is formed by four building blocks, adenine, cytosine, guanine and thymine, whose alterations can give rise to point mutations such as β-thalassemia that can be potentially deadly. Thanks to base editing those minimal variations of DNA can be precisely corrected in the patient's primary cells. In 2017 the feasibility of curing genetic disease causing point mutations in human somatic cells and embryos by base editor system has been demonstrated by a Chinse team of Sun Yat-sen University, in Guangzhou, China, where protocols and limitations of European ethical committees are less strong (Liang et al., 2017). Cured human embryos with β-thalassemia by this technique were not be implanted. These studies have already raised some concerns from the scientific community since embryos can occasionally start to develop from maternal healthy DNA alone.[12] These studies therefore would have not proven the efficacy of the method. However, even supposing that their safety and efficiency, these advances cannot completely overcome the existing challenge

---

[12]http://www.sciencemag.org/news/2017/08/skepticism-surfaces-over-crispr-human-embryo-editing-claims. Accessed 5 October 2017.

for the notion of dignity where this technique surpasses the limits of therapy and leads, instead, to eugenics.

Another possibility of genetic modification is the creation of human–animal chimeras, namely organisms made out of the cells of two or more different zygotes.[13] This technique could pave the way to several useful applications on the human being in particular in the field of transplantation. Recently, thanks to the CRISPR-Cas9 techniques a team of the University of California at Davis created a human–pig chimera, which might pave the way to 'the possibility of xeno-generating transplantable human tissues and organs towards addressing the worldwide shortage of organ donors' (Wu et al., 2017). A human–animal chimera should be considered as an interspecies chimera resulting from a human and an animal zygote. As one can easily imagine, this technique raises several concerns related to the dignity of human beings because it cannot be completely equated to genetic engineering from the legal standpoint. Therefore, it is not certain that solutions offered by the ECHR to genetic engineering in general are so also for the human–animal hybridization.

4.3 The State's Discretion in Ethical Controversial Issues

4.3.1 Since the ECHR is inseparable from the application of the Strasbourg Court of its provisions, in order build the notion of dignity in legal terms, first we need to analyse the occurrence of this concept in its jurisprudence, especially considering those instances where human dignity was used to solve the case. In these cases, we need to take into account also the internal practices of the judges at Strasbourg such as those that focus on the state of uncertainty in a given field. Emerging technologies in general are an ethically controversial issue. In this context opinions, are much polarized and this impedes the formation of a single and clear position, not only

---

[13] Recently a US team of scientists has injected human stem cells into pig embryos to produce human-pig embryos known as chimeras in order to obtain a human pancreas which could be transplanted. See http://www.bbc.com/news/health-36437428.

in the scientific world. In the same sense also genetic enhancement is an ethically controversial issue. Easily, positions in its favour face contrary opinions with the arsenal of different arguments, which are often equally plausible. In this regard the Court has developed the margin of appreciation doctrine.

The ECtHR has affirmed the doctrine of the margin of appreciation enabling the State a degree of discretion in multiple circumstances. This margin of appreciation should not be confused with two further situations. These are first that according to the subsidiary principle, the State has to try to apply the ECHR norms on its own before facing the Strasbourg Court (Kratochvìl, 2011, 335), and second, that in a state of emergency which threatens the life of the nation (e.g. civil war, terrorism), Article 15 ECHR permits the suspension of the application of the ECHR (Donati and Milazzo, 2003). In this latter situation the derogation of the ECHR norms cannot however involve the compromising Articles 2 (life), 3 (torture), 4,1 (slavery), 7 (the principle of no punishment without law) of the ECHR – they are therefore called 'absolute rights'.[14] Instead, the margin of appreciation is generally used in the application of Articles 8-11 ECHR (respect for private and family life, freedom of thought, freedom of expression, freedom of association), as well as of Article 1 Protocol 1 (propriety).[15] Moreover, a margin of appreciation is acknowledged in immigration controls (Art. 2,3 Protocol 4), and, as said, in morally sensitive questions (such as the status of embryo, medically assisted procreation, surrogacy, end-life decisions, assisted suicide) on which States parties in Europe do not have a single position.[16] A space of discretion for a State is therefore likely to be recognized in questions raised by the development of emerging technologies, including human genetic modification.

Contrary to what one could think, the established margin of appreciation is not completely free (the State's decision is not unfettered) and it does not entail a lack of control by the Strasbourg Court.[17] The ECtHR normally scrutinizes these matters

---

[14] The logic of the application of Article 15 is the 'better position rationale' by which the State can better assess the state of emergency. On this see Geer, 2000, 24.

[15] Ibid., 9–11.

[16] Ibid., 24–26.

[17] See e.g. *Lambert and others v. France* (Appl. 46043/14) judgement of 5 June 2015 *Reports of Judgements and Decisions*, 2015, par. 148: 'this margin of appreciation is

closely,[18] assessing (in the case of Article 8 ECHR for example) whether the State's interference with ECHR rights was made in accordance with the law (principle of legality), pursuing a legitimate aim (provided for by the ECHR), necessary in a democratic society, and through the adoption of a measure that is proportionate to the pursued legitimate aim (the 'proportionality' test).[19] This intense scrutiny by the Court has been criticized by scholars,[20] as well as by States (who should have benefited from it), leading to the adoption of a specific Protocol in 2013 which has not entered into force yet. In this sense, the margin of discretion does not alleviate the need for the State to prove they have respected the rights and freedoms set forth by the Convention.[21] This is especially true in the domain of health, where human genetic modification is likely to fall. This premise has to be taken into account when considering challenges of emerging technologies to the human rights system existing in Europe.

4.4 The Case Law of the Strasbourg Court on the Life of the Unborn and Reproductive Rights

4.4.1 With regard to genetic enhancement in general, we should basically distinguish two different situations whether a person decides to modify their own genome and whether s/he decides to change the genetic heritage of their descendants. The first case

---

not unlimited [. . .] and the Court reserves the power to review whether or not the State has complied with its obligations'.

[18] *Costa and Pavan v. Italia* (Appl. 54270/10) judgement of 28 August 2012, par. 68.

[19] E.g. *Lambert and others v. France* (Appl. 46043/14) judgement of 5 June 2015 *Reports of Judgements and Decisions*, 2015. While the State enjoys a broader discretion in the environmental matter, where a nation's strategic choices is at stake (e.g. nuclear, transport, industry), this margin is quite strict and generally subject to a deeper scrutiny of the Court in the immigration and health matters, where individuals are directly affected. See e.g. *Costa and Pavan v. Italia* (Appl. 54270/10) judgement of 28 August 2012, par. 77; *Open Door Counselling Ltd and Dublin Well Woman v. Ireland* (Appl. 14234/88, 14235/88) judgement of 29 October 1992 *Series A No. 246-A*, par. 68.

[20] See, e.g. Kratochvìl 2011, 324–357.

[21] Council of Europe (2013) *Additional Protocol No. 15 amending the Convention for the Protection on Human Rights and Fundamental Freedoms* (CETS n. 213), adopted in Strasbourg 24 June 2013 (it has not entered into force yet).

mainly falls under the issue of freedom to be enhanced as regards the legitimacy of the consent and its restrictions.[22] The second appears instead more complicated and mainly involves dignity and, at a lower level, the consent. In this second instance, the Oviedo Convention appears to be decisive.

The current jurisprudence of the Strasbourg Court might provide a set of solutions to ethical controversial issues related to genetic modification of offspring. In this context, we need to consider two aspects: the fact that these forms of enhancement involve human embryos, as well as the parents' decision to modify genetic features of offspring. Therefore, in order to assess genetic enhancement under the ECHR we need to consider the ECtHR jurisprudence on the legal status of the embryo (as the possible holder of the right to life) and that on reproductive rights.

With regard to the status of the embryo the ECtHR in its early jurisprudence first dealt with the question whether the unborn is or is not a person. This aspect is crucial when we tackle the issue of the genetic modification of offspring.

First of all, the wording of the ECHR leaves the question as to whether the terms 'everyone' or 'person' also apply to the unborn open (Bernat, 2009, p. 93). In the *Paton case*,[23] the Commission noted that the term 'everyone' which is used in several articles of the ECHR, in particular Article 2 (right to life), is not defined by the Convention. However, in almost all these cases, it is used as it can apply only *postnatally*, whereas none of these instances clearly indicates that it can also apply *prenatally*, even though this outcome cannot completely be excluded (Wescha, 2009, p. 102). In this case the Commission excluded that the term 'life' in Article 2 would grant the human foetus an absolute right to life, while it left the question whether it does not cover the human foetus at all or it recognizes a right to life only in certain circumstances open. This leads not to equate the position of the unborn to that of the human being already born as recognized by the Strasbourg Court. In other words, the ECHR protect the unborn's potentiality of life, not its right to life, since potentiality cannot be equated to the notion of existence

---

[22] see Chapter 6 Part II on Consent.
[23] *X v. the United Kingdom* (Appl. 8416/78) decision of the Commission of 13 May 1980 *Decision and Reports*, 19, also cited as *Paton v. UK.*

tout court. For example, the ECtHR has clarified that the concept of embryo and also foetus cannot be included in the same category of 'child', following the opinion of the Committee on Bioethics of the Council of Europe (DH-BIO), in the *Costa and Pavan case* on medically assisted procreation for an Italian couple suffering from a genetically transmissible disease.[24]

Basically this openness of the term 'everyone' was maintained also by the subsequent jurisprudence of the Strasbourg Court, since it is used to state that the question of the beginning of life is covered by a margin of State's appreciation, as well as the question of the status of the embryo or the foetus.[25] For example, in the *Evans case*,[26] where the applicant wanted to implant the fertilized embryos notwithstanding the subsequent withdrawal of the consent by her partner, the Court had to decide for the first time whether the right to life was granted to the embryo in vitro under the ECHR provisions. Also in this case, the Court concluded that the question whether the unborn enjoys the right to life or not falls under the margin of State's appreciation, by depending on the national legislation which did not provide for any right to life. However, it is to say that essentially this limitation of the interest of the unborn is recognized only in front of the protection of the right to health of another (living) subject such as the mother. In this regard, according to the ECtHR (*Paton case*) the protection of the right to life of the foetus can be limited at any time of its development only by virtue of a 'medical indication', namely to protect the life and health of the pregnant woman according to an empowerment conception of dignity. Conversely, it cannot be limited for any other type of indication (ethics, *eugenics*, social, time).[27] This

---

[24] *Costa and Pavan v. Italia* (Appl. 54270/10) judgement of 28 August 2012.

[25] *Vo v. France* (App. 53924/00) judgement of 8 July 2004, par. 18: 'The Convention does not define the term 'everyone' (in French '*toute personne*'). These two terms are equivalent and found in the English and French versions of the European Convention on Human Rights, which however does not define them. In the absence of a unanimous agreement on the definition of these terms among member States of the Council of Europe, it was decided to allow domestic law to define them for the purposes of the application of the present Convention'.

[26] *Evans v. the United-Kingdom [GC]* (App. 6339/05) judgement of 7 March 2006 *Reports Judgements and Decisions*, 2007-I, par. 54.

[27] *X v. the United Kingdom* (Appl. 8416/78) decision of the Commission of 13 May 1980 *Decision and Reports*, 19, also cited as *Paton v. UK*, par. 22.

should be considered a limit for the manipulation of descendants' genome in genetic enhancement.

Lastly, in this regard we cannot exclude that the recent jurisprudence of the Court of Justice of the European Union (CJEU) which has defined the human embryo as an ovum, fertilized or not, 'capable of commencing the process of development of a human being'[28] could influence the ECtHR in the future by consolidating this jurisprudential trend (Puppnick, 2013, p. 4). The *Brüstle case* is the first case where dignity was at stake as individual right (Ruggiu, 2013). In 1997, Mr. Oliver Brüstle, a stem cell researcher, filed a patent related to isolated and purified neural progenitor cells for the treatment of neural defects. The transplantation of brain cells into the nervous system can be a promising method of treatment for numerous neurological diseases. The same techniques could be used also in brain enhancement. Greenpeace filed an appeal before the Bundespatentgericht (the Federal Patent Court in Germany) seeking a ruling of invalidity of the above-mentioned patent. The Bundespatentgericht decided to stay the proceeding and to refer the question of stem cells patentability to the CJEU. However, it is worthwhile noting that the Court did not refer to the right to dignity as protected by the EU Charter (Art. 1), but to the notion of internal market of the European Union, which would have been threatened by the existence of differences in the legal protection of biotechnological inventions in Europe. Notwithstanding the lacking reference to the EU Charter the outcome was similar. Therefore, lacking a uniform definition of human embryo within the EU, the CJEU provided one including: (a) any human ovum as soon as fertilized; (b) any non-fertilized human ovum into which the cell nucleus from a mature human cell has been transplanted; (c) any non-fertilized human ovum whose division and further development have been stimulated by parthenogenesis, including stem cells obtained from a human embryo at the blastocyst stage. Then, it excluded from patentability some uses of human embryos

---

[28] *Oliver Brüstle v Greenpeace eV* (Case C-34/10) judgement of 18 October 2011, *OJ C 362*, p. 5. See also *International Stem Cell Corporation v. Comptroller General of Patents and designs and Trade Marks* (C-364/13) judgement of 18 December 2014, not published yet. On the embryonic research within the ECHR see *Parrillo v. Italia* (Appl. 46470/11) judgement of 27 August 2015.

(with industrial, commercial purposes, or for scientific research). Although the aim of the scientific research must be distinguished from industrial and commercial purposes, the use of human embryos for purposes of scientific research cannot be separated from the patent itself and the rights attached to it. According to the Court, techniques for producing stem cells that require the prior destruction of human embryos are also unpatentable.[29] The framework as expressed by the CJEU might further reduce chances of genetically manipulating human embryo also under the ECHR.

In the case of eugenics, reproductive rights of parents are at stake as well and they could lead to an opposite outcome.

Article 8 ECHR (respect for private and family life) covers several aspects under the concept of 'private life', among others the right to respect the decision whether to become or parents or not.[30] This right, however, is not limit-free since, as seen, the State usually enjoys a margin of discretion,[31] which falls under the strict scrutiny of the ECtHR.

In this framework, the Menneson case[32] has expressed a principle, which gains a particular relevance in the field of genetic enhancement. In this case a couple of French nationals went in the U.S. in order to have surrogacy, which was forbidden in France. Subsequently, after the birth of a couple of twins, the French Cassation Court annulled the recognition of the biological father's paternity of the children born thanks to surrogacy. Notwithstanding the Menneson family could live together in France without any interference of the State, the children could not acquire French nationality and enjoy normal hereditary rights by breaching thus their private life (Art. 8). In this instance the ECtHR has affirmed the principle that 'where a particularly important facet of an individual's existence or identity is at stake, the margin allowed to the State will

---

[29] However, although he did not obtain the patent in Europe, Oliver Brüstle patented his invention in U.S.

[30] *Costa and Pavan v. Italia* (Appl. 54270/10) judgement of 28 August 2012, par. 62.

[31] *Evans v. the United-Kingdom [GC]* (App. 6339/05) judgement of 7 March 2006 *Reports Judgements and Decisions*, 2007-I, par. 54.

[32] *Menneson v. France* (Appl. 65192/11) judgement of 26 June 2014 *Reports of Judgements and Decisions*, 2014.

normally be restricted'.[33] This principle is valid especially for those whose identity is or can be mutated due to human enhancement technologies according to unregulated techniques. Like, for example, in the case of genetic enhancement or human–animal hybridization. In other words, the fact that a technique is forbidden does not exclude the protection of its outcomes under the ECHR.

4.5 The Framework of the Council of Europe as a Reference Point of the Judges in Strasbourg

4.5.1 The framework of the Strasbourg Court decisions in the field of status of embryo and reproductive rights appears however not decisive with regard to genetic enhancement. We can draw out only an incomplete and non-conclusive guidance here. In this regard we can find a decisive help in the context of the Oviedo Convention provisions and other documents of both the Council of Europe and human rights international law. Whenever the Strasbourg Court jurisprudence is not able to solve a hard case, it can refer to the entire framework of the Council of Europe instruments in order to integrate the interpretation of the ECHR provisions.

Therefore, notwithstanding States enjoy a margin of appreciation in morally sensitive issues, according to the Strasbourg Court jurisprudence, this must be used in accordance with the ECHR provisions, including soft law documents of the Council of Europe such as the Oviedo Convention, the opinions of the Council of Europe Committee on Bioethics (DH-BIO)[34] or any other formal statement of the Parliamentary Assembly of the Council of Europe.[35] This is likely the case of the genetic enhancement of human being.

With regard to genetic enhancement the Oviedo Convention provisions on dignity must be framed in a net of several other norms

---

[33] Ibid., 77.

[34] https://www.coe.int/en/web/bioethics/dh-bio.

[35] *Costa and Pavan v. Italia* (Appl. 54270/10) judgement of 28 August 2012, par. 68: '[w]hile acknowledging that the question of access to PGD raises sensitive moral and ethical questions, the Court notes that the solutions reached by the legislature are not beyond the scrutiny of the Court'.

ranging from the human rights international law to the Convention itself.

Firstly, the UNESCO Universal Declaration on Bioethics and Human Rights when it establishes a set of norms to guide biomedical practice, assigns the first place to the principle of respect for 'human dignity, human rights and fundamental freedoms' (Art. 3.1) (Andorno, 2009, p. 224). Secondly, the Universal Declaration on the Human Genome and Human Rights[36] solemnly proclaims human genome as heritage of humanity and postpones the interest of science 'over respect for the human rights, fundamental freedoms and human dignity of individuals' (Art. 10). Further, the Oviedo Convention provisions on dignity must be connected to the entire framework of norms of the ECHR on the self-determination principle (Art. 8 and its associate case law), as well as those provisions of the Convention itself related to the consent in biomedicine (Art. 5-9). This conclusion is motivated by the fact that the offspring subject to genetic enhancement would be deprived thus of their possibility to express their consent to the medical treatment without any legal justification (existence of a genetic disease).

In this regard, given this framework, the protection of self-determination together with personal dignity under the ECHR should also preclude the possibility of applying these techniques whenever these genetic mutations can be transmitted to offspring beyond the therapeutic use.

This solution appears confirmed by the set of norms of the Oviedo Convention on the protection of the human genome. While Article 21 of the Oviedo Convention prohibits the economic exploitation of the human genome, according to Article 13 any modification of the human genome (germline genetic modification) is prohibited unless it is preventive, diagnostic or therapeutic, and provided that it introduces any modification in the genome of any descendant. In this sense, the genetic modification of mankind is not prohibited as such, but only when it leads to the mere enhancement of the specie (eugenics) in accordance with what *the Paton case* has

---

[36]UNESCO (1997) *Universal Declaration of the Human Genome and Human Rights,* adopted in 11 November 1997.

affirmed.[37] Instead, it is allowed when it has preventive, diagnostic and therapeutic purposes.

Recently the DH-BIO has arguably reaffirmed the value of the Oviedo Convention as a reference point for issues in the biomedical field, in particular genetic enhancement.

In this regard the DH-BIO has recalled that 'Article 13 of the Oviedo Convention addresses these concerns about genetic enhancement or germline genetic engineering by limiting the purposes of any intervention on the human genome, including in the field of research, [...] prevention, diagnosis or therapy. Furthermore, it prohibits any intervention with the aim of introducing a modification in the genome of any descendants'.[38]

Given this normative framework, human genetic modification should be deemed allowed only in limited cases where those techniques aim at the prevention, diagnosis or therapy (to enhance immune system, for example).[39]

As regards experimentation on embryos, instead, Article 18 of the Oviedo Convention provides that where laws allow research on embryos in vitro, they shall ensure adequate protection of embryos, whereas the creation of embryos for research purposes is prohibited.

It must be added, finally, that, according to the principles set by the DH-BIO, '[t]he techniques of human artificial procreation must not be used for obtaining particular characteristics in the future child' (Principle 1, 2).[40]

With regard to techniques of human–animal hybridization the answer is a little more complex. According to Marion Weschka,

---

[37] *X v. the United Kingdom* (Appl. 8416/78) decision of the Commission of 13 May 1980 *Decision and Reports*, 19, also cited as *Paton v. UK*, par. 22.

[38] Committee on Bioethics (DH-BIO) (2015) *Statement on Genome Editing Technologies*, adopted in Strasburg 1–4 December 2015.

[39] Recently, on 21 June, an advisory committee at the US National Institutes of Health (NIH) approved a proposal to use CRISPR–Cas9 to modify Human T cells (blue) and enhance the body's immune system of 18 patients with several types of cancers. On this see Reardon (2016).

[40] Committee on Bioethics of the Council of Europe (DH-BIO) (1989) *Report on Human Artificial Procreation. Principles set out in the report of the Ad Hoc Committee of Experts on Progress in the Biomedical Sciences (CAHBI)*, 1989 http://www.coe.int/en/web/bioethics/embryo/cloning.

the only conclusions one could draw from the set of provisions of the Oviedo Convention as well as the Protocol on human cloning with regard to Chimbrids research are the following: it should be 'prohibited to create human embryos by fertilisation for any purposes, including Chimbrids research (Art. 18,2). Interspecies cell somatic transfer is prohibited for reproductive purposes at least where a somatic cell nucleus is taken from a born human being and transferred into an animal ovum for the purpose of creating another human being genetically identical to the first one. As mentioned above, this requires that the resulting clone is considered human despite the animal mitochondrial DNA' (Weschka, 2009, p. 109).

However, in the Council of Europe, there is a quite dated recommendation of the Parliamentary Assembly on the Use of Human Embryos and Foetuses for Diagnostic, Therapeutic, Scientific, Industrial and Commercial Purposes, which lays down a set of guidelines on the use of embryos also for Chimbrids research.[41] In particular, it recommends forbidding 'the implantation of a human embryo in the uterus of another animal or the reverse' and 'the fusion of embryos or any other operation which might produce chimeras' (Point A, iv). The context of research in this field has mutated and there is probably the need of a more updated document today. However, the meaning of this this guideline does not appear controversial, at least up to now.

References

Andorno, R. (2009). Human Dignity and Human Rights as a Common Ground for a Global Bioethics, *Journal of Medicine and Philosophy*, **34**(3), pp. 223–240.

Brownsword, R. (2009). Regulating Human Enhancement: Things Can Only Get Better, *Law Innovation & Technology*, **1**, pp. 125–152.

---

[41]Council of Europe (1986) Recommendation 1046(1986) *on the Use of Human Embryos and Foetuses for Diagnostic, Therapeutic, Scientific, Industrial and Commercial Purposes*, adopted on 19 and 24 1986 (13th and 14th Sittings) http://www.coe.int/t/dg3/healthbioethic/Texts_and_documents/INF_2014_5_vol_II_textes_%20CoE_%20bioéthique_E%20(2).pdf

Bernat, E. (2009). Which Being Should Be Entitled to Human Rights, in: Taupitz, J., and Weschka, M. (eds.), *CHIMBRIDS – Chimeras and Hybrids in Comparative European and International Research: Scientific, Ethical, Philosophical and Legal Aspects*, Berlin, Heidelberg: Springer, Verlag, pp. 88–97.

Beyleveld, D., and Brownsword, R. (2001). *Human Dignity in Bioethics and Biolaw*, Oxford: Oxford University Press.

Connor, S. (26 July 2017). First Human Embryos Edited in U.S., *MIT Technology Review* https://www.technologyreview.com/s/608350/first-human-embryos-edited-in-us/. Accessed 25 September 2017.

Cyranoski, C. (15 November 2016). CRISPR Gene-Editing Tested in a Person for the First Time, *Nature*, http://www.nature.com/news/crispr-gene-editing-tested-in-a-person-for-the-first-time-1.20988. Accessed 25 September 2017.

Cyranoski, D., and Reardon, S. (22 April 2015). Chinese Scientists Genetically Modify Human Embryos. Rumours of Germline Modification Prove True — and Look Set to Reignite an Ethical Debate, *Nature*, doi: 10.1038/nature.2015.17378, https://www.nature.com/news/chinese-scientists-genetically-modify-human-embryos-1.17378. Accessed 25 Spetember 2017.

Donati, F., and Milazzo, P. (2002). La dottrina del margine di apprezzamento nella giurisprudenza della Corte europea dei diritti dell'uomo, in: P. Falzea, A. Spadaro and L. Ventura (eds.) *La Corte costituzionale e le Corti d'Europa. Atti del seminario svoltosi a Capannello (CZ) il 31 maggio-1 giugno 2002*, Torino: Giappichelli, pp. 65–117.

Geer, S. (2000). *The Margin of Appreciation: Interpretation and Discretion under the European Convention on Human Rights*, Council of Europe Publishing.

Jacobson, N. (2007). Human Dignity and Health: A Review, *Social Science and Medicine*, **64**, pp. 292–302.

Kratochvìl, J. (2011). The Inflation of the Margin of Appreciation Doctrine of the European Court of Human Rights, *Netherlands Quarterly of Human Rights*, **29**(3), pp. 324–357.

Liang, P., Ding, C., Sun, H. Xie, X., Xu, Y., Zhang, X., Sun, Y., Xiong, Y., Ma, W., Liu, Y., Wang, Y., Fang, J., Liu, D., Songyang, Z., Zhou, C., and Huang, J. (2017). Correction of B-Thalassemia Mutant by Base Editor in Human Embryos, *Protein & Cell*, https://doi.org/10.1007/s13238-017-0475-6.

Ma, H., Marti-Gutierrez, N., Park, S.-W., Wu, J., Lee, Y., Suzuki, K., Koski, A., Ji, D., Hayama, T., Ahmed, R., Darby, H., Van Dyken, C., Li, Y., Kang, E., Park,

A.-R., Kim, D., Kim, S., Gong, J., Gu, Y., Xu, X., Battaglia, D., Krieg, S.A., Lee, D.M., Wu, D.H., Wolf, D.P., Heitner, S.B., Izpisua Belmonte, J.C.I., Amato, P., Kim, J.K., Kau, S., and Mitalipov, S. (2017). Correction of a Pathogenic Gene Mutation in Human Embryos, *Nature*, **548**(413), https://www.nature.com/nature/journal/vaop/ncurrent/pdf/nature23305.pdf. Accessed 25 September 2017.

Pollack, R. (2015) Eugenics Lurk in the Shadow of CRISPR, *Science*, **346**(6237), p. 871.

Puping, L., Yanwen, X., Xiya, Z., Chenhui, D., Rui, H., Zhen, Z., Jie, L., Xiaowei, X., Yuxi, C., Yujing, L., Ying, S., Yaofu, B., Zhou, S., Wenbin, M., Canquan, Z., Junjiu, H. (2015). CRISPR/Cas9-Mediated Gene Editing in Human Tripronuclear Zygotes, *Protein & Cell*, **6**(5), pp. 363–372, http://dx.doi.org/10.1007/s13238-015-0153-5. Accessed 25 September 2017.

Puppnick, G. (2013). *Synthetic Analysis of the ECJ Case C-34/10 Oliver Brüstle v Greenpeace e.V. and Its Ethical Consequences*, European Centre for Law and Justice, https://7676076fde29cb34e26d-759f611b127203e9f2a0021aa1b7da05.ssl.cf2.rackcdn.com/eclj/Synthetic%20analysis%20of%20the%20ECJ%20case%20of%20Br%C3%BCstle%20v%20Greenpeace%20and%20its%20ethical%20consequences.pdf. Accessed 25 September 2017.

Reardon, S. (2016). First CRISPR Clinical Trial Gets Green Light from US Panel, *Nature* (22 June) http://www.nature.com/news/first-crispr-clinical-trial-gets-green-light-from-us-panel-1.20137. Accessed 25 September 2017.

Regalado, A. (2017). One Man's Quest to Hack His Own Genes. When Brian Hanley Set Out to Test a Gene Therapy, He Started with Himself, *MIT Technology Review*, https://www.technologyreview.com/s/603217/one-mans-quest-to-hack-his-own-genes/. Accessed 25 September 2017.

Ruggiu, D. (2013). A Rights-Based Model of Governance: The Case of Human Enhancement and the Role of Ethics in Europe, in: K. Konrad, C. Coenen, A. Dijkstra, C. Milburn, and H. Van Lente (eds.), *Shaping Emerging Technologies: Governance, Innovation, Discourse*, Berlin: IOS Press/AKA, pp. 103–115.

Ruggiu, D. (2018). Implementing a Responsible, Research and Innovation Framework for Human Enhancement According to Human Rights: The Right to Bodily Integrity and the Rise of 'Enhanced Societies', *Law, Innovation & Technology*, **10**(1), pp. 1–40.

Weschka, M. (2009). Chimbrids and International Law, in: J. Taupitz, and M. Weschka (eds.), *CHIMBRIDS – Chimeras and Hybrids in Comparative European and International Research: Scientific, Ethical, Philosophical and Legal Aspects,* Berlin, Heidelberg: Springer, Verlag, pp. 98–115.

Wu, J., Platero-Luengo, A., Sakurai, M., Martinez, E.A., Ross, P.J., and Izpisua Belmonte, J.C. (2017). Interspecies Chimerism with Mammalian Pluripotent Stem Cells, *Cell,* **168**, pp. 473–486.

Chapter 5

Right to Health

5.1 Introduction

5.1.1 The application of emerging technologies often also raises issues of safety, which directly involve the right to health. Nanotechnologies, example, are clearly illustrative of this complexity. Despite the incredible number of opportunities, to which nanotechnologies have paved the way,[1] studies have revealed the existence of a criticality regarding the toxicology and the ecotoxicology of some engineered nanomaterials. For example, there are studies addressing the allegation that carbon nanotubes (both single and multi-walled) can cause lung inflammations (Shvedova et al., 2005), as well as granulomas once insert into abdomen of mice (Poland et al., 2008)[2];

---

[1] For example, there are already nanotechnological drugs used in anti-cancer therapy in nanomedicine (e.g. Myocet) that are commercialized within the EU (Hafner et al., 2014).

[2] With regard to carbon nanotube toxicology, it can be added that there is also a promising study in nanomedicine, which has identified an enzyme produced by some kinds of white blood cells (myeloperoxidase) that can biodegrade them by decomposing the carbon nanotubes into two innocuous elements, water and carbon dioxide (Kagan et al., 2010).

Human Rights and Emerging Technologies: Analysis and Perspectives in Europe
Daniele Ruggiu
Copyright © 2018 Pan Stanford Publishing Pte. Ltd.
ISBN 978-981-4774-93-2 (Hardcover), 978-0-429-49059-0 (eBook)
www.panstanford.com

other studies address dangers to health and the environment with regard to silver nanospheres (Mwilu et al., 2013), titanium dioxide nanoparticles included in several sun creams (Zhang et al., 2007), or the use of nickel nanoparticles in the working environment which could cause allergic reactions in a setting without any specific respiratory protection or control measures (Journeay and Goldman, 2014). Therefore, in these cases the protection of the right to health is crucial.

The right to health is a cluster of rights which covers several aspects such as the right to life, access to healthcare services, self-determination, the right to information, privacy and the protection of the confidentiality and, finally, the right to bodily integrity. These rights are all closely intertwined and can together be seen as having two essential aspects: a negative aspect (the protection of the physical and mental state of wellness of the individual) and a positive one (the right to access to healthcare) (Foà, 1998, p. 58). In order to understand exactly the content of the right to health we need therefore to make an accurate act of mapping that involves all these aspects.

First, the right to health is a human right. It is established by Article 25 of the Universal Declaration of Human Rights[3] which states that '[e]veryone has the right to a standard of living adequate for the health and well-being of himself and of his family, including ... medical care'.[4] Then, in more precise terms, the International Covenant on Economic, Social and Cultural Rights[5]

---

[3] ONU (1948) *Universal Declaration on Human Rights* (UDHR), adopted in New York on 10 December 1948 by the General Assembly of the United Nations.

[4] Article 25
1. Everyone has the right to a standard of living adequate for the health and well-being of himself and of his family, including food, clothing, housing and medical care and necessary social services, and the right to security in the event of unemployment, sickness, disability, widowhood, old age or other lack of livelihood in circumstances beyond his control.
2. Motherhood and childhood are entitled to special care and assistance. All children, whether born in or out of wedlock, shall enjoy the same social protection.

[5] ONU (1966) *International Covenant on Economic, Social and Cultural Rights* (ICESCR), adopted in New York on 16 December 1966 by the General Assembly of the United Nations (entered into force on 3 January 1976).

recognizes the right to health in Article 12[6] ensuring 'the enjoyment of the highest attainable standard of physical and mental health' to everyone. In this regard the right to health covers both health as an individual right and health as a policy which governments have to implement in order to enhance health conditions of the population. The UNESCO Universal Declaration on Bioethics and Human Rights[7] recalls health as a privileged field of public policy (Art. 14).[8] But what are the boundaries of the right to health in Europe?

It is worth noting that the right to health is not specifically recognized by the ECHR. The ECHR sets out only civil and political rights, not social rights such as the right to health. The right to health therefore should not have any formal legal basis in the ECHR. This represents a serious obstacle to the recognition of the right to health

---

[6] Article 12

1. The States Parties to the present Covenant recognize the right of everyone to the enjoyment of the highest attainable standard of physical and mental health.

2. The steps to be taken by the States Parties to the present Covenant to achieve the full realization of this right shall include those necessary for:

(a) The provision for the reduction of the still birth-rate and of infant mortality and for the healthy development of the child;

(b) The improvement of all aspects of environmental and industrial hygiene;

(c) The prevention, treatment and control of epidemic, endemic, occupational and other diseases;

(d) The creation of conditions which would assure to all medical service and medical attention in the event of sickness.

[7] UNESCO (2005) *Universal Declaration on Bioethics and Human rights*, adopted by acclamation by the 33rd session of the General Conference of the UNESCO in 19 October 2005.

[8] Article 14 Social responsibility and health

1. The promotion of health and social development for their people is a central purpose of governments that all sectors of society share.

2. Taking into account that the enjoyment of the highest attainable standard of health is one of the fundamental rights of every human being without distinction of race, religion, political belief, economic or social condition, progress in science and technology should advance:

(a) access to quality health care and essential medicines, especially for the health of women and children, because health is essential to life itself and must be considered to be a social and human good;

(b) access to adequate nutrition and water;

(c) improvement of living conditions and the environment;

(d) elimination of the marginalization and the exclusion of persons on the basis of any grounds;

(e) reduction of poverty and illiteracy.

in this context, as well as to the act of mapping which we need in this case. However, within the Council of Europe framework the European Social Charter (ESC)[9] and the Oviedo Convention[10] have several provisions with regard to the right to health which could be recalled.

The ESC was brought into existence in order to protect social rights such as the right to health. It was signed in Turin in 1961 and subsequently followed by three protocols (in 1988, 1991, 1995) before being revised in 1996. According to it, States undertake to be bound by at least some of the articles of the Charter (six specific articles, as well as an additional set of articles or numbered paragraphs[11]). These provisions, at least some of them, might be relevant to emerging technologies, such as nanotechnologies. For example, the ESC prescribes a right to safe and healthy working conditions (Art. 3 Part II), which might be affected by the use of nanotechnologies (e.g. nickel nanoparticles) by workers without due safeguards. As known, in 2008 two female workers in a paint factory in China died presumably due to exposure to nanoparticles without any precautionary measure (Song et al., 2008). The improvement of working conditions to the new challenges of the use of nanotechnologies in the workplace is therefore crucial in this regard. Nanotechnologies can also enhance the level of healthcare assistance. In this instance, the ESC sets out a right to benefit from any measures which enable individuals to enjoy the highest possible standard of health (Art. 10 Part II), as well as the right to social and medical assistance (Art. 13 Part II), both of which could be pertinent if nanomedicine was to become further spread in healthcare. The effectiveness of the ESC is however impaired by some shortcomings that can directly reduce the possibilities of implementing the right to health. The impact of the Charter, indeed, is limited by the lack of

---

[9]Council of Europe (1961/1996) *European Social Charter* (ESC) (CETS n. 35), adopted in Turin on 3 October 1961 (entered into force on 26 February 1965) revised in Strasbourg on 3 May 1996 (CETS n. 163) (entered into force on 1 July 1999).

[10]Council of Europe (1997) *Convention for the Protection of Human Rights and Dignity of the Human Being with regard to the Application of Biology and Medicine* (Convention on Human Rights and Biomedicine or the Oviedo Convention) (CETS n. 164), adopted in Oviedo on 4 April 1997 (entered into force on 1 December 1999).

[11]This possibility to choose which article should be implemented by a State led some to talk of *menu à la carte*, as reported by Zaghì (2002, p. 217).

means of an enforcement mechanism. While the additional Protocol in 1995 established a system of collective complaints to be lodged to a Committee of Independent Experts that is then responsible for drawing up a report (Art. D Part IV),[12] no judicial organ, even the Strasbourg Court, is able to apply any rule of the Charter. This might leave the right to health without protection in Europe.

The Oviedo Convention meanwhile, safeguards health in the field of biomedicine. It is the main legally binding document in international law for the protection of human dignity and personal integrity in the application of biomedicine in Europe. This makes its provisions particularly interesting with regard, for example, to nanotechnologies. These additional protocols on several matters (namely, those on the prohibition of the human cloning,[13] on human transplantation,[14] and on biomedical research,[15] on genetic testing for health purposes[16]) form a galaxy around the Convention, which deepens several aspects that are crucial for the protection of health of individuals and that can be recalled by the Strasbourg Court when interpreting the ECHR. Moreover, there are also documents different from additional protocols such as recommendations of the Committee of Ministers or the Parliamentary Assembly of the Council of Europe (e.g. xenotransplantation[17] and nanotechnology[18]) which can be refereed to when the individual health is at stake. This

---

[12] Ibid.

[13] Council of Europe (1998) *Additional Protocol to the Convention on Human Rights and Biomedicine on the Prohibition of Cloning Human Beings* (CETS n. 168), adopted in Paris on 12 January 1998 (entered into force on 1 March 2001).

[14] Council of Europe (2002) *Additional Protocol to the Convention on Human Rights and Biomedicine concerning Transplantation of Organs and Tissues of Human Origins* (CETS n. 186), adopted in Strasbourg on 24 January 2002 (entered into force on 1 May 2006).

[15] Council of Europe (2005) *Additional Protocol to the Convention on Human Rights and Biomedicine concerning Biomedical Research* (CETS n. 195), adopted in Strasbourg on 25 January 2005 (entered into force on 1 September 2007).

[16] Council of Europe (2008) *Additional Protocol to the Convention on Human Rights and Biomedicine concerning the genetic testing for health purposes* (CETS n. 203), adopted in Strasbourg on 11 November 2008 (it has not entered into force yet).

[17] Council of Europe (1997) Recommendation RE (97) 15 of the Committee of Ministers of the Council of Europe *on Xenotransplantations*, adopted on 1997; Council of Europe (2003); Council of Europe (2003) Recommendation RE (2003)10 of the Committee of Ministers of the Council of Europe *on Xenotransplantations*, adopted on 19 June 2003.

[18] Council of Europe (2013) Recommendation 2017 (2013) of the Parliamentary Assembly of the Council of Europe Nanotechnology: *Balancing Benefits and Risks*

promising framework can be of interest, therefore, for the right to health.

While the Oviedo Convention only sets out general principles, the protocols deal with specific matters in more detail. In general, the Oviedo Convention lays down three levels of protection (de Salvia, 2000, p. 107ff.). The first protects the individual from the misuse of scientific progress. The second protects the individual as a member of mankind. The third protects the individual by ensuring the involvement of the public on questions of scientific progress.

As noted, the Oviedo Convention has several aspects that are relevant to the right to health. First, Article 1 protects the dignity and identity of the individual, which might be called into question by any medical treatment performed without the consent of the individual, especially when it has experimental nature such as in the case of nanotechnologies. Article 2 meanwhile is noteworthy as it affirms the prevalence of the individual over the *sole* interest of society or science,[19] meaning the interests of a human being can only be subjugated in certain conditions (e.g. the coexistence of a plurality of public interests, such as the protection of public health and of the right to health of the individual in vaccination). Further, given the strong connection between a therapy and notions of individual choice, the Convention's emphasis on the importance of consent in biomedical interventions is pertinent (Arts. 5, 6, 7, 9, 17). The protection of personal data concerning health (Art. 10) may also be important, with special care required to ensure confidentiality for patients. The improvement of diagnostics thanks to nanotechnologies (Acimovic Srdjan et al., 2014) may greatly improve the quality of data collected but also pose new threats to privacy of patients. The nearly infinite applications of nanotechnologies can also be incorporated in the study of DNA by improving gene therapy, as well as diagnostics. New DNA tests, for

---

to *Public Health and the Environment,* adopted by the Assembly on 26 April 2013 (18th sitting).

[19]Article 2 – Primacy of the human being
The interests and welfare of the human being shall prevail over the *sole* interest of society or science.

example, use nanotechnologies to find early signs of cancer (Johns Hopkins University, 2009). In general, the Convention restricts predictive genetic testing only to those being tested for health purposes or for the use in scientific research linked to health purposes, and always subject to appropriate genetic counselling (Art. 12). In addition, biomedical research also raises questions about the application of the non-discrimination principle for (Art. 11), since individuals could be discriminated against as a result of their own characteristics, even genetic which can depend by the use of nanotechnologies (Kumar et al., 2016). Nanotechnologies are used for the cryopreservation of embryos in *in vitro* fertilization and other assisted reproductive technologies (Khosla et al., 2017), as well as in embryonic stem cell research (Chen, Qiu and Li, 2014) and in gene therapy (Liu and Zhang, 2011). These promising studies also raise important ethical questions. In this regard, Article 13 limits interventions on the human genome to preventive, diagnostic or therapeutic purposes. Similarly, Article 14 prohibits using assisted reproductive technologies in order to select the sex of descendants, except where it is to avoid serious hereditary sex-related diseases. Furthermore, the Convention prescribes some limits on human embryo research (Art. 18), in particular on creating human embryos for research purposes. In this regard, Article 4[20] highlights the imperative that any health related interventions in biomedical research are carried out in accordance with professional standards and obligations which might be improved by the applications of nanotechnology. This provision must be linked to that which provides for equitable access to health care (Art. 3),[21] and which raises the question of distributive justice with the spread of nanotechnologies in healthcare.[22]

---

[20] Article 4 –Professional standards
  Any intervention in the health field, including research, must be carried out in accordance with relevant professional obligations and standards.
[21] Article 3 –Equitable access to health care
  Parties, taking into account health needs and available resources, shall take appropriate measures with a view to providing, within their jurisdiction, equitable access to health care of appropriate quality.
[22] On this see Lev (2011) 25(4) Brownsword (2009, p. 136).

Although the ESC and the Oviedo Convention can be pertinent when nanotechnologies are at stake, they both have some shortcomings.

First, although the Oviedo Convention has entered into force, many key European States either did not sign it (for example, Austria, Belgium, Germany, Ireland, Liechtenstein, Russia, the United Kingdom), or, when they signed it, have not ratified the Convention yet (for example, France, Italy, The Netherlands, Poland, Sweden, and Ukraine).[23]

Second, there is no enforcement mechanism provided for both instruments. While the ESC provides only the measure of collective complaints, in the framework of the Oviedo Convention, the Strasbourg Court has a mere interpretative role – to clarify the meaning of its provisions (Art. 29). The Court cannot apply its provisions for censuring their violations. Therefore, although the Oviedo Convention is ratified by a State, there is no consequence for breaching it, and no sanction can be imposed on any State by a court. Since both the ESC and the Oviedo Convention lack any mechanism of enforcement, the right to health is at risk of being left unprotected by any judicial body. Fortunately, this outcome is avoided thanks to the jurisprudence of the Strasbourg Court.[24]

5.2 The Right to Health's Entry to the Sphere of the ECHR

5.2.1 While both the ESC and the Oviedo Convention cannot be the legal basis for the application of the right to health by the Strasbourg Court, the right to health has however entered into the ECHR in two ways.

First, they may have an indirect impact on the case law of the Strasbourg Court by affecting the interpretation of the ECHR provisions by reflex (*par ricochet* in French). The judges at Strasbourg stated that 'the Court took account, in interpreting [...] the Convention, of the standards enshrined in the Oviedo

---

[23] http://www.coe.int/en/web/conventions/full-list/-/conventions/treaty/164.
[24] See e.g. Jurisconsult's Department (June 2015).

Convention on Human Rights and Biomedicine of 4 April 1997, even though that instrument had not been ratified by all the States parties to the Convention'.[25] This happened especially in ethically controversial cases, as is likely in the case of emerging technologies such as nanotechnologies.[26] Therefore, where there is no ECHR provision for solving a hard case, soft law instruments such as the Oviedo Convention can play a decisive role. This is the case, for example, when human dignity is at stake in the instance of human genetic modification and when future generations rights are concerned. According to the Court, the Convention must be read as a whole and interpreted (as well as implemented) in such a way as to promote internal consistency and harmony between its various provisions and the various values enshrined therein.[27] This implies not only that its provisions have to be put in the context of all the other ECHR norms, but also that even other soft law instruments can be used, if necessary, to help interpret the ECHR. In addition, the fact that a given instrument has not been ratified by a State might be not sufficient to render any part of the ECHR inapplicable. Since in the view of the Court the essence of the Convention must be preserved, the lack of ratification cannot be opposed to avoid the application of the ECHR provisions.[28]

Secondly, when the violation of the rights and freedoms guaranteed by the ECHR result in economic and social consequences, as in the case of the violation of the right to health, the Strasbourg Court

---

[25] *Demir and Baykara v. Turkey* (Appl. 34503/97) judgement of 12 November 2008 *Reports of Judgements and Decisions*, 2008, par. 81.

[26] E.g. *Lambert and others v. France* (Appl. 46043/14) judgement of 5 June 2015 *Reports of Judgements and Decisions*, 2015.

[27] *Ivi*, par. 142. See here also the separate opinion of judges Hajiyev, Šikuta, Tsotsoria, De Gaetano and Griţco.

[28] See e.g. *Abdulaziz, Cabales and Balkandai v. the United Kingdom* (App. 914/80, 9473/81, 9474/81) judgement of 28 May 1985 *Series A*, No. 94, par. 60: 'the Court recalls that the Convention and its Protocols must be read as a whole; consequently a matter dealt with mainly by one of their provisions may also, in some of its aspects, be subject to other provisions there of [. . .]. Thus, although some aspects of the right to enter a country are governed by Protocol No. 4 as regards States bound by that instrument, it is not to be excluded that measures taken in the field of immigration may affect the right to respect for family life under Article 8 (Art. 8)'. In this case the Protocol 4 was not ratified by the United Kingdom. Nevertheless, it was condemned for the violation of Article 8 ECHR.

has subject-matter jurisdiction. Starting from the Airey case the Strasbourg Court stated that '[w]hilst the Convention sets forth what essentially civil and political rights, many of them have implications of a social or economic nature'.[29] In this instance, the Court can directly recall the right to health through the ECHR provisions. As repeated by the judges at Strasbourg, the 'interpretation of the Convention may extend into the sphere of social and economic rights'. Therefore, although the ECHR formally protects only civil and political rights, thanks to the ECtHR jurisprudence the ECHR provisions also cover some social rights such as the right to education, the right to the environment and, crucially, the right to health. This is the main way through which the right to health has been protected within the ECHR, and thus also where the right to bodily integrity derives from.

In these instances the Strasbourg Court started to refer to Article 2 of the ECHR (which protects the right to life),[30] Article 3 (prohibition of inhuman and degrading treatment),[31] and Article 8 (respect for private life).[32] First, when a person is forced to undergo medical treatment against his will, it can be considered either as

---

[29]*Airey v. Ireland* (Appl. 6289/73) judgement of 9 October 1979 *Series A*, No. 32, par. 26. See also Foà (1998, pp. 57–93); Olivieri (2008, pp. 509–539); Gitti (1998, pp. 719–735).

[30]Article 2 – Right to life

1. Everyone's right to life shall be protected by law. No one shall be deprived of his life intentionally save in the execution of a sentence of a court following his conviction of a crime for which this penalty is provided by law.

2. Deprivation of life shall not be regarded as inflicted in contravention of this article when it results from the use of force which is no more than absolutely necessary: a. in defence of any person from unlawful violence; b. in order to effect a lawful arrest or to prevent the escape of a person lawfully detained; c. in action lawfully taken for the purpose of quelling a riot or insurrection.

[31]Article 3 – Prohibition of torture

No one shall be subjected to torture or to inhuman or degrading treatment or punishment.

[32]Article 8 – Right to respect for private and family life

1. Everyone has the right to respect for his private and family life, his home and his correspondence.

2. There shall be no interference by a public authority with the exercise of this right except such as is in accordance with the law and is necessary in a democratic society in the interests of national security, public safety or the economic well-being of the country, for the prevention of disorder or crime, for the protection of health or morals, or for the protection of the rights and freedoms of others.

inhuman or degrading treatment (Art. 3), depending on its severity. Then, when a State's measure, on which basis the medical treatment is performed, is not an inhuman treatment, it might however be considered an illegal interference with the private life of the individual (Art. 8).[33] In this regard it is to be noted that Article 8 not only protects the individual's privacy, home, correspondences and personal data, but also private life much more broadly construed, including their autonomy, self-determination, and control over their own body.

When confronted with the challenges posed by the society, the ECHR may be forced to extend its provisions further. As the Strasbourg Court has reiterated, the ECHR is a 'living instrument' evolving according to changes in society.[34] The impact of emerging technologies will greatly transform our society and the needed flexibility, which the act of the Strasbourg Court gives to the ECHR provisions, appears even more crucial today.

5.3 The Strasbourg Court Jurisprudence Related to Health

5.3.1 Since the right to health is a bunch of rights covering several different aspects, which will be singularly deepened in following chapters, here I will only propose a brief overview of the different dimensions forming the right to health. Therefore, in subsequent chapters the reader will find further in-depth analyses about those dimensions that I mean to rapidly sketch here.

The first aspect that comes into play when we analyse the right to health is the consequences of a State's measure concerning the health on the personal integrity of the individual. In this instance one can refer to both Article 2 of the ECHR, which protects the right to life, Article 3 (prohibition of inhuman and degrading treatment), and Article 8 (respect for private life). When a medical treatment,

---

[33]*Juhnke v. Turkey* (Appl. 1620/03) judgement of 23 September 2010 *Reports of Judgements and Decisions*, 2010.
[34]*Soering v. the United Kingdom* (Appl. 14038/88) judgement of 7 July 1989 *Series A*, No. 161, par. 105.

even innovative due to the application of ultimate results of scientific research, is performed against the will of the person, the right to bodily integrity is at stake and the *Yhunke against Turkey case*[35] is pertinent. This is a case of a female detainee whom was charged of aiding an illegal organization, the Workers' Party of Kurdistan (PKK). During her detention in a Turkish prison she was precluded to go before a judge for nine days and finally she was subjected to a forced gynaecological exam in front of the male military personnel. The Turkish government argue that the examination was performed on voluntary basis the applicant stated the opposite and complained therefore the violation of Article 3 ECHR since this treatment was an instance of ill-treatment and degrading. The Strasbourg Court ruled that, according to documentary evidence submitted by both parties, on charges of ill-treatment (being threatened with death, left waiting up and blindfolded for long periods of time) covered by Article 3 should be archived because charges were unproven while the issue of the gynaecological examination had to be thorough in this regard. Nevertheless, when a given measure falls outside the application of Article 3, this can fall anyway under Article 8 (respect to private and family life which covers also the right to bodily integrity). In this sense, a medical treatment conducted against of the subject's will constitutes an interference with regard to the respect for private and family life of the individual and, in particular, with regard to their integrity. In the opinion of the Court under Turkish legislation a forced gynaecological examination was a violation of personal integrity except in cases of need for medical intervention where expressly provided by law. The case shows that the disputed gynaecological examination does not meet the standard procedures for medical examinations of those arrested and detained, and that it represents a discretionary decision not subject to any procedural requirement taken by the authority at the sole protection of the military personnel who arrested the applicant against any false allegations of violence and sexual harassment. Moreover, if it ever was in accordance with the law, it would have also been disproportionate to the aim provided by the law. In this sense,

---

[35]*Juhnke v. Turkey* (Appl. 1620/03) judgement of 23 September 2010 *Reports of Judgements and Decisions*, 2010.

the measure was not necessary in a democratic society. The medical treatment performed against the will of the patient was therefore deemed as a violation of the Convention (Art. 8) by a Court's decision, which expresses an essential principle in biomedical field.

5.3.2 The refusal of medical treatment, especially in the case of moral objection, appears a significant issue in the liberal-democratic society, since it goes to the core of the relation which join together consent and personal integrity. The defence of the will of the patient can find though a limit in some prevalent public interests. In *Acmanne and others against Belgium*[36] the applicants were sentenced for refusing to undergo, and to let their children to undergo, methods for tuberculosis screening authorized by Belgian law, namely the tuberculin test and a radiological examination of the throat (a chest x-ray). In this case they complained an interference with their private and family life (Art. 8), especially since these tests were not required outside Belgium. In dismissing the complaint as manifestly ill-founded the Commission first recognized that the medical treatment was an interference with the private life of the applicants. Therefore, it analysed in more detail the nature of this interference in order to see whether it was according to the domestic law, as well as necessary in a democratic society. In fact, the interference was taken in accordance with Belgian law, namely the Medical School Inspection Act of 1964 and the Royal Order of the same year. Furthermore, it was also necessary in a democratic society because there was proportionality between the legitimate aim pursued, the protection of public health, as well as health of the applicants, and the measure to strengthen the social duty of each individual not to endanger the health of others, where his life was not in danger. Moreover, while it is true that the same tests are not compulsory in several countries, this fact does not exclude that Belgium is entitled to do so, especially in the face of the amendments adopted in the Belgian regulation during and after in order to make less frequent tests against tuberculosis and increase the flexibility of regulation according to the present-day conditions of life. In this

---

[36] *Roger Acmanne and others v. Belgium* (Appl. 10435/83) decision of the Commission of 10 December 1984.

sense, the measure was assessed as necessary for the protection of public health in a democratic society.

5.3.3 As seen in the *Juhnke against Turkey case*[37] the consent to the medical treatment is crucial, especially in those treatment having experimental nature. In its initial jurisprudence, the Commission acknowledged that a medical treatment of experimental nature may constitute, in certain circumstances, an inhuman and degrading treatment under Article 3 of the Convention,[38] since the consensus is the *condicio sine qua* non of any clinical trial (Byk, 1999, p. 107). In this case, a Danish lady underwent an operation of sterilization by electric cauterization of ovarian ducts. The technique was in use for a couple of years. Therefore, it was quite old. In that occasion, however, the surgeon used a new model of pincers introduced just three months before which would have to avoid damage to the affected area. Once discharged, the patient was informed that it would be neither necessary preventive measures, or a check. A few weeks later, instead, the patient noticed that she was pregnant. A subsequent investigation showed that on 72 operations with these newly designed pincers 10 failed and for this reason the applicant sued the State for damages.

In the opinion of the applicant she underwent a surgical operation, when she was not previously told the changed instrument and for that reason she complained a violation of Article 3 (prohibition of inhuman and degrading treatment). The Commission, recalling that both for the inhuman treatment and to those degrading is required to pass a certain threshold of severity, noted that the applicant agreeing to undergo the operation of sterilization had been informed that there was a 1 or 2% of possibility that it could go wrong. Besides, the operation was also conducted in accordance with the methodology established until 1973. Furthermore, the replacing of the model of pincers changed only slightly the technique in order to minimize their negative effects. The new model of pincers therefore did not change the agreed methodology. In these circumstances according to the Commission, the intervention of an experimental

---

[37]*Juhnke v. Turkey* (Appl. 1620/03) judgement of 23 September 2010 *Reports of Judgements and Decisions*, 2010.

[38]*X v. Denmark* (Appl. 9974/82) decision of the Commission of 2 March 1983 *Decision and Reports*, 32.

nature as such, even if conducted without the consent of the patient, fell under the scope of treatment previously authorized and thus it could not be considered in violation of Article 3. In this sense, the decision of the Commission affirmed the principle that any variation in the medical treatment request the consent of the patient, unless the introduced novelty aligns in the same methodology already subjected to information.

5.3.4 The protection of health is closely linked to the quality and quantity of information available to the subject. Information becomes more crucial in front of the innovative nature of a treatment. The use of nanotechnologies, such as carbon nanotubes, is quite promising, especially in the biomedical field. For example, Alan G. MacDiarmid of the NanoTech Institute of University of Texas, Dallas, developed an aerogel made of carbon-nanotubes which is able to enhance performances of artificial muscles over 100 times heavier than can human muscle of the same length and weight, and generate 5.3 kilowatts of mechanical work per kilogram of muscle weight, similar to that produced by a jet engine (Haines et al., 2014). This application could be used in prosthetic technologies leading to forms of human enhancement of disabled people. At the same time, there are studies stating that carbon nanotubes (both single and multi-walled) can cause lung inflammations (Shvedova et al., 2005), as well as granulomas once insert into abdomen of mice (Poland et al., 2008). This imposes great prudence in the development of these applications, as well as in the correct communication with patients. The protection of health therefore strictly depends on the amount of the information available.

This principle is evident in the whole jurisprudence of the ECtHR, starting from that related to the termination of pregnancy. Termination of pregnancy is a typical ethical controversial question. Therefore, according to the case law of the Court on the abortion in these instances there is a wide discretion for the State, but cannot sacrifice or endanger the health of the person. In the *Open Door case*,[39] the Irish authorities, in view of protecting the lives of unborn children as provided for in the Constitution of Ireland, adopted some restrictive measures of healthcare of pregnant women that

---

[39] *Open Door Counselling Ltd and Dublin Well Woman v. Ireland* (Appl. 14234/88, 14235/88) judgement of 29 October 1992 *Series A*, No. 246-A.

was carried out by both public and private clinics. While in Ireland termination of pregnancy was prohibited (although with some exceptions), the assistance and counselling of pregnant women was legal. In addition, it should be noted that to go abroad in order to terminate an unwanted pregnancy was not a crime. In this case the limitation of information about the availability of abortion outside the national borders, when the pregnancy was at an advanced stage, could seriously endanger women's health. For this reason, the applicants raised the question of its compatibility with Article 10 (freedom of expression which includes the freedom to give and receive information without interference by public authority and regardless of frontiers), as well as with Article 8 (respect for private and family life) and Article 14 (prohibition of discrimination regarding to the health of women and men in similar matters concerning the reproductive life of the people). The Irish Supreme Court granted an injunction restraining non-profit making organizations (Open Door Counselling Ltd and Dublin Well Woman), which provided information on abortion, came to affect from their activities. However, this directly affected the welfare of women in Ireland, by restricting the possibility of giving and receiving basic health information (Art. 10). Although, the restrictive measure was taken accordance with a domestic law, by pursuing a legitimate aim (namely the moral belief, protected by the Constitution, that the life of the unborn was sacred), in the opinion of the judges in Strasbourg the permanent restriction of information carried out by counselling organizations appeared disproportionate because it affected the essential health information for women who were so compelled to get this information otherwise, without any control and, ultimately, by creating a serious threat to their health. Therefore, the measure breached Article 8 of the Convention.

Vaccination is a field where information plays a crucial role. Scientific research in the field of vaccination certainly paves the way to new possibilities, but also may generate unprecedented risks. This field attracts the contribution of several emerging technologies ranging from nanotechnology to synthetic biology in order to make vaccines safer and more effective. Nanoscale size materials, for example, such as virus-like particles, liposomes, polymeric, and non-degradable nanospheres can be used as delivery vehicles for vaccine antigens. Some of these nanoparticles are able to enter antigen-presenting cells modulating thus the immune response to the disease. Several delivery systems have been studied for the development of new vaccines against a number of infectious diseases and today new great opportunities have been paved the way by this research field (Gregory, Titball, and Williamson, 2013). This, however, requests a further effort in the right communication of risks implied by these routes of research. Vaccination is therefore a very delicate field that involves both directly health of those who submit to it and information about the risks involved. In this regard, the Court acknowledged in a quite famous case,[40] that a mass vaccination campaign carried out in the UK, which had had tragic consequences for children with different damage and some deaths, fell under a positive State's obligation of correctly informing the population about the potential risks to health (Arts. 2 and 8 of the ECHR). This principle was affirmed although the application was rejected as manifestly ill-founded.

5.3.5 Within the technoscientific ambit it may be hard to determine the causal link between the cause and the damage on health, especially when the latest technology developments are at stake. In these instances, the scarcity of data and the increasing uncertainty related to a rising research field may increase the risks. Scarcity of information means that there is not enough knowledge to determine whether a certain application involves risks and what they exactly are. In these instances, it would be better to talk of scientific uncertainty tout court which must not be shifted on the responsibility of whom undergoes to a treatment, especially with experimental nature. Therefore, in these contexts the right to be informed plays a crucial role. The case discussed below is illustrative of role of information with regard to the etiological nexus and highlight the principle according to which public authorities have however a positive obligation to inform whom undergoes to experimental tests about the implied risks.

---

[40]*Association X v. The United Kingdom* (Appl. 7154/75) decision of the Commission of 12 July 1978 *Decision and Reports*, 14, p. 35.

In L.C.B.[41] the appellant, whose father in the sixties had done the service assistant catering to the army of His Majesty in the Pacific during a series of nuclear tests, had been diagnosed with leukaemia. First, it is worthwhile noting the centrality of the issue, raised by judges in Strasbourg, of the causal link between the risk exposure of the father and the onset of leukaemia in the applicant. In this case the application was entirely focused on the violation of Article 2 (right to life) that requires to be realized quite severe terms (e.g. murder of a person in custody by the authority) and requests therefore a high margin of proof. For, this complaint was rejected for the lack of proof for establishing a cause/effect relationship between radiation exposure of the father and the occurrence of leukaemia in his daughter.[42] The circumstances of the case appeared quite nuanced and they were mainly focused on the causal link between the onset of the disease and the exposure of the applicant's father. This causation nexus then was clearly difficult to determine. In addition, the applicant complained that the absence of warning by the British authorities of the danger to her parents, had prevented a timely diagnosis of leukaemia in violation of her right to life. The survey, however, concluded that it was highly uncertain that an early diagnosis in utero could monitor the health of the foetus or that a medical intervention at that stage would have lessened the severity of the disease then contracted by the applicant.[43] For this reason, the application with regard to the violation of Article 2 (right to life) was immediately rejected. However, the Court recognized that, though the applicant's belief that there was a causal link between the significant exposure of the

---

[41] *L.C.B. v. The United Kingdom* (App. 23413/94) judgement of 9 June 1998 *Reports*, 1998-III.

[42] Ibid., par. 37, 39.

[43] Ibid., par. 40: 'in the light of the conflicting evidence of Dr Bross and Professor Eden (see paragraphs 29 and 33 above), and as the Commission also found (see paragraph 34 above), it is clearly uncertain whether monitoring of the applicant's health in utero and from birth would have led to earlier diagnosis and medical intervention such as to diminish the severity of her disease. It is perhaps arguable that, had there been reason to believe that she was in danger of contracting a life threatening disease owing to her father's presence on Christmas Island, the State authorities would have been under a duty to have made this known to her parents whether or not they considered that the information would assist the applicant'.

father and her leukaemia was justifiable, there was for the British authorities a positive obligation to inform about the risks arising from participation testing in the Pacific. In theory this latter aspect could fall under Article 8 (respect for private life), but not having been raised it earlier during the proceedings of first instance in front of the Commission, the Court was not competent to examine it since the proceeding before the Commission determines the scope of the competence of the Court. For this reason, the application was finally rejected. However, notwithstanding it was not originally included in the applicant's claim, the Court ruled the principle according to which the State authorities were under the duty to make her parents known about the danger for the offspring of contracting health threatening diseases arising from the father's participation in testing in the Pacific.

5.3.6 The need of further information due to the experimental nature of a biomedical trials is the object of a right. The use of nanotechnologies has gain the attention of large internet companies which wants develop new services in the healthcare field. For example, large companies, such as Google, are researching systems of self-screening which use nanoparticles (e.g. tiny iron oxide particles) in the human body that can be monitored through a wearable device such as a watch or a sports tracker (Barr and Wilson, 2014). This search raises concerns not only with regard to their safety, which must be fully tested, but also with regard to the information delivered to potential patients. The development of science and technology requires therefore the predisposition of a series of tests that must take place in respect of certain protocols and some fundamental principles of the ECHR.

The *Roche case*[44] is perhaps the main case of clinical trials decided by the judges in Strasbourg. Michael Roche was a man who between 1953 and 1968 during his military service participated in the chemical experiments in the Chemical and Biological Defence Establishment at the city of Porton Down. These experiments, which entailed tests on both animals and men, and the exposure

---

[44] *Roche v. The United Kingdom (App. 32555/96)* judgement of 19 October 2005 *Reports of Judgements and Decisions*, 2005-IX.

to 'mustard' gas (sulfide dicloroetile), and nerve gas, caused him several pathologies.

The 1962 tests served to try some protective materials against mustard gas in view of manufacturing suits. Once discharged, the medical tests done on him witnessed the onset of complications such as bronchial asthma, bronchitis, hypertension and high blood pressure, so much so that in 1988 he was recognized as invalid. The firm belief that there was a relationship of cause and effect between experiments '62 and '63 and the worsening of his health led him to request all personal medical reports available by the Ministry of Defence relating to that period. But they were not given to him. Besides, in 1991, the request for a disability pension was denied due to the lack of proof of a causal link between the tests with 'mustard gas' and the nerve gas and his health state. Therefore, those medical reports were of crucial importance for Michael Roche and is application for a disability pension. During the appeal process the records on tests at Porton Down were made public, though only in part. Also this process, where the medical records were issued, did not detect any cause/effect relationship between the experiments and the state of disability. For this reason, the appeal was rejected. This led Michael Roche to complain that the impossibility of bringing an action for damages against the Crown under Section 10 of the Crown Proceedings Act was a violation of paragraph 1 of Article 6 of the Convention (fair trial) and the stubborn refusal of medical information about his health state and the consequent impossibility of knowing whether the experiments which had undergone had or not long-term effects, was a violation of Article 8 (respect for private life). While the first instance of his application (the violation of Art. 6) was dismissed, since there already was a Law establishing a kind of pension for him, the question of the violation of the Convention under Article 8 was accepted because, if the causal link between the state of his disability and the experiments at Porton Down lacked of proof, it was certain that the continued denial of information on the health risks associated the tests caused a state of anxiety and suffering that certainly represented a restriction of his private life.[45] Therefore, from the Roche case we can derive the

---

[45] Ibid., par. 166.

principle that beyond the issues of safety and health on a given material or procedure a complete information is always requested when emerging technologies are at stake. This informative duty falls within the positive obligation of the State.

5.3.7 As said, the right to health entail both a negative aspect (the protection of the bodily integrity of the individual) and a positive aspect. In this latter case there is a positive obligation of the State to take action through appropriate legislative solutions, even of constitutional nature, so that the protection of the individual's good of health is effective. The lack of an adequate legislation could be therefore a source of responsibility for the State in certain circumstances. The rise of emerging technologies always entails a regulatory uncertainty, which, though, does not free the State to adopt concrete steps in order to minimize risks. In *Tysiac against Poland*[46] a woman who had already had two children and suffered from a severe form of myopia, intended to terminate her third pregnancy because, due to the high risk to her eyesight if the pregnancy were to continue. Three ophthalmologists had diagnosed a serious risk to her health, but the doctor of the state hospital where she was admitted, in spite of this dissenting opinion, forced her to give birth by caesarean section. So much so that after her delivery, her eyesight was rapidly deteriorating. On this occasion, the Court ruled that the right to respect for private life of the applicant (Art. 8 ECHR) had been violated when the public authorities had rejected the diagnosis of the three ophthalmologists (obligation to abstain from acting) and when they were not foreseen the appropriate decision-making processes that would protect its interest in the health (obligation to act).

5.3.8 The use of nanotechnologies in gene diagnosis and medically assisted procreation (MAP) can further improve the precision and effectiveness of current techniques (Khosla et al., 2017). These advances, however, entre the middle of a hot discussion which is open from years. The advance of techniques of MAP and preimplantation genetic diagnosis (PGD) led Italy to adopt a highly

---

[46] *Tysiąc v. Poland* (Appl. 5410/03) judgement of 3 March 2007 *Reports of Judgements and Decisions*, 2007-I.

restrictive law, Law 40, which was variously involved by several decisions of unconstitutionality by the Italian Constitutional Court. Its Article 4 limited the use of assisted reproductive technology (ART) 'only where proof is adduced that it is otherwise impossible to eliminate the causes of inability to procreate, and, in any event, [said access] shall be limited to medically certified inexplicable cases of sterility or infertility and to cases of sterility or infertility [deriving] from a medically certified and verified cause'. A subsequent decree of the Minister of Health in 2008 (No. 15165) extended practices medically assisted procreation even when the man is a carrier of sexually transmissible viral diseases by infection of HIV and hepatitis B and C that could infect both mother the foetus. This instance was therefore equated as those of male infertility.

In 2010, a married couple, Rosetta Costa and Walter Pavan,[47] healthy carriers of cystic fibrosis (or mucoviscidosis), a genetic disease that causes breathing difficulties and can be fatal, went before the Court complaining of not being able to access preimplantation genetic diagnosis for the purposes of selecting an embryo unaffected by the disease and alleged that the technique was available to categories of persons to which they did not belong. Therefore, they complained about the violation of Article 8 (respect for private and family life) and the principle of non-discrimination (Art. 14). Basically, the couple, who already had a daughter in 2006 suffering from cystic fibrosis and had to terminate a second pregnancy in 2010 for the same reason, for having a child would have to start a pregnancy by natural means and then terminate it if the prenatal test shows that the foetus is unhealthy. For this reason, the two applicants wanted access to the techniques of MAP and of PGD, but these techniques (those of MAP) were only available to sterile or infertile couples and, following the 2008 decree, also to fertile couples, whose the man is suffering from a sexually transmissible viral diseases that could access to artificial insemination after a process of 'sperm washing' prior to the *in vitro* fertilization (IVF) of the embryo, while the PGD was banned for any category of people. According to the Court, the Italian legal system apparently was lacking of coherence since, as also the report of the European

---

[47] *Costa and Pavan* (Appl. 54270/10) judgement of 28 August 2012.

Commission recalled, on the one hand prohibits the implantation limited to embryos unaffected by the disease of which the applicants are healthy carriers, on the other authorizes recurring to abort a foetus with the same disease.[48] The consequence for the private and family life of the applicants was that 'to have a child unaffected by the disease of which they are healthy carriers, the only possibility available to them is to start a pregnancy by natural means and then terminate it if the prenatal test shows that the foetus is unhealthy' (like it already happened in February, 2010).[49] In this context, one cannot but take into account the state of anxiety that the applicant experienced, since she was unable to access the preimplantation genetic diagnosis, by facing the risk that 'the child will be born with the disease or the suffering inherent in the painful decision to undergo, as the case may be, an abortion on medical grounds'.[50] Notwithstanding in the field of heterologous fertilization, due to the technology advance, States enjoy a margin of appreciation, this case is clearly about homologous fertilization. Therefore, the Court has to criticize the proportionality of the measure in the face of the fact that the applicants can only take the way of therapeutic abortion. This case involves among the 32 States Parties to the Convention only two other countries apart from Italy (namely Switzerland and Austria). For this reason, given the inconsistency of the Italian legal system and the consequences of its application, the interference with the applicants' right was disproportionate and there was a violation of the Convention. Instead, the application was dismissed as clearly ill-founded with regard to the violation of the prohibition of discrimination under the combined provisions of Articles 8 and 14 since, if it is true that the applicants were not sterile and infertile, though suffering from a genetic disease transmissible to the offspring, differently from sterile and infertile couples, following the 2008 decree, also from couples of which the man is a carrier of sexually transmissible viral diseases, the ban on preimplantation genetic diagnosis (PGD) affect all without discrimination, both the non-sterile couples and infertile, and suffering from a genetic

---

[48] Ibid., par. 59.
[49] Ibid., par. 57.
[50] Ibid., par. 59.

disease transmissible, and couples of which the man is a carrier of sexually transmissible viral diseases. Therefore, where there was no discrimination regarding access to PGD they would, however, have challenged the discrimination in the techniques of MAP. Following the *Costa and Pavan case*, we understand that although there are matters where States enjoy a space of discretion, even great, the limitation of the use of technoscientific progress in reproductive technologies has to pay a great attention to the rights concerned since it cannot lead to completely sacrifice reproductive rights of people by neglecting the proportionality test in accordance with to Article 8.

5.3.9 This principle is also confirmed in other fields of reproductive technologies. The progress of biomedical sciences, as in the case of surrogacy, also leads to a radical redefinition of social relations. In *Menneson against France*[51] the couple of Menneson family suffering from infertility went to the United States to procreate through surrogacy. In surrogacy, the pregnancy results from the transfer of an embryo created by *in vitro* fertilization (IVF). The embryos, produced by using the sperm of Mr. Menneson, were implanted in the uterus of another woman (who also provided for the ova) that, after nine months, gave birth to twins. A subsequent trial in California ruled that Menneson couple are parents. Mr. Menneson was therefore both the biological father and the social father, while the wife was only the social mother. This family relationship, however, was not recognized by the French law. The French authorities, suspecting the existence of a case of surrogacy arrangement, contested the birth certificate, which had previously been entered in the register of births, marriages and deaths at the French consulate in Los Angeles and the French Court of Cassation rejected the appeal of the spouses on the basis of the fact that in the Civil Code the contract of surrogacy is void. In addition, it claimed that there was not a violation of their private and family life under Article 8 of the ECHR because the act of invalidation of the birth certificate would have had no effect

---

[51]*Menneson v. France* (Appl. 65192/11) judgement of 26 June 2014 *Reports of Judgements and Decisions*, 2014.

regarding parental relationships of filiation of the twins with their natural parents under the laws of California. The applicants, for their part, complained about the violation of Article 8 (private and family life) because the cancellation, causing the detriment of the interests of minors, would have prevented them from obtaining, in France, the recognition of the parent-child relationship recognized abroad. In addition, they complained the discriminatory treatment between their children and other children (combined provisions of Articles 14 and 8) when their right to respect for their family life came to exist. The Strasbourg Court considered relevant Article 8 in two aspects. Firstly, with regard to the 'family life', since the Menneson couple created a social relationship with the children from the outset, completely indistinguishable from the biological family life as such. Secondly, with regard to the 'private life', since the identity is an integral part of the right in question and there is a direct relationship between the private life of children born through surrogacy techniques and the determination of their legal relationship. According to the judges in Strasbourg the interference of the French authorities which refused to recognize the legal relationship between parents and children took place in accordance with a national law which pursued the legitimate aim of protecting the health and the rights and freedoms of others. The authorities' refusal to recognize the legal relationship between the children born with the techniques of surrogacy and intended parents which is subjected to the techniques of artificial procreation was aimed at discouraging French citizens to undergo abroad in those techniques that are forbidden at home in order to protect both children and the surrogate mother. While the measure did not violate the right to the respect for family life of Mr. and Mrs. Menneson, with regard to the right of twins to respect for their private life, it should be noted that they are in a situation of uncertainty. In fact, although the French authorities were aware that the twins were identified as children of the Menneson family, did not recognize them the status under French law. This contradiction was to undermine the very identity of the children in the French society. Furthermore, although their biological father was French, they were facing a worrying uncertainty regarding the possibility of obtaining French citizenship, a situation which could have negative repercussions in terms of their identity. Moreover, in the current situation they could inherit from Menneson only by legate, meaning that their hereditary rights are less favourable than other French children. This fact deprives them of an additional component of their identity in relation to their parenthood. Thus, the refusal to recognize the family relationship with the twins affected not only the couple, but also the children. In this sense, their right to respect for private life also implied that everyone should be able to establish the essence of their identity, including family relationships, had been significantly affected. The fact that the intended father is also the biological father of the twins makes the dimension of the parental relationships particularly relevant in the context of their personal identity. In this case it could not be said that the best interests of those children was to deprive them of their family relationship with the surrogate parents when the biological reality of that bond had been legally established abroad and the children and the parents had sought the recognition at home. Article 8 was deemed violated since not only the full recognition had been refused by the cancellation of the registration of the certificate in the register of births, but according to the French law and the case law of the Court of Cassation was not possible both the declaration of paternity and the adoption, by preventing to establish any legal relationship with their biological father. This case therefore establishes the principle according to which the limitations by the law for a given technology cannot produce consequences that may affect the right to identity of the persons even in the instance where they resulted by its illegitimate use.

5.3.10 The will is at the centre of biomedical field and requests its respect in front of any medical treatment over one's own body. This emerges, in particular, in the field of end-of-life decisions in situations where the individual can have a reduced form of control over their own body.

On the euthanasia and assisted suicide there is the *Pretty case*[52] which represents a significant precedent in the ECtHR

---

[52] *Pretty v. the United Kingdom* (Appl. 2346/02) judgement of 29 April 2002 *Reports of Judgements and Decisions*, 2002-III.

jurisprudence. Dianne Pretty was a woman of 43 years' old who suffered from a neurodegenerative disease that affects the motor cells of the central nervous system (involving progressively all muscles: in particular, arts, respiratory system) and that paralyzed her from the neck down. Since there is no cure for the disease, which inevitably leads to death after onset of respiratory complications following a total decline in her speech and swallowing abilities, she wanted to end her sufferings and a condition that she deemed unworthy before the disease had total course.

However, due to her current situation, Dianne was not able to commit suicide by herself (fact that was not punishable by law British), requiring thus the assistance of others (fact that was, instead, punishable in the United Kingdom). She would have required the help of her husband, but he would have run the risk of being sentenced. For this goal, Dianne experienced several proceedings without any success. She therefore considered violated her rights guaranteed under the ECHR, in particular Article 2 (right to life). According to Diane Article 2 would not protect life in itself, but the right to life of the individual against any interference by public and private entities, which would also include the right to self-determination by choosing whether to live or not. Under Article 2, therefore, it would fall even the right to die and choose when to die without pain and with dignity that is merely the corollary to the right to life. In this sense, assisted suicide would be permitted under the ECHR, otherwise Switzerland, where it is regulated, would violate the Convention. All other instances were around the application for the recognition of a right to die.

By confuting the argument of Diane Pretty, in particular that there exists a right to die protected under the Convention, the judges in Strasbourg affirmed two principles. Firstly, that the State has a positive obligation to protect the life of the individual, especially they are in situations of great vulnerability. Secondly, that the individual has a right to take decisions over their life and their body. Therefore, notwithstanding the application was dismissed and the Diane Pretty's argument was rejected, the heart of her argument was acknowledged by the Court.

Dianne Pretty died on 11 May 2002 for respiratory failure just thirteen days after the verdict of the Court.

5.3.11 The application of these principles can be found in the *Lambert case*,[53] a case on the interruption of artificial hydration and nutrition of a terminally ill patient. In the field of end-of-life decisions, science and technology are much advancing by using deep thalamic stimulation for treating patients lying in a vegetative state present severe impairments of consciousness (Corazzol et al., 2017). These studies showed that it is possible to activate the thalamocortical network based on vagus nerve stimulation and to restore a minimum of consciousness in the patient. As known, the effectiveness of deep thalamic stimulation can be enhanced by using magneto-electric (ME) nanoparticles to artificially stimulate the neural activity deep in the brain (Yue et al., 2012). If these studies further advance, the ethical framework of end-of-life decisions will be, if possible, more challenging. In general terms, when the individual is not able to express their consent to the medical treatment, their will can be reconstructed through evidences, or it can be substituted with the will of legal representative (relatives, public authorities, etc.). However, it is in critical situations, where the patient is not able to express the consent, that the self-determination principle is reaffirmed.

In 2008, Vincent Lambert became quadriplegic after a blow to the head received in an automobile accident. Following the consultation procedure on the basis of the 2005 law 'Leonetti' on patients' rights and end of life, on 11 January 2014 the doctor who was treating Mr. Lambert decides to discontinue, on 13th January Vincent Lambert's artificial nutrition and hydration. In favour of this decision, his wife and one of his nephews expressed on the basis of his previous willingness expressed before the accident. His parents, his brother and his sister went before the courts to order the hospital to not discontinuing the nutrition and hydration and to transfer the patient to another facility. The court initially gave reason to parents and suspended the execution of the decision of the doctor while refusing the application for transfer of the patient to another facility.

Against this decision, the wife and the nephew appealed to the Conseil d'État that on the basis of a new medical report established

---

[53]*Lambert and others v. France* (Appl. 46043/14) judgement of 5 June 2015 *Reports of Judgements and Decisions*, 2015.

the legitimacy of the choice to stop the procedure of feeding and hydration of the patient. Parents and siblings went so urgently before the Strasbourg Court claiming that the discontinuance of the Vincent Lambert's nutrition and hydration would be contrary to the duties of the State to protect its citizens in accordance with Article 2 (right to life). Furthermore, by depriving him of nutrients and of 'hydration needed would violate also Article 3 (prohibition of inhuman and degrading treatment), and breach his personal integrity against Article 8 (privacy). Based on these considerations, the appellants requested the application of urgent provisional measures (cd. Interim Measures) under Article 39 of the Rules of Court.

In this case the Strasbourg Court acknowledged that the State adopted all needed measures, even legislative (i.e. a clear legal framework, such as the law 'Leonetti', for the process of decision-making), in order to comply with its positive obligation to protect the Vincent Lambert's life. Furthermore, it recalled the consent principle which falls under the protective shield of Article 8. In this contest, it is fully legitimate any effort to reconstruct the will of the patient by taking into account any evidence of the patient's previously expressed wishes.

5.3.12 Even in health matter the responsibility of the State can cover not only the activities of all its bodies (executive, legislative, and judiciary), but also the activities of private actors, who have gained increasingly importance in the healthcare organization of contemporary societies (Lear, Mossialos and Karl, 2010). This can be seen also in the recent evolution of the market with the spread of several high-tech products related to health for the large public. The increasing interest of private subjects, like companies, for health sphere of individuals can pave the way to new business opportunities, as well as a new generation of risks which must be considered by the State. In *Öneryildiz*[54] the Court examined the preliminary question of whether the State can be held responsible for dangerous activity conducted by private industry that managed

---

[54]*Öneryildiz v. Turkey* (App. 48939/99) judgement of 30 November 2004 *Reports of Judgement and Decisions*, 2004-XII.

a hazardous waste dump. On this occasion the Court observed that the State 'must govern the licensing, setting up, operation, security and supervision of the activity and must make it compulsory for all those concerned to take practical measures to ensure the effective protection of citizens whose lives might be endangered by the inherent risks'.[55] Among these measures, the right to information (also protected under Art. 8 ECHR, namely the right to a private and family life) has great relevance.[56] The Court, which had to judge whether Turkey was responsible for the death of thirty-nine persons caused by a methane explosion occurred in the landfill without the necessary security conditions, concluded that it had been infringed Article 2 of the Convention (right to life) (Xenos, 2003).

References

Acimovic Srdjan, S., Ortega, M.A., Sanz, V., Berthelot, J., Garcia-Cordero, J.L., Renger, J., Maerkl, S.J., Kreuzer, M.P., and Quidan, R. (2014). LSPR Chip for Parallel, Rapid, and Sensitive Detection of Cancer Markers in Serum, *Nano Letters*, **14**, pp. 2636–2641.

Barr, A., and Wilsonm, R. (2014). Google's Newest Search: Cancer Cells, *The Wall Street Journal*, 29 October 2014, http://www.wsj.com/articles/google-designing-nanoparticles-to-patrol-human-body-for-disease-1414515602. Accessed 2 October 2017.

Brownsword, R. (2009). Regulating Human Enhancement, *Law Innovation & Technology*, **1**, p. 136.

Byk, C. (1999). Bioéthique et Convention européenne des droits de l'homme, in: Pettiti, L.E., Decaux, E., and Imbert, P.H. (eds.), *La Convention européenne des droits de l'homme. Commetraire article par article*, Paris: Economica, pp. 101–121.

de Salvia, C. (2000) La Convenzione del Consiglio d'Europa sui diritti dell'uomo e la biomedicina, *I diritti dell'Uomo. Cronache e battaglie*, **11**(1–2), pp. 99–109.

---

[55] Ibid., par. 90.

[56] In this regard the Court judges have long recognized that there is a right to receive early warning information as a positive measure related to dangerous activity of both public and private industry. See, *Guerra and others v. Italy* (App. 14967/89) judgement of 19 February 1998 *Reports*, 1998-I.

Chen, L., Qiu, R., and Li, L. (2014). The Role of Nanotechnology in Induced Pluripotent and Embryonic Stem Cells Research, *Journal of Biomedical Nanotechnology*, **10**(12), pp. 3431–3461.

Corazzol, M., Lio, G., Lefevre, A., Deiana, G., Tell, L., André-Obadia, N., Pierre Bourdillon, P., Guenot, M., Desmurget, M., Luauté, J., and Angela Sirigu, A. (2017). Restoring Consciousness with Vagus Nerve Stimulation, *Current Biology*, **27**, R979–R1001.

Foà, S. (1998). Il fondamento europeo del diritto alla salute. Competenze istituzionali e profili di tutela, in: C.E. Gallo and B. Pezzini (eds), *Profili attuali del diritto alla salute*, Milano: Giuffrè, pp. 57–93.

Gitti, A. (1998). La Corte europea dei diritti dell'uomo e la Convenzione sulla biomedicina, *Rivista internazionale dei diritti dell'uomo*, **11**(3), pp. 719–735.

Gregory, A.E., Titball, R., and Williamson D. (2013). Vaccine Delivery Using Nanoparticles, *Front Cell Infect Microbiology*, **3**, p. 13.

Hafner, A., Lovric, J., Lakos, G.P., and Pepic, I. (2014). Nanotherapeutics in the EU: An Overview of the Current State and Future Directions, *International Journal of Nanomedicine*, **9**, pp. 1005–1023.

Haines, C.S., Lima, M.D., Li, N., Spinks, G.S., Foroughi, J., Madden, J.D.W., Kim, S.H., Fang, S., de Andrade, M.J., Göktepe, F., Göktepe, Ö., Mirvakili, S.M., Naficy, S., Lepró, X., Oh, J., Kozlov, M.E., Kim, S.J., Xu, X., Swedlove, B.J., Wallace, G.G., and Baughman, R.H. (2014). Artificial Muscles from Fishing Line and Sewing Thread, *Science*, **343**, pp. 868–872.

Johns Hopkins University (19 August 2009). New DNA Test Uses Nanotechnology to Find Early Signs of Cancer, *ScienceDaily*, www.sciencedaily.com/releases/2009/08/090817142847.htm. Accessed 28 September 2017.

Journeay, W.S., and Goldman, R.H. (2014). Occupational Handling of Nickel Nanoparticles: A Case Report, *American Journal of Industrial Medicine*, **57**(9), pp. 1073–1076.

Jurisconsult's Department (June 2015). *Health-related Issues in the Case-law of the European Court of Human Rights*, Council of Europe/European Court of Human Rights.

Kagan, V.E., Konduru, N.V., Feng, W., Allen, B.L., Conroy J., Volkov Y., Vlasova, I.I., Belikova, N.A., Yanamala, N., Kapralov, A., Tyurina, Y.Y., Shi, J., Kisin, E.R., Murray, A.R., Franks, J., Stolz, D., Gou, P., Klein-Seetharaman, J., Fadeel, B., Star, A., and Shvedova, A.A. (2010). Carbon Nanotubes Degraded by Neutrophil Myeloperoxidase Induce Less Pulmonary Inflammation, *Nature Nanotechnology*, **5**, pp. 354–359.

Khosla, K., Wang, Y., Hagedorn, Qin, M.Z., and Bischof, J. (2017). Gold Nanorod Induced Warming of Frozen Embryos Enhances Viability, *ACS Nano*, **11**(8), pp. 7869–7878.

Kumar, V., Palazzolo, S., Bayda, S., Corona, G., Toffoli, G., and Rizzolio, F. (2016). DNA Nanotechnology for Cancer Therapy, *Theranostics*, **6**(5), pp. 710–725.

Lear, J., Mossialos, E., and Karl, B. (2010). EU Competition Law and Health Policy, in: E. Mossialos, G. Permanand, R. Beaten, and T.K. Hervey (eds.), *Health Systems Governance in Europe: The Role of European Union Law and Policy*, Cambridge: Cambridge University Press, pp. 337–378.

Lev, O. (2011). Will Biomedical Enhancements Undermine Solidarity, Responsibility, Equality and Autonomy? *Bioethics*, **25**(4), pp. 177–184.

Liu, C. and Zhang, Na (2011). Nanoparticles in Gene Therapy: Principles, Prospects, and Challenges, in: G. Legname and S. Vanni (eds.), *Progress in Molecular Biology and Translational Science*, **101**, Cambridge, Ma, S. Diedo, Oxford, London, Elsevier, pp. 509–562.

Olivieri, F. (2008). La Carta sociale europea tra enunciazioni dei diritti, meccanismi di controllo e applicazioni delle corti nazionali. La lunga marcia verso l'effettività', *RDSS. Rivista del diritto della sicurezza sociale*, **8**(3), pp. 509–539.

Poland, C.A., Duffin, R., Kinloch, I., Mynard, A., Wallace, W.A.C., Seaton, A., Stone, V., Brown, S., Mac Nee, W., and Donaldson, K. (2008). Carbon Nanotubes Introduced into the Abdominal Cavity of Mice Show Asbestos-Like Pathogenicity in a Pilot Study, *Nature Nanotechnology*, **17**, pp. 423–428.

Shvedova, A.A., Kisin, E.R., Mercer, R., Murray, A.R., Johnson, V.J., Potapovich, A.I., Tyurina, Y.Y., Gorelik, O., Arepalli, S., Schwegler-Berry, D., Hubbs, A.F., Antonini, J., Evans, D.E., Ku, B.K., Ramsey, D., Maynard, A., Kagan, V.E., Castranova, V., and Baron, P. (2005). Unusual Inflammatory and Fibrogenic Pulmonary Responses to Single-Walled Carbon Nanotubes in Mice, *American Journal of Physiology Lung Cellular Molecular Physiology*, **289**(5), pp. L698–L708.

Song, Y., Xue, L., and Xuqin, D. (2009). Exposure to Nanoparticles is Related to Pleural Effusion, Pulmonary Fibrosis and Granuloma, *European Respiratory Journal*, **34**(3), pp. 1–31.

Xenos, D. (2003). Asserting the Right to Life (Art. 2 ECHR) in the Context of Industry, *German Law Journal*, **8**, pp. 231–254.

Yue, K., Guduru, R., Hong, J., Liang, P., Nair, M., and Khizroev, S. (2012). Magneto-Electric Nano-Particles for Non-Invasive Brain Stimulation, *PLoS One*, **7**(9), pp. e44040.

Zaghì, C. (2002). *La protezione internazionale dei diritti dell'uomo*, Torino: Giappichelli.

Zhang, X., Sun, H., Zhang, Z., Niu, Q., Chen, Y., and Crittenden, J.C. (2007). Enhanced Bioaccumulation of Cadmium in Carp in the Presence of Titanium Dioxide Nanoparticles, *Chemosphere*, **67**, pp. 160–166.

Chapter 6

Consent

6.1 Introduction

6.1.1 Consent is at the heart of biomedicine and of the right to health. In general, any medical intervention may only be carried out after the person concerned has given free and informed consent to it. This also entails the possibility of withdrawing the expressed consent at any time. This is because it is presumed that the individual has a special relation with their own body, which represents a sort of 'forbidden ground' for public powers. If the individual has a special sphere of personal nature (the private sphere), at least their body physically delimitates this sphere, which is, obviously, wider than their mere physicality. This is also linked to the dignity principle, which prohibits using another person merely as a means to an end, especially when someone commodifies the body of another individual even for a good aim (Jacobson, 2007, p. 293). This is the reason why the organs and tissues donation finds several limits even when it rests on voluntary basis.

This principle of bioethics strictly joins the principle of autonomy which is the architrave of our societies. In liberal-democratic societies autonomy has a special position and can be questioned by some developments of science and technology. Human enhancement

Human Rights and Emerging Technologies: Analysis and Perspectives in Europe
Daniele Ruggiu
Copyright © 2018 Pan Stanford Publishing Pte. Ltd.
ISBN 978-981-4774-93-2 (Hardcover), 978-0-429-49059-0 (eBook)
www.panstanford.com

(HE), for instance, is a field where autonomy has a special relevance since the individual can autonomously decide to enhance themselves. This makes this field an excellent starting point for analysing the topic of consent.

HE is a multidisciplinary field, which covers neuroscience, robotics, nanotechnologies, biotechnology, and synthetic biology, etc. (Roco and Brainbridge, 2003, p. 1ff.), that can give rise to provisional (e.g. pharmaceutical drugs) or permanent modifications of human performances (e.g. genetic enhancement) (Allhoff et al., 2010, p. 6). This makes the implementation of easy solutions difficult, as each technological sector carries different concerns and risks (as well as opportunities), and raises different questions for the protection of individual rights, ranging from the right to bodily integrity, to self-determination, to the right to information, the right to non-discrimination and the rights of future generations, etc.

We have already seen the impact of human genetic modifications over the principle of dignity.[1] Here we analyse the impact of HE with regard to the self-determination principle.

In liberal-democratic countries, where autonomy has a special relevance, human enhancement technologies (HETs) will likely become goods which can be purchased and sold (Brey, 2009, p. 175). Since there is no ban at the international level impeding the diffusion of HETs as such, but only limits for some specific forms of enhancement (human cloning,[2] human–animal hybridization,[3] genetic enhancement[4]), they will slowly conquer our markets and, together with them, our societies. The consumers' freedom of choice, which is strictly linked to autonomy, could therefore facilitate the settlement of this phenomenon at any level.

---

[1] See Chapter 4 Part II on Human Dignity.

[2] Council of Europe (1998) *Additional Protocol to the Convention on Human Rights and Biomedicine on the Prohibition of Cloning Human Beings* (CETS n. 168), adopted in Paris on 12 January 1998 (entered into force on 1 March 2001).

[3] Council of Europe (1986) Recommendation 1046(1986) *on the Use of Human Embryos and Foetuses for Diagnostic, Therapeutic, Scientific, Industrial and Commercial Purposes*, adopted on 19 and 24 1986 (13th and 14th Sittings) http://www.coe. int/t/dg3/healthbioethic/Texts_and_documents/INF_2014_5_vol_II_textes_%20Co E_%20bioéthique_E%20(2).pdf. Accessed 26 November 2017.

[4] Article 13 of the Oviedo Convention.

Furthermore, the development of assistive technologies, from which many HETs are derived, will greatly boost this sector. The Air Force Research Laboratory (AFRL), for example, is testing devices such as tDCS that is able to improve human performances in the field of aviation, aerospace and cyberspace.[5] These devices, engineered for disabled people but tested on healthy, give rise concerns since these people might hardly refuse the treatment due to their particular position and role such as in the case of military personnel. Assistive technologies show that drawing a clear-cut line here is likely to become increasingly difficult while progress advances. Indeed it may be even possible for assistive technologies to become so effective that therapeutic treatments for disabled people can enhance them beyond even that of a healthy person.[6] For example, in the field of prosthetic technology, the convergence of robotics and nanotechnology have enabled the production of an artificial skin, which offers disabled people a sense of touch.[7] This will allow people to interact with the environment in even more humanoid ways, bringing with them new and unimaginable opportunities for people with disabilities in the future (Wu et al., 2013). In a close future persons suffering from neuromuscular disorders could use smart exoskeleton technology machines,[8] power-assisted gauntlets[9] or other devices. Soon these prosthetic technologies, which are already on the market, may be more effective and widespread so that they will become indispensable to disabled for performing ordinary tasks one day. For instance, if wheelchairs are generally replaced by exoskeletons which are able to raise one's state of health to almost that of a healthy person, there will be the need of new adaptations

---

[5] http://www.bbc.com/future/story/20140603-brain-zapping-the-future-of-war. Accessed 26 November 2017.

[6] According to Brey (2009, p. 174) the advance in neuroprosthetics and robotics and the growth of artificial organs could lead prosthesis to perform better than normally functioning organs in future.

[7] The Defense Advanced Research Projects Agency (DARPA), for example, is working on an upper-limb prosthetic technology that is able to give the sense of touch to amputees, 'which offers increased range of motion, dexterity and control options'. See http://www.darpa.mil/program/revolutionizing-prosthetics.

[8] http://eksobionics.com/. Accessed 26 January 2017.

[9] https://www.wired.com/2016/07/gm-nasa-roboglove-robonaut/. Accessed 26 January 2017.

to accommodate these people. This will lead to a consequence: once the effectiveness of prosthetic technology enables them to surpass healthy people's physical performances (prosthetic enhancement), the relationship between health and disability may reverse, at least in certain, though exceptional, circumstances.

Therefore, the concomitant effect of autonomy and the improvement of conditions of disabled people will likely contribute to the entry of HETs into the market of mass consumer products and to the transformation of our current habits and behaviour. Without realizing it, ours will become societies where performance-enhancing technologies are commonly available for anyone. We could call them 'enhanced societies'. This will happen without the support of any form of ideology such as that of posthumanism or transhumanism, at least in the large public, although it is likely that this kind of philosophical support will be diffused in some groups of population.

Right now several firms, engaged in the field of HE, propose new products which offer new possibilities for consumers. For example, a firm called Cyborg Nest, which works in the field of HETs, engineers performance-enhancing products such as chips implanted under the skin.[10] Its first product into the market, 'North sense', is able to turn the human body into a compass by making the individual able to feel what direction she/he is facing.[11] There are already some groups of biohackers who have modified their body in order to cure some handicaps by improving their performances at the same time.[12] This is an emblematic case of coincidence of therapy for disability and enhancement. The artist Neil Harbisson, for example, suffering for an extreme form of colour blindness called achromatopsia decided to implant on himself a neural implant (an 'eyeborg' antennae)

---

[10]See http://www.cyborgnest.net/. Accessed 5 October 2017.

[11]Some scholars talked of 'enhancement societies' in this regard. See, e.g. Coenen et al. (2009, p. 6). Differently from Coenen and his collegues, I do not think necessary that the diffusion of performance-enhancing technologies in daily life would be accompanied by the spread of any particular vision (transhumanism, posthumanism or similar). I think that is more verisimilar that HETs could be common in several sections of the society such as current social practices like piercing or tattoo are today.

[12]https://www.wired.com/2016/07/gm-nasa-roboglove-robonaut/ Accessed 26 January 2017.

which allows him to 'hear colours' thanks to skull vibrations that it transmits to his ears. In this way, differently from any other human being, he can hear all light frequencies of the spectrum including invisible colours such as infrared and ultraviolet. It is therefore plausible that when the availability of these possibilities, as well as their efficiency, is improved, they will encounter the large public easier than today.

In this context of progressive transformation of our societies, to have a clearer idea of what are the boundaries of autonomy in biomedical field within the ECHR is crucial.

6.2 The Self-Determination Principle

6.2.1 Autonomy principle is enshrined by the UNESCO Universal Declaration on Bioethics and Human Rights[13] (Art. 5) which emphasizes that '[t]he autonomy of persons to make decisions, while taking responsibility for those decisions and respecting the autonomy of others, is to be respected'. On the consent is dedicated Article 6[14] which requests the expression of the free and informed

---

[13]UNESCO (2005) *Universal Declaration on Bioethics and Human Rights*, adopted by acclamation by the 33rd session of the General Conference of the UNESCO in 19 October 2005.

[14]Article 6 – Consent

1. Any preventive, diagnostic and therapeutic medical intervention is only to be carried out with the prior, free and informed consent of the person concerned, based on adequate information. The consent should, where appropriate, be express and may be withdrawn by the person concerned at any time and for any reason without disadvantage or prejudice.

2. Scientific research should only be carried out with the prior, free, express and informed consent of the person concerned. The information should be adequate, provided in a comprehensible form and should include modalities for withdrawal of consent. Consent may be withdrawn by the person concerned at any time and for any reason without any disadvantage or prejudice. Exceptions to this principle should be made only in accordance with ethical and legal standards adopted by States, consistent with the principles and provisions set out in this Declaration, in particular in Article 27, and international human rights law.

3. In appropriate cases of research carried out on a group of persons or a community, additional agreement of the legal representatives of the group or community concerned may be sought. In no case should a collective community agreement or the consent of a community leader or other authority substitute for an individual's informed consent.

consent to any medical treatment, even of experimental nature. How is this choice therefore protected under the ECHR?

Within the ECHR there is no specific article aimed at protecting the autonomy of the individual in general. With regard to the self-determination principle we could mention Article 9 (freedom of thought), which protects the freedom of conscience of the individual and therefore the freedom of determining what they feel it is just for their life, and Article 10 (freedom of expression), which protects the expression of the decisions taken by the individual in accordance with their conscience that could be those aimed a modifying their body.

However, although there is no specific article for this, the individual's choice on one's own body normally falls within Article 8 ECHR (respect for private and family life)[15] or Article 3 (inhuman and degrading treatment)[16] depending on circumstances. Article 8, in fact, protects the private life of the individual in general. Therefore, it protects not only correspondence and privacy in general, but also the choices of the individual with regard to their own life. Article 3, in the meantime, protects the individual from ill-treatments of the public powers (first torture), therefore from those forced treatments which coerce the individual's will, for example, in biomedical field.

In this field there is also the Oviedo Convention with some non-binding provisions related to biomedicine, which can be extremely important in the interpretation of Articles 8 and 3 of the ECHR. The Oviedo Convention emphasizes the subtlety of the nexus between the will and the body. It is expressly aimed at the

---

[15] Article 8 – Right to respect for private and family life

1. Everyone has the right to respect for his private and family life, his home and his correspondence.

2. There shall be no interference by a public authority with the exercise of this right except such as is in accordance with the law and is necessary in a democratic society in the interests of national security, public safety or the economic well-being of the country, for the prevention of disorder or crime, for the protection of health or morals, or for the protection of the rights and freedoms of others.

[16] Article 3 – Prohibition of torture

No one shall be subjected to torture or to inhuman or degrading treatment or punishment.

protection of the dignity of all persons related to the applications of biomedicine (Art. 1), which would be violated where the individual is forced against their will to a treatment. The Oviedo Convention, therefore, recalls that every medical intervention requires the free and informed consent of the person, which can be withdrawn at any time (Art. 5). The centrality of the individual's will emerge in particular in the face of the vulnerability of the subject (e.g. disabled, ill, pregnant). In the case of people who do not have the capacity to give their consent (because of a mental disability, a disease or similar reasons), medical intervention may be authorized by their legal representative, by an authority or a person or an organ authorized by the law (Art. 6). The incapacitated person may however participate in the decision-making process and even the authorization to the process, which replaces the subject's consent (Art. 6, 3). This centrality of the will to intervention on their own body is confirmed, in particular, by Article 9 of the Oviedo Convention, which states that the doctor must also consider wishes previously expressed by the patient if one cannot express their will at the time of intervention. This special relevance of the individual's consent in front of the medical treatment, can be deemed as central even when interventions have an enhancing nature. In this regard, the Strasbourg Court jurisprudence is outmost clear. According to the ECtHR 'the primacy of the consent reflected in the relevant instruments concerned with medical interventions'.[17] Therefore, with regard to the decision to enhance one's self in general, since there is no rule impeding HE, from the entire framework of the documents of the Council of Europe we can argue that within the ECHR an individual is free to enhance themselves if s/he wants.

But what are the boundaries of this choice? To answer to this question we have to consider the case law of the Strasbourg Court related to decisions to undergo to a medical treatment more in depth.

---

[17] *Evans v. the United-Kingdom [GC]* (App. 6339/05) judgement of 7 March 2006 *Reports Judgements and Decisions*, 2007-I, par. 68.

6.3 The Strasbourg Court Jurisprudence Related to the Consent to Medical Treatments Aimed at Enhancing Human Performances[18]

6.3.1 The ECtHR jurisprudence on the limits of the will of the patient is crucial vis-à-vis HE. This set of decisions is essential in order to tackle the issue whether the individual can legitimately decide to enhance themselves and on the basis of what limits. First of all: to which limits can the will of the individual pushes? What about the possibility of doing harm to one's own body, or to the possibility of renouncing even one's own life?

In general terms, on the consent over medical treatments it is worth referring to the *Juhnke case*.[19] *Juhnke against Turkey* is a case of a female detainee whom was charged of aiding an illegal organization, the Workers' Party of Kurdistan, or PKK, and was then subjected to a forced gynaecological exam. At the centre of the case there were thus personal integrity and the free consent of the individual. This case gives us a clear idea of the functioning of Article 3 of the ECHR as regards the prohibition of inhuman and degrading treatment, and Article 8 (respect of private life). The applicant was arrested in 1997 in northern Iraq by the Turkish soldiers, with a medical kit and a set of documents concerning the organization of PKK. During the detention the woman was subjected to gynaecological examination carried out by a male doctor to avoid false accusations of sexual harassment and violence against the military personnel. The applicant also complained that the examination was conducted in the presence of the male military personnel. She added that only 9 days after her arrest, she was brought before a military judge, thus not a civil judge, for the formalization of the allegations. Among the various alleged violations of the Convention in addition to violation of the principles of a fair trial (Art. 6), there was the inhuman and degrading treatment (Art. 3), for being threatened with death, left waiting up and blindfolded for long periods of time, and the non-respect of her

---

[18]This paragraph partially retakes and deepens the analysis of Ruggiu (2018).
[19]*Juhnke v. Turkey* (Appl. 1620/03) judgement of 23 September 2010 *Reports of Judgements and Decisions*, 2010.

private life by being subjected to a so invasive medical treatment without her consent (Art. 8). The charges of ill-treatment were contested by the Turkish government and there was no evidence able to prove the alleged facts.

Therefore, Court decided that, according to documentary evidence submitted by both parties, the charges of ill-treatment covered by Article 3 should be archived. Instead, the issue of the gynaecological examination had to be thorough in this regard.

According to the Court Article 3 (prohibition of torture and inhuman and degrading treatment) is absolute and requires to be applied when the severity of the ill-treatment must exceed a certain threshold. The assessment of the minimum level, however, is purely relative and 'depends on all the circumstances of the case, such as the duration of the treatment, its physical and mental effects and, in some cases, the sex, age and state of health of the victim'.[20] To be considered inhuman, the treatment presupposes a certain degree of premeditation, must be applied for hours and cause either actual bodily injury or intense physical and mental suffering. To be degrading, the treatment should arise 'in its victims feelings of fear, anguish and inferiority capable of humiliating and debasing them and possibly breaking their physical or moral resistance or [...] drive the victim to act against his will or conscience'.[21] In this context the Court has to consider whether the aims was to humiliate and degrade the victim.[22] In order to assess whether a punishment or a treatment are inhuman or degrading 'the suffering or humiliation involved must in any event go beyond that inevitable element of suffering or humiliation connected with a given form of legitimate treatment or punishment'.[23]

With regard to more specifically the charge of forced medical treatment, we have to point out that the State bears the obligation to provide medical care to detained persons. A measure that appears 'therapeutically necessary from the point of view of established

---

[20] Ibid., par. 69.
[21] Ibid., par. 70.
[22] *Raninen v. Finland* (Appl. 20972/92) judgement of 16 December 1997 *Reports of Judgements and Decisions*, 1997-VIII, pp. 2821–2822, par. 55.
[23] *Juhnke v. Turkey* (Appl. 1620/03) judgement of 23 September 2010 *Reports of Judgements and Decisions*, 2010, par. 70.

principles of medicine cannot in principle be regarded as inhuman and degrading'.[24] However, the intervention and the existence of appropriate procedural guarantees for the patient must be proven. In any case, even if the medical treatment is not necessary, nevertheless it can be legitimized by the need to obtain evidences of a crime (through the examination of DNA, fingerprints, etc.). In this instance, it must be justified and must not exceed the minimum level required by the case law the Court for the application of Article 3.

However, the ill-treatment under Article 3 must be proved and the Court follows the principle of proof 'beyond reasonable doubt', but the test may also come from inferences that are sufficiently strong, clear and concordant or of similar unrebutted presumptions of fact. In this case since there was a disagreement between the parties on how the gynaecological examination was conducted (the government maintained that the applicant had voluntarily submitted, the applicant argued that she had been threatened and forced to do the exam), the Court recognized that there was no evidence to substantiate the applicant's version. Besides, in similar cases it was found that when the victims refused to undergo the gynaecological examination, this was not followed. In this case it was found, instead, that the victim has resisted the gynaecological examination until she was persuaded to undergo it. For this reason, the appeal did not appear founded with regard to Article 3.

Nevertheless, when a given measure falls outside the application of Article 3, this can fall anyway under Article 8 (right to privacy which also protects personal integrity). In this sense, a medical treatment conducted against of the subject's will constitute an interference with regard to the respect for private and family life of the individual and in particular with regard to his integrity.[25] According to the second paragraph of Article 8 an interference with the private life of a person can be justified if: (i) it is taken in accordance with the law (the principle of legality), (ii) it is necessary in a democratic society, (iii) the social need to which it

---

[24] Ibid., par. 71.
[25] Ibid., par. 71.

responds appears proportionate to the legitimate aim pursued (the 'proportionality' test).

The Court noted that in this case the applicant had been held incommunicado with about nine days before being submitted to gynaecological exam. Furthermore, the documentation available to the Court did not reveal any medical reason for such an examination 'or that it was carried out in response to a complaint of sexual assault lodged by the applicant'.[26] The examination was not therefore required by the applicant. Conversely, it is proved that the applicant resisted the examination until she was persuaded by the physician who conducted the examination. It's not clear whether she was informed of the nature and reasons of the examination, while it is possible that the consent was induced by the false belief that it was compulsory. Therefore, there is no certainty that the consent of the applicant was free and informed.

In order to assess whether the case fulfils the criteria of the second paragraph of Article 8, the Court checked first whether the measure was based on domestic law, second whether the measure was necessary in a democratic society, finally whether it was proportionate.

As regards the first point, the measure must be taken under a law which the Court can criticize with regard to whether it was knowable by the recipient, so that he can predict its consequences, as well as whether it was in accordance with the Rule of Law.[27] If the law provides for discretionary acts by public authorities, they are required to indicate precisely the scope. The Court also notes that under Turkish legislation a forced gynaecological examination was a violation of personal integrity except in cases of need for medical intervention where expressly provided by law. The case shows that the disputed gynaecological examination does not meet the standard procedures for medical examinations of those arrested and detained, and that it represents a discretionary decision not subject to any procedural requirement taken by the authority at the

---

[26] Ibid., par. 71.
[27] *Narinen v. Finland* (Appl. 45027/98) judgement of 1 June 2004, par. 34, http://hudoc.echr.coe.int/sites/eng/pages/search.aspx?i=001-61798#{'itemid': ['001-61798']}.

sole protection of the military personnel who arrested the applicant against any false allegations of violence and sexual harassment.[28] In this sense, the interference with the applicant's right cannot be considered in accordance with the law. Yet, a measure that did not pursue any legitimate aim, must also be framed under the second aspect, namely with regard to the question whether it was necessary in a democratic society. Even assuming that the aim (namely to exonerate the gendarmes and police from false accusations) was legitimate, it could not be considered proportionate to that aim. In fact, '[w]hile in a situation where a female detainee complains of a sexual assault and requests a gynaecological examination, the obligation of the authorities to carry out a thorough and effective investigation into the complaint would include the duty promptly to carry out the examination, a detainee may not be compelled or subjected to pressure to such an examination against her wishes'.[29] In this case the applicant had not filed any charge or was willing to do it. The protection of gendarmes against false accusations of rape 'is, in any event, not such as to justify overriding the refusal of a detainee to undergo such an intrusive and serious interference with her physical integrity'.[30] Since it is neither in accordance with the law nor necessary in a democratic society the measure violated Article 8 and the government had to be condemned also to pay a fine for moral damages of 4000 euro.

From this case we infer that no medical treatment can be performed without the free and informed consent of the patient, therefore that a performance-enhancing treatment requests the free decision of the individual on the basis of complete information.

However, which limits does the individual's will find in the field of HE?

6.3.2 As known, the issue whether self-determination over one's own body may include even renouncing their life was at the centre of

---

[28]*Juhnke v. Turkey* (Appl. 1620/03) judgement of 23 September 2010 *Reports of Judgements and Decisions*, 2010, par. 79.

[29]Ibid., par. 81.

[30]Ibid., par. 81.

the *Pretty case*[31] which shacked the jurisprudence of the Strasbourg Court in the years 2000s.

Dianne Pretty suffered from a neurodegenerative disease. This affected the motor cells of the central nervous system (involving progressively all muscles, in particular, arts, respiratory system, and paralyzed her from the neck down. Unfortunately, there was no cure for it today like then. Its course inevitably leads to death after onset of respiratory complications following a total decline in speech and swallowing abilities. She therefore wanted to die before the disease had total course.

Given her physical condition, Dianne was not able to commit suicide by herself, requiring therefore the assistance of others. However, her clear will was to terminate this state of suffering that she considered unworthy. However, the regulatory framework of the United Kingdom, though, limited this possibility. The attempt of suicide was not punishable by the British law. Instead, the assistance to the suicide was punishable in the United Kingdom. She would have required the help of her husband, who then would have run the risk of the jail time. To be assisted in the suicide, Dianne experienced several proceedings without any success. She considered therefore violated her rights guaranteed under the ECHR, in particular Article 2 (right to life),[32] which would entail a right to die. According to the Diane's interpretation Article 2 would not protect life in itself, but the right to life of the individual against any interference by public and private entities, which would also include the right to self-determination by choosing whether to live or not. Under Article 2, therefore, it would fall even the right to die and choose when to die without pain and with dignity that is merely the corollary to the right

---

[31] *Pretty v. the United Kingdom* (Appl. 2346/02) judgement of 29 April 2002 *Reports of Judgements and Decisions*, 2002-III.

[32] Article 2 – Right to life

1. Everyone's right to life shall be protected by law. No one shall be deprived of his life intentionally save in the execution of a sentence of a court following his conviction of a crime for which this penalty is provided by law.

2. Deprivation of life shall not be regarded as inflicted in contravention of this article when it results from the use of force which is no more than absolutely necessary:

a. in defence of any person from unlawful violence; b. in order to effect a lawful arrest or to prevent the escape of a person lawfully detained; c. in action lawfully taken for the purpose of quelling a riot or insurrection.

to life. In this sense, assisted suicide would be permitted, otherwise other countries where it is regulated (e.g. Switzerland) would violate the Convention.

All other instances are around the application for the recognition of a right to die.

First of all, the suffering, which she would have incurred for the refusal to recognize the legality of the assisted suicide in her case, should have been qualified in this regard as inhuman and degrading treatment under Article 3 (prohibition of inhuman and degrading treatment). In her opinion, according to the case law of the Court, Article 3 would have not implied only a negative obligation for the State, but also a positive obligation to prevent its citizens to suffer much that can be considered inhuman and degrading.

The refusal of not interrupting the prosecution against her husband would then also have infringed Article 8 (respect for private life) because it would also include the right to self-determination by establishing a special relationship between the will of the subject and his own body, which would also include the right to choose how to live and how to die which would be denied by the decision of the Director of Public Prosecution not to want to stop the criminal prosecution against her husband if he had helped her in the suicide.

Finally, it would have also violated the right to equality (Art. 14) since she would have been treated in the same way as other people who are in situations much different from her, being able, differently from her, to commit suicide.

The reject of the Diane's argument for the protection of a right to die under Article 2 of the Convention caused the dismissal of all other instances of the applicant.

According to the judges at Strasbourg Article 2 of the Convention guarantees no right to die, whether with the assistance of a third party or of the State. In this sense, differently from Article 11 which identifies freedom of association by including also a freedom not to associate, the right to life has no corresponding negative freedom. Article 2 rules the legitimate use of force by the State and indicates some exceptional situations where the deprivation of life may be justified (in defence of a person from unlawful violence; in the execution of a warrant of arrest or to prevent an escape of a person lawfully detained; in order to quell a riot or insurrection). It enucleates not only the negative obligation not to take life except in the cases listed, but also to take in certain circumstances, typified by the legislator, all necessary measures of preventive nature to protect individuals whose life is endangered by a crime or by the action of others (for example, in the case to avoid the suicidal tendencies of a mentally ill detainee). For the judges at Strasbourg it is clear that Article 2 implies a positive obligation to protect the human life, and it does not seem, due to even the manner in which it is formulated, involve a negative aspect that includes, for example, quality of life or what the person does or does not want to do with his life. The protection of these aspects against State interference, alluded to by the applicant, would instead fall under other articles of the Convention (and here the allusion of the Court is clearly to Article 8). Protection that only through a significant distortion of the literal meaning of the Article could have drew under Article 2. It therefore does not include neither its opposite, the right to die, nor the right to self-determination in the sense of giving the individual the right to choose death instead of life.[33]

Article 3 does not in principle oblige the State to guarantee criminal impunity for assisting a person to commit suicide or to create a legal basis for another form of assistance with that act. However, the State must not sanction actions intended to terminate life (such as suicide or its attempt). In this case there was no question of ill-treatment put in place by the State against the applicant, or whether the health service did or not provide for necessary medical care, but the fact that, following the refusal to block the prosecution of those who assist her suicide, by doing so, it would lead to a Mrs. Pretty an inhuman and degrading treatment by failing to protect her from the inevitable suffering of the last stage of her illness. According to the Court, in fact, this interpretation would have introduced a meaning entirely new and extensive in the concept of 'treatment'. Article 3 is, in fact, to be interpreted in harmony with Article 2 which discloses a prohibition of the use of lethal force and of those acts that could lead to the death of a person without giving any right to die or facilitating the death of some. It is true that under Article 3 there is also a positive obligation, but this

---

[33] Ibid., par. 39.

cannot extend to where exactly to Article 2 stops, by providing what this clearly is not included (i.e. a right to die).

Even the interpretation of Article 8 should then be rejected, because if it is true that the concept of 'private life' is rather extensive since it includes the possibility of undertaking activities that may be perceived physically or morally dangerous for the individual (e.g. sadomasochistic activities), or to refuse medical treatments necessary to prevent a fatal outcome. However, in this case medical treatments provided for to the applicant were not pertinent,[34] but the interference with her private life that the refusal not to prosecute the husband's behaves implied. If that refusal appeared legitimate because it was adopted on the basis of a law, it also appeared necessary in a democratic society to protect the fundamental value of human life, which is based on the Convention itself. The notion of 'necessity in a democratic society' requires that the interference serves to pursue a legitimate aim, such as the protection of human life, and that it is proportionate to the legitimate aim pursued. The State has the task of regulating the criminal behaviour that can affect human life and personal safety. In this case assisted suicide was ruled by Section 2 of the 1961 Act which was designed to protect the terminally ill persons because of their condition they are in a particularly vulnerable situation where there are clear risks of abuse. Therefore, the refusal not to stop the prosecution was, in this sense, not arbitrary nor unreasonable and, for this reason, Article 8 should not be considered violated. However, with regard to Article 8, the Court found that – without in any way negating the principle of sanctity of life – the quality of life and, in consequence, the question of the individual's autonomy play a role under this provision. The Court considered that an individual's right to decide by what means and at what point his or her life will end, provided he or she is capable of freely reaching a decision on this question and acting in consequence, is one of the aspects of the right to respect for private life within the meaning of Article 8 of the Convention.

Finally, the Court rejected the argument according to which there was a violation of the right to equality since the law, prohibiting assisted suicide and allowing people to commit suicide, where

---

[34]Ibid., par. 62.

the terminally ill patients would not be able to end their life by themselves, would not have considered the peculiarities of her case. In fact, Article 14 requires that for discriminating between different situations there should be an objective and reasonable justification, that is, precisely in this case, the protection of human life in particularly vulnerable people such as the terminally ill patients. This is why the application was rejected also on this point.

However, the acknowledgment of the freedom of deciding by what means and at what point his or her life will end can be deemed the major success of Diane Pretty.

Soon Dianne Pretty died on 11 May 2002 for respiratory failure just thirteen days after the verdict of the Court.

6.3.3 The *Pretty case* leads us to highlight the limits of the individual's will in the end of life decision cases. The will of the patient is at the centre of anticipated statements about end of life wishes. The *Vincent Lambert case*[35] shows that the Convention protects of the will of the individual, even when they are in a condition of extreme vulnerability which could definitely undermine their capacity to express it.

In 2008, Vincent Lambert became quadriplegic after a blow to the head received in an automobile accident. Following the consultation procedure on the basis of the 2005 law 'Leonetti' on patients' rights and end of life, on 11 January 2014 the doctor who was treating Mr. Lambert decides to discontinue, on January 13th, Vincent Lambert's artificial nutrition and hydration. In favour of this decision, his wife and one of his nephews expressed on the basis of his previous willingness expressed before the accident. His parents, his brother and his sister went before the courts to order the hospital to not discontinuing the nutrition and hydration and to transfer the patient to another facility. The court initially gave reason to parents and suspended the execution of the decision of the doctor while refusing the application for transfer of the patient to another facility.

---

[35] *Lambert and others v. France* (Appl. 46043/14) judgement of 5 June 2015 *Reports of Judgements and Decisions*, 2015.

Against this decision, the wife and the nephew appealed to the Conseil d'État that on the basis of a new medical report established the legitimacy of the choice to stop the procedure of feeding and hydration of the patient. Parents and siblings went so urgently before the Strasbourg Court claiming that the discontinuance of the Vincent Lambert's nutrition and hydration would be contrary to the duties of the State to protect its citizens in accordance with Article 2 (right to life). Furthermore, by depriving him of nutrients and of hydration needed would violate also Article 3 (prohibition of inhuman and degrading treatment), and breach his personal integrity against Article 8 (privacy). Based on these considerations, the appellants requested the application of urgent provisional measures (cd. Interim Measures) under Article 39 of the Rules of Court. On 24 June 2014, the Court in taking the case as non-manifestly ill-founded proceeded in the interest of the parties to the eventual suspension of the execution of the decision of the Conseil d'Etat to discontinue the Vincent Lambert's artificial nutrition and hydration for the duration of the process before it, also giving it the highest priority. In the second stage it ordered that the patient was not transferred to other structure, most of all outside France. In addition to questions concerning the compatibility with Article 2, 3, 8, the Court questioned whether artificial feeding and hydration constitute or not medical treatment.[36]

On 25 April 2015 the Grand Chamber issued the decision on the *Lambert case.*

At the preliminary level, the Court noted that according to the Oviedo Convention any medical intervention needs the free informed consent which can be withdrawn at any time (Art. 5). In the instance of persons with limited capacity of consenting (mental disorder or illness, etc.) the intervention may only be carried out with the authorization of his or her representative or an authority or a person or body provided for by law (Art. 6). The person concerned can take part to the procedure of authorization which can be withdrawn at any time (Art. 6). Crucially, Article 9 provides that

---

[36] *Pierre Lambert et autres contre la France* (Requête no 46043/14) introduite le 23 juin 2014, http://www.revuedlf.com/wp-content/uploads/2015/09/Affaire-Lambert-et-autres-c.-France-Intention-de-dessaisissement-de-la-chambre-en-faveur-de-la-Grande-Chambre2.pdf. Accessed 24 October 2017.

the previously expressed wishes relating to a medical intervention by a patient who is not, at the time of the intervention, in a state to express his or her wishes shall be taken into account. This framework must be considered while applying Article 8 ECHR and other ECHR provisions.

First the Court ruled that Vincent Lambert's parents did not have standing to act on behalf of Vincent Lambert but on their own behalf since Vincent Lambert is not dead although the applicants are seeking to raise complaints on his behalf. In fact, in this case, the Court did not discern any risk, firstly, that Vincent Lambert would have been deprived of effective protection of his rights since, in accordance with its consistent case-law, it was open to the applicants, as Vincent Lambert's close relatives, to invoke before the Court on their own behalf the right to life protected by Article 2.[37]

Then, the Court considered how the France complied with the positive obligation enshrined by Article 2 of the Convention. Article 2, in fact, 'enjoins the State not only to refrain from the "intentional" taking of life (negative obligations), but also to take appropriate steps to safeguard the lives of those within its jurisdiction (positive obligations)'.[38] In this regard, France adopted the law 'Leonetti' in 2005, which did not authorize neither euthanasia, nor assisted suicide, but it fixed a precise decision-making procedure. This is in three phases: an individual phase, where each part could express their view point; a collective phase, where many consultations must occur (with the care team, at least one other doctor, the person of trust, the family or those close to the patient); and finally a deliberative phase, where it was the doctor in charge of the patient who alone could take the decision. According to this legal framework a treatment amounted to unreasonable obstinacy if it was futile or disproportionate or has 'no other effect than to sustain life artificially'.

Where there is the absence of consensus on a given subject (such as this) in Europe, States enjoy a wide margin of appreciation. However, the State's margin of discretion must be used under the

---

[37] *Lambert and others v. France* (Appl. 46043/14) judgement of 5 June 2015 *Reports of Judgements and Decisions*, 2015, par. 103.

[38] Ibid., par. 117.

boundaries left free by the ECHR. In this regard, 'the organisation of the decision-making process, including the designation of the person who takes the final decision to withdraw treatment and the detailed arrangements for the taking of the decision, [fell] within the State's margin of appreciation of France'.[39] Furthermore, the Vincent Lambert's parents had the availability of several remedies in the present case, including the appeal to the Conseil d'Etat, which could adopt interim measures (i.e. provisional measures in order to preserve the *status quo*). In this context the French Court had the opportunity to order a new and complete medical report of a group of experts, which examined Vincent Lambert on nine occasions. According to the report findings the 'Vincent Lambert's clinical condition corresponded to a chronic vegetative state'. On its basis it was clear that 'he had sustained serious and extensive damage whose severity, coupled with the period of five and a half years that had passed since the accident, led to the conclusion that it was irreversible and that there was a "poor clinical prognosis"'.[40]

Since according to the judges at Strasbourg the patient's consent must remain at the centre of the decision-making process, also in accordance with wishes the Council of Europe's 'Guide on the decision-making process regarding medical treatment in end-of-life situations', in the absence of advance directives or of a 'living will', the Vincent Lambert's will was, however, clearly proven in the proceeding before the Court. On the basis of several testimonies Conseil d'État could precisely reconstruct the will of Vincent Lambert during the process of decision making. Vincent Lambert and his wife, Rachel Lambert, in fact, were both nurses with experience of patients in resuscitation and those with multiple disabilities. They had often discussed their professional experiences and on several such occasions Vincent Lambert had voiced the wish not to be kept alive artificially in a highly dependent state.[41] Since 'it was primarily for the domestic authorities to verify whether the decision to withdraw treatment was compatible with the domestic legislation and the Convention, and to establish the patient's wishes

---

[39] Ibid., par. 168.
[40] Ibid., par. 175.
[41] Ibid., par. 176.

in accordance with national law', according to the Court the State had fulfilled its positive obligations under Article 2 of the Convention, *inter alia*, of protecting the Vincent Lambert's life. This conclusion was also consistent with the *Pretty case* (§69), which 'recognised the right of each individual to decline to consent to treatment which might have the effect of prolonging his or her life'.[42]

From the *Vincent Lambert case*, therefore, we can derive the principle according to which the wish related to a medical treatment must always be protected by the State, especially in consideration to the state of vulnerability of the individual that can undermine its expression. This protection can imply the reconstruction of the will of the individual even when it aims at withdrawing a life-saving treatment when it is disproportionate.

6.3.4 With regard to the protection of the will to enhance oneself, gender re-assignment cases are of outmost importance. Gender re-assignment cases are instances of HE, which belong to *aesthetic* or *cosmetic enhancement* (Brey, 2009, p. 173) that are quite diffused in liberal-democratic countries. There is already a relevant and consolidated jurisprudence on this, in particular from the Strasbourg Court. Normally this case law is on discrimination on grounds of gender, for example on the lack of legal recognition of changed gender in the birth certificate[43] In the case of *Y.Y. versus Turkey*[44] the ECtHR directly tackled the limits and conditions of the personal choice to undergo to a re-assignment for the first time. Y.Y. is a Turkish transsexual who wanted to undergo to a gender re-assignment treatment. The refusal of authorization by public authorities was justified according to Article 40 of the civil code establishing particularly restrictive preconditions (infertility and sterility). This restrictive legal framework, however, easily led many persons to go abroad or to undergo to the delivery of medicaments and to treatments beyond the control of the judge and medicine with a consequent risk for health. For this reason, as Article 8 ECHR

---

[42] Ibid., par. 180.
[43] See e.g. *Christine Goodwin v. The United Kingdom* (App. 28957/95) judgement of 11 July 2002 *Reports of Judgements and Decisions*, 2002-VI.
[44] *Affaire Y.Y. c. Turquie* (Requête n° 14793/08) arrêt 10 mars 2015. On this judgement, when it was still pending, see also Ruggiu (2013, p. 109).

was concerned, although the interference with the right of self-determination of the applicant was in accordance with the law, the ECtHR considered the restrictive conditions regarded in the civil code as non-necessary in a democratic society, in particular in the light of the set of recommendations of the Council of Europe in this field.[45]

The principle in defence of one's self-determination has been affirmed when such choice is grounded on one's sexual and gender identity, namely on deep motivations upon which the person builds their own identity. However, the application of HETs to some forms of disability could find the same basis of justification. In particular, where those devices (e.g. prosthetic arms or legs, implants) are increasingly integrated and perceived as an extension of the individual's personhood since they are used in daily routine.

The State positive obligation to protect the free will of the individual vis-à-vis any medical treatment can imply also to preserve the conditions in which it is expressed. Since the State is responsible not only for the action of its organs, but also for private actors[46] acting within its territory (Xenos, 2003), we have to conclude that it has an obligation to ensure that the choice to be enhanced is free and there is not undue pressure being placed, even indirectly, on the individual's choice, as could be the case in clinical trials, in the workplace, or in the military. For example, recently 'employees at Three Square Market, a technology company in Wisconsin, can choose to have a chip the size of a grain of rice injected between their thumb and index finger' in order to accomplish with a wave of the hand 'any task involving RFID technology—swiping into the office building, paying for food in the cafeteria'.[47]

---

[45] Ibid., par. 118.

[46] *Öneryildiz v. Turkey* (App. 48939/99) judgement of 30 November 2004 *Reports of Judgement and Decisions*, 2004-XII, par. 71: '[t]he Court considers that this obligation must be construed as applying in the context of any activity, whether public or not, in which the right to life may be at stake, and a fortiori in the case of industrial activities, which by their very nature are dangerous, such as the operation of waste-collection sites'.

[47] See, https://www.nytimes.com/2017/07/25/technology/microchips-wisconsin-company-employees.html. Accessed 25 October 2017.

In these instances, Article 3 of ESC, which lays down the right to safe and healthy working conditions, might also be relevant in the interpretation of the ECHR provisions such as Article 8.

In most controversial cases such as 'the engineering of humans so that either a moral disposition is designed in or immoral acts are designed out' (the case of the novel *A Clockwork Orange* (Burgess, 1962) or moral enhancement), this space for the free choice of the individual has thus to be ensured by the State. This implies a set of guarantees, first of all by the law, for protecting the freedom of choice that can lead to even a general prohibition (when the risk of abuses can hardly be overcome through legislative measures). A similar obligation, therefore, should be deemed to exist for the State as well as where employers put improper pressure on their employees to agree to some form of enhancement (for example, as to their ability to work for longer hours or with greater attention). And it can be seen also in the school environment where students could be led to enhance their learning performances (so-called 'academic doping'), provided that these HETs are safe as regards health. In other words, it should be always granted the possibility not to enhance in any context. Moreover, particular guarantees might be requested in consideration of the age of the person concerned, and other personal conditions. Therefore, following the ECtHR jurisprudence these guarantees exerted by public authorities have to be modulated according to the state of vulnerability of the person (detention, young age, inequality of bargaining power, etc.) that can lead to a ban according to the State's margin of appreciation subject to strict scrutiny of the Court.

References

Allhoff, F., Lin, P., Moor, J., and Weckert, J. (2010). Ethics of Human Enhancement: 25 Questions & Answers, *Studies in Ethics, Law, and Technology*, **4**(1), Art. 4.

Brey, P. (2009). Human Enhancement and Personal Identity, in: Berg, J.K., Friis, O., Selinger, E., and Riis, S. (eds.), *New Waves in Philosophy of Technology. New Waves in Philosophy*, London: Palgrave Macmillan, pp. 169–185.

Burgess, A. (1962). *A Clockwork Orange*, London: William Heinemann.

Coenen, Ch., Schuijff, M., Smits, M., Klaassen, P., Hennen, L., Rader, M., and Wolbring, G. (2009) *Human Enhancement* (IP/A/STOA/FWC/2005-28/SC32 & 39), Brussels: European Parliament, http://www.itas.kit.edu/pub/v/2009/coua09a.pdf. Accessed 4 October 2017.

Jacobson, N. (2007). Human Dignity and Health: A Review, *Social Science and Medicine*, **64**, pp. 292–302.

Roco, M., and Brainbridge, W.S. (eds.) (2003). *Converging Technologies for Improving Human Performance. Nanotechnology, Biotechnology, Information Technology and Cognitive Science,* Dordrecht, The Netherlands: Kluwer Academic Publishers (currently Springer).

Ruggiu, D. (2013). A Rights-Based Model of Governance: The Case of Human Enhancement and the Role of Ethics in Europe, in: K. Konrad, C. Coenen, A. Dijkstra, C. Milburn, and H. Van Lente (eds.), *Shaping Emerging Technologies: Governance, Innovation, Discourse*, Berlin: IOS Press/AKA, pp. 103–115.

Ruggiu, D. (2018). Implementing a Responsible, Research and Innovation Framework for Human Enhancement According to Human Rights: The Right to Bodily Integrity and the Rise of 'Enhanced Societies', *Law, Innovation & Technology*, **10**(1), pp. 1–40.

Wu, W., Wen, X., and Wang, Z.L. (2013). Taxel-Addressable Matrix of Vertical-Nanowire Piezotronic, *Science*, **340**(6135), pp. 952–957.

Xenos, D. (2003). Asserting the Right to Life (Art. 2 ECHR) in the Context of Industry, *German Law Journal*, **8**, pp. 231–254.

Chapter 7

Right to Bodily Integrity[1]

7.1 Introduction

7.1.1 The right to bodily integrity protects the physical integrity of the individual against the interferences of public (Roagna, 2012, p. 24) and private powers (ibid., p. 65).[2] It creates a sphere of inviolability for the subject, which is strictly linked to the consent of the individual. We could call it their *habeas corpus*. It protects the psychophysical integrity of the subject, that covers the not only the physicality of the individual but also their mental state which is strictly linked to their own well-being.

Biomedical activity must respect the mental and physical integrity of the person. Pursuing the good of the patient (beneficence) and the harm principle (*primum non nocere* – according to which the actions of individuals should only be limited to preventing harm being caused to others) are the two essential goals of clinical medicine, lying at the core of the Hippocratic Oath.

As seen, respecting the individual's bodily integrity is an essential element of the right to health. The right to health is a cluster of rights

---

[1] This Chapter partially retakes and deepens the analysis of Ruggiu (2018).
[2] On this see also Jurisconsult's Department (June 2015, p. 5).

Human Rights and Emerging Technologies: Analysis and Perspectives in Europe
Daniele Ruggiu
Copyright © 2018 Pan Stanford Publishing Pte. Ltd.
ISBN 978-981-4774-93-2 (Hardcover), 978-0-429-49059-0 (eBook)
www.panstanford.com

which covers, *inter alia*, the right to bodily integrity and mainly expresses two aspects: a negative one (as negative liberty) and a positive one (as positive liberty) (Foà, 1998, p. 58). Like the right to health also the right to bodily integrity implies both a negative obligation not to interfere with the mental and physical integrity of the person, and a positive obligation to adopt preventive measures to protect the health state of the individual.

In the field of biomedicine the UNESCO Universal Declaration on Bioethics and Human Rights[3] linked personal integrity to the *state of vulnerability* of the individual (Art. 8)[4] and consent (Art. 6).[5] The will of the individual represents therefore a barrier against the interferences of others on the sphere of the physicality of the subject. This protection is ensured both when a treatment has medical nature and when it is driven with purposes of scientific research. This makes this aspect crucial when emerging technologies are at stake.

---

[3]UNESCO (2005) *Universal Declaration on Bioethics and Human rights*, adopted by acclamation by the 33rd session of the General Conference of the UNESCO in 19 October 2005.

[4]Article 8 – Respect for human vulnerability and personal integrity
In applying and advancing scientific knowledge, medical practice and associated technologies, human vulnerability should be taken into account. Individuals and groups of special vulnerability should be protected and the personal integrity of such individuals respected.

[5]Article 6 – Consent
1. Any preventive, diagnostic and therapeutic medical intervention is only to be carried out with the prior, free and informed consent of the person concerned, based on adequate information. The consent should, where appropriate, be express and may be withdrawn by the person concerned at any time and for any reason without disadvantage or prejudice.
2. Scientific research should only be carried out with the prior, free, express and informed consent of the person concerned. The information should be adequate, provided in a comprehensible form and should include modalities for withdrawal of consent. Consent may be withdrawn by the person concerned at any time and for any reason without any disadvantage or prejudice. Exceptions to this principle should be made only in accordance with ethical and legal standards adopted by States, consistent with the principles and provisions set out in this Declaration, in particular in Article 27, and international human rights law.
3. In appropriate cases of research carried out on a group of persons or a community, additional agreement of the legal representatives of the group or community concerned may be sought. In no case should a collective community agreement or the consent of a community leader or other authority substitute for an individual's informed consent.

The relevance of the state of vulnerability, as well as of the will of the person, is equally reflected also within the ECHR and its associate jurisprudence on personal integrity. Under the ECHR the right to psychophysical integrity is protected by Article 2 of the ECHR[6] (which protects the right to life), Article 3 (prohibition of inhuman and degrading treatment),[7] and Article 8 (respect for private life).[8] First, when a person is forced to undergo medical treatment against his will, it can be considered either as inhuman or degrading treatment (Art. 3), depending on its severity. Then, when a State's measure, on which basis the medical treatment is performed, is not an inhuman treatment, it might however be considered an illegal interference with the private sphere of the individual (Art. 8).[9] In this regard it is to be noted that Article 8 not only protects the individual's privacy, home, correspondences and personal data, but also private life much more broadly construed, including their autonomy, self-determination, and control over their own body. Moreover, it is worth noting that there is a strict link between the psychophysical integrity of a person and their free will.

---

[6] Article 2 – Right to life
1. Everyone's right to life shall be protected by law. No one shall be deprived of his life intentionally save in the execution of a sentence of a court following his conviction of a crime for which this penalty is provided by law.
2. Deprivation of life shall not be regarded as inflicted in contravention of this article when it results from the use of force which is no more than absolutely necessary: a. in defence of any person from unlawful violence; b. in order to effect a lawful arrest or to prevent the escape of a person lawfully detained; c. in action lawfully taken for the purpose of quelling a riot or insurrection.
[7] Article 3 - Prohibition of torture
No one shall be subjected to torture or to inhuman or degrading treatment or punishment.
[8] Article 8 – Right to respect for private and family life
1. Everyone has the right to respect for his private and family life, his home and his correspondence.
2. There shall be no interference by a public authority with the exercise of this right except such as is in accordance with the law and is necessary in a democratic society in the interests of national security, public safety or the economic well-being of the country, for the prevention of disorder or crime, for the protection of health or morals, or for the protection of the rights and freedoms of others.
[9] *Juhnke v. Turkey* (Appl. 1620/03) judgement of 23 September 2010 *Reports of Judgements and Decisions*, 2010.

Like consent, the right to bodily integrity is always at stake in human enhancement (HE), since HE necessarily occurs with the physical modification of the individual's body even when it aims at enhancing the mental performance. No alteration of cognitive performance of a subject, in fact, may occur without the concomitant modification of the human body, for example through the supply of a psychotropic substance. The spread of human enhancements technologies (HETs) in societal practices and the rise of 'enhanced societies', namely of societies where HE is a concrete possibility available for all people, can create an increasing pressure over the integrity of persons. It is therefore crucial knowing limits and possibility of intervention over the body of individual within the European framework. The case of HE therefore can be illustrative of the potential of the jurisprudence of the Strasbourg Court on the right to bodily integrity.

7.2 Assistive Technologies and the Decision of Modifying One's Own Body

7.2.1 The field of assistive technology represents a great opportunity for HETs. Brain-computer interfaces (BCI), neuroprosthetics, exoskeletons, neurostimulators, and transcranial direct current stimulation (tDCS) are increasingly being trialled on healthy people. We have already mentioned the study of the Air Force Research Laboratory (AFRL), which is testing special devices of tDCS able to improve human performances in the fields of aviation, aerospace and cyberspace.[10] These devices can treat forms of disabilities but once used on healthy can enhance their reactions in particularly stressful conditions, finding several applications in the civil field such as gaming. This double range of application of assistive technology is not uncommon, like the off-label use of drugs clearly illustrates. This raises the issue of the needed research on their effects on healthy. As argued by Henry Greely and his colleagues, in this field there is still the need for an adequate 'assessment of both risks and benefits for enhancement uses of drugs and devices,

---

[10]http://www.wpafb.af.mil/shared/media/document/AFD-100614-041.pdf.

with special attention to long-term effects on development and to the possibility of new types of side effects unique to enhancement' (Greely et al., 2008).

The off-label use of pharmaceutical drugs, though, is also an example of the development of therapeutic medicine furthering advancements in HE (Larriviere et al., 2009; Racine and Forlini, 2010).

Analogously, the extension of assistive technology to other fields raises the issue of the boundaries between therapy and enhancement. Especially in this context therefore has come the question: where does the therapy end and the enhancement begin? This question is far from being abstract. The distinction between treatment and enhancement, in fact, is relevant since it 'determines who pays, and thus who has access to certain medicines and services' (Colleton, 2008). For example, insurance companies in the United States establish whether medical treatment is therapeutic for their insurance holders. If so, they cover the expense. Otherwise they do not.

Drawing a clear-cut line for assistive technologies is likely to become increasingly difficult. As technologies develop, they are able both to cure and to enhance at the same time. Such as in the case of laser eye surgery to treat myopia that sometimes can enhance eyesight of the patient. The same might occur with prosthetic technologies. According to Brey (2009, p. 174) the advance in neuroprosthetics and robotics and the growth of artificial organs could lead prosthesis to perform better than normally functioning organs one day. For example, the convergence of robotics and nanotechnology have enabled the production of an artificial skin, which offers disabled people prostheses with the sense of touch (Wu et al., 2013). In this regard, the Defense Advanced Research Projects Agency (DARPA), for example, is working on an upper-limb prosthetic technology that is able to give the sense of touch to amputees, 'which offers increased range of motion, dexterity and control options'.[11] Assistive technologies are trying to solve many

---

[11]The Defense Advanced Research Projects Agency (DARPA), for example, is working on an upper-limb prosthetic technology that is able to give the sense of touch to amputees, 'which offers increased range of motion, dexterity and

handicaps of disabled people thanks to the engineering of smart exoskeleton technology machines,[12] power-assisted gauntlets[13] or other devices for persons suffering from neuromuscular disorders. These HETs are already commercialized but thanks to the technological development can become common and even indispensable to disabled persons for performing ordinary tasks. It is likely that smart exoskeletons may replace wheelchairs for daily life in the future. This will need new adaptations of our cities to accommodate the use of these technologies by the people. The State, therefore, might be requested to adopt specific measures for these new needs in our societies.

As performance-enhancing technologies develop and become more widespread, the impact of HE on human rights will likewise increase. In particular, with regard to the right to bodily integrity which is at the centre of HE. HE will generate new, unforeseen forms of vulnerability which the State has to respond to. In the future we cannot exclude that even enhanced people might be, in certain particular circumstances, in a condition of vulnerability since the removal of a prosthesis, for example, might leave them in a condition which is worse of non-enhanced persons. This concern will be particularly evident in the case of application of HETs to disabled persons where modifications are merely temporary and can therefore be easily removed, exactly like today in the case of wheelchairs or other prostheses. In this sense, the case of disabled people is particularly interesting for HE.

7.2.2 The development of assistive technology is also fostering cognitive enhancement (CE) with an impact on the personal integrity of persons.

Robotic applications are developed not only on the human body, but also in the human body becoming increasingly integrated to our physicality. Nowadays these applications are starting to become more common, ranging from implants to hybrid bionic systems such

---

control options'. See http://www.darpa.mil/program/revolutionizing-prosthetics. Accessed 2 November 2017.

[12]http://eksobionics.com/. Accessed 26 January 2017.

[13]https://www.wired.com/2016/07/gm-nasa-roboglove-robonaut/. Accessed 26 January 2017.

as artificial limbs connected to the nervous system (Koops and Pirni, 2013). In order to cure neuromuscular disorders scientists developed devices such as special headphones equipped with electrodes that use pulses of energy to increase the action of motor neurons and send stronger, more synchronous signals to muscles.[14] The insertion of chips can be used for curing severe forms of headache, epilepsy, or even restoring functional movements in people with quadriplegia (Bout et al., 2016). Some of these applications are already to the lay public. A Dutch biohacker, Sander Pleji, for instance, decided to insert a hand remote-controlled neurostimulator into his back for example, which sends electrical impulses to the nerves in his head to calm headaches when they hit. In the military field meanwhile, the DARPA is testing microchips implanted between the brain and skull of brain-wounded warfighters to treat post-traumatic stress disorder (PTSD).[15] These same techniques could be also used for connecting technology-enabled people with the things of our interconnected world. Recently an Irish firm has engineered a device of tDCS (called BrainWaveBank) able to detect symptoms of Alzheimer and other mental disease at early stage, such as for dementia.[16] This device work together with an app downloaded on a common mobile phone recording patient's health data. As known, the same devices are tested in gaming for improving gamers' performance in order to introduce the experience of augmented reality to the large public.

7.2.3 The spread of the Internet of Things (IoT) could lead us towards more integrated forms of living with our devices and even our body (Internet of Me). Companies increasingly offer to implant their workers and startup members with microchips the size of grains of rice working as swipe cards for opening doors, operating printers, buying a coffee from the vending machine with a wave of the hand[17] or even gaining a sixth sense. A firm called Cyborg Nest,

---

[14] https://www.haloneuro.com/get-halo-sport.
[15] http://www.darpa.mil/program/detection-and-computational-analysis-of-psychological-signals.
[16] https://www.brainwavebank.com/. Accessed 2 November 2017.
[17] http://www.latimes.com/business/technology/la-fi-tn-microchip-employees-20170403-story.html. Accessed 21 November 2017.

for example, works in the field of HETs, such as chips implanted under the skin. Its first product into the market, 'North sense', is able to turn the human body into a compass by making the individual able to feel what direction she/he is facing.[18]

We have already mentioned the case of Neil Harbisson, the Spanish artist suffering for colour blindness, a disease called achromatopsia, that decided to implant on himself an 'eyeborg' antennae which allows him to 'hear colours' thanks to particular skull vibrations which it transmits to his ears. this solution allows him to hear all light frequencies of the spectrum including invisible colours such as infrared and ultraviolet.

In the future there will be further opportunities of integrating technology into the human body. The Finnish computer programmer, Jerry Jalava, for instance, replaced half a finger he had lost in an accident with a USB stick for example.[19]

The spread of the communities of biohackers addresses the potential of these developments for our society. When these applications are part of our common opportunities and our behaviour starts to be shaped by them, the scenario of 'enhanced societies', which we attempted to advance in Chapter 3, will become more realistic and to have a clear regulatory framework on the right to psychophysical integrity will become crucial. In this framework, therefore, we need to think of the challenge of HE for the right to personal integrity.

7.3 The Strasbourg Court Jurisprudence Related to the Right to Bodily Integrity

7.3.1 With the right to psychophysical integrity we are at the centre of the right to health.

The ECtHR jurisprudence associated to it covers several aspects which can be relevant for HE both negative, as negative liberty, and positive, as positive liberty.

---

[18] See http://www.cyborgnest.net/. Accessed 02 November 2017.
[19] http://www.theguardian.com/artanddesign/architecture-design-blog/2015/aug/14/body-hackers-the-people-who-turn-themselves-into-cyborgs. Accessed 2 November 2017.

With regard to the first aspect, the respect of physicality of the person in biomedical field obliges any medical intervention over the body of the individual to be performed with their consent. Therefore, the first bastion against the violation of the bodily integrity of the person is the protection of their will. This exclude in principle that any performance-enhancing treatment might be carried out without gaining the prior free and informed consent of the patient.[20]

In this regard, as we know, there is the *Juhnke case*[21] where a detainee was subject to a medical treatment against her will for the preservation of a supposed higher public interest (the protection of the military personnel against false allegations of rape).

The protection of this special link between the will of the person and their body is confirmed by the jurisprudence on gender re-assignment cases such as *Y.Y. versus Turkey*[22] where the Strasbourg Court directly tackled the issue of the limits and conditions of the decision of modify their own gender. This case related to forms of aesthetic enhancement affecting the (sexual) identity of the individual shows that it is likely legitimate, as well as protected, the choice to enhance themselves within the ECHR.

As known from *Roche*[23] the expressed consent must also be manifested on the basis of complete information covering also the subsequent health state of the individual.

This State responsibility to ptotect the individuals' will over their own body can extend itself beyond the limited boundaries of the public sphere.

The State is responsible, in fact, not only for the action of its organs, but also for private actors[24] acting within its territory.

---

[20] See Chapter 6 Part II on Consent.

[21] Juhnke v. Turkey (Appl. 1620/03) judgement of 23 September 2010 *Reports of Judgements and Decisions*, 2010.

[22] *Affaire Y.Y. c. Turquie* (Requête n° 14793/08) arrêt 10 mars 2015.

[23] *Roche v. The United Kingdom*, (App. 32555/96) judgement of 19 October 2005, *Reports of Judgements and Decisions*, 2005-IX.

[24] *Öneryildiz v. Turkey* (App. 48939/99) judgement of 30 November 2004 *Reports of Judgement and Decisions*, 2004-XII, par. 71: '[t]he Court considers that this obligation must be construed as applying in the context of any activity, whether public or not, in which the right to life may be at stake, and a fortiori in the case of industrial activities, which by their very nature are dangerous, such as the operation of waste-collection sites'.

Therefore, we have to conclude that it has an obligation to ensure that the choice to be enhanced is free and there is not undue pressure being placed, even indirectly, on the individual's choice, as could be the case in clinical trials, in the workplace, or in the military. In these instances, Article 3 of ESC, which lays down the right to safe and healthy working conditions, might also be relevant. In most controversial cases such as 'the engineering of humans so that either a moral disposition is designed in or immoral acts are designed out' (the case of the 1962 novel of Anthony Burgess, *A Clockwork Orange*, or moral enhancement), this space for the free choice of the individual has thus to be ensured by the State. This implies a set of guarantees, first of all by the law, for protecting the freedom of choice that can lead to even a general prohibition (when the risk of abuses can hardly be overcome). Therefore, a similar obligation should be deemed to exist for the State as well as where employers put improper pressure on their employees to agree to some form of enhancement (for example, as to their ability to work for longer hours or with greater attention). And it can be seen also in the school environment where students could be led to enhance their learning performances (so-called 'academic doping'), provided that these HETs are safe as regards health. However, particular guarantees might be requested in consideration of the age of the person concerned. Therefore, following the ECtHR jurisprudence these guarantees exerted by public authorities have to be modulated according to the *state of vulnerability* of the person (detention, young age, inequality of bargaining power, etc.) that can lead to a ban according to the State's margin of appreciation subject to strict scrutiny of the Court.

However, this principle is not without restrictions. They may be some cases, in fact, where the will of the subject might not be decisive in decision about a medical intervention since the State has an obligation to protect human life under the Convention. The positive obligation to take appropriate steps to safeguard the lives of those within its own jurisdiction is significantly relevant in the consideration of the state of vulnerability of particular category of persons, for example due to their mental state. In this instance a person who is suffering from a psychiatric disorder can be at risk of causing harm to themselves or others.

With regard to harm to others the *Cocaign case*[25] is pertinent. Nicolas Cocaign was an insane detainee who was sentenced to solitary confinement for more than four years for having killed a cellmate in the overcrowded jail at Bonne Nouvelle in Rouen and then having eaten a part of his lungs. The Court held that the solitary confinement of a detainee who was a danger for himself and other cellmates with the support of appropriate medical supervision during his detention was not in breach of Article 3 ECHR since he was not being subject to hardship of an intensity exceeding the unavoidable level of suffering inherent in detention.

As regards the harm to themselves, the case of *Ketreb* is pertinent.[26] *Ketreb* concerns a drug addict who was suffering from a psychiatric disorder, who committed suicide while detained in prison, after both the prison authorities and physicians failed to notice the gravity of his illness. As well as being deemed a violation of Article 2 ECHR (right to life), it was also held that placing an individual with a mental illness in isolation for two weeks constituted inhumane treatment according to Article 3 ECHR. This limitation on the right to self-determination could perhaps exceptionally be extended to some instances where the choice to have performance-enhancing treatments puts the health and safety of the person at risk, as well as the safety of others. Indeed, the Court has been clear that Article 2 ECHR (right to life) extends beyond merely a negative obligation to refrain from taking a person's life (outside the cases where the legitimate use of force by the State is justified), but also to a positive obligation to protect the subject's life and integrity.[27] The right to self-determination protected under Article 8 ECHR (respect for private life), must therefore be read in accordance with the other ECHR articles such as Article 2 (right to life),[28] which places a positive obligation on the State to take any

---

[25] *Affaire Cocaign c. France* (Requête no. 32010/07) 3 November 2011.

[26] *Affaire Cocaign c. France* (Requête no. 32010/07) 3 November 2011.

[27] *Osman v. the United Kingdom* (Appl. 23454/94) judgement of 28 October 1998 *Reports* 1998-VIII, p. 3159, par. 115; *Pretty v. the United Kingdom* (Appl. 2346/02) judgement of 29 April 2002 *Reports of Judgements and Decisions*, 2002-III, par. 38; *Lambert and others v. France* (Appl. 46043/14) judgement of 5 June 2015 *Reports of Judgements and Decisions*, 2015, par. 117.

[28] *Haas v. Switzerland* (Appl. 31322/07) judgement of 20 January 2011 *Reports of Judgements and Decisions*, 2011, par. 54.

steps required to protect the life of the individual, even from their own actions. These principles apply to everyone, even enhanced persons, as we cannot be sure that HE practices will not pose risks for oneself or others.

7.3.2 The development of assistive, adaptive, and rehabilitative technologies for people with disabilities, which can enhance human performance, can be more or less integrated into human body. In this case, we should wonder not only whether a prosthesis is part of the body of the disabled person, but also if a performance-enhancing device is part of the physical identity of the individual.

As recognized by Article 1 of the Oviedo Convention, a central aspect of the human being is not only his dignity but also his identity. Especially non-permanent modifications of the human body raise the question whether the physical identity of the individual also covers a performance-enhancing technology and must therefore be protected like their own body. Connected to this, there is also the question whether a violation related to a performance-enhancing device could be deemed as a violation of one's integrity. These aspects are not protected by ECHR as such (the ECHR does not recognize dignity on the whole), but as part of the Convention rights, since they fall within the spectrum of application of Article 8 ECHR (respect for private and family life), Article 3 ECHR (inhuman and degrading treatment), and (more rarely) Article 14 ECHR (non-discrimination).

In the *Price case*,[29] the instance of a device with performance-enhancing features has been at the centre of the attention of the judges at Strasbourg for the first time. In this case, a paraplegic on a wheelchair was imprisoned for debts. When she was in detention, she was denied the use of the charger to the wheelchair as it was considered a luxury. The decision of the prison authorities forced her to rely only on the help of the prison staff and of her cellmates for all her needs. The Court considered this a violation of Article 3 ECHR (inhuman and degrading treatment), being it a degrading

---

[29] *Price v. the United Kingdom* (Appl. 33394/96) judgement of 7 July 2001 *Reports of Judgements and Decisions*, 2001-VII, par. 37.

treatment in detriment to her dignity.[30] The statement that the denial of devices improving the prosthesis performance (such as a charger to a wheelchair) affects the personal dignity of the person seems to hint at the fact that both the prosthesis and the device improving it fall within the individual's bodily integrity.

This point was clearly addressed in her separate opinion by Judge Greve. As known, Article 3 regulates three different situations: torture, inhuman treatments and degrading treatments. Judge Greve held that the behaviour of the prison authorities in preventing Ms. Price from bringing the battery charger for her wheelchair to prison constitutes not only a degrading treatment (in accordance with Article 3), but it also directly impacted upon her personal integrity (in accordance with Article 3, again, but under a different aspect, namely inhuman treatment). Her dignity was at the centre of the case. In this instance, in fact, her life largely depended on the functioning of the wheelchair from which she could not be separated. To separate her from the wheelchair in order to treat her like the other inmates was therefore deemed discriminatory, since degrading treatment covered by Article 3 also includes the general prohibition against discrimination in accordance with Article 14 ECHR. This idea was confirmed, for example, in the *Thlimmenos case*,[31] which stated: '[t]he right not to be discriminated against in the enjoyment of the rights guaranteed under the Convention is also violated when States without an objective and reasonable justification fail to treat [. . .] persons whose situations are significantly different' in a different manner, like in Price case. According to Judge Greve, however, in this case the State not only failed to take the necessary measures to safeguard the dignity of the victim but also—by preventing her from using her wheelchair with the battery charger—directly affected the sphere of her bodily integrity as a disabled person. She held therefore that 'treating

---

[30] Ivi, par. 30: 'the Court considers that to detain a severely disabled person in conditions where she is dangerously cold, risks developing sores because her bed is too hard or unreachable, and is unable to go to the toilet or keep clean without the greatest of difficulty, constitutes degrading treatment contrary to Article 3 of the Convention'.

[31] *Thlimmenos v. Grece* (Appl. 34369/97) judgement of 6 April 2000 *Reports of Judgements and Decisions*, 2000-IV, par. 38.

[Ms. Price] like others is not only discrimination but [it] brings about a violation of Article 3' ECHR, namely inhuman treatment.[32]

7.3.3   While people are free to be enhanced, there is no right to be enhanced as such in the ECHR framework. No right to be enhanced can be found in the ECHR nor in the ECtHR case law. However, the relevance of the dignity of the human being, which is at the centre of the Oviedo Convention (Art. 1) and can be read at the basis of several ECHR provisions such as Article 3 (prohibition of inhuman and degrading treatment), Article 14 (non-discrimination principle), can lead to a different conclusion.

In the *Price case*[33] we can infer a further relevant consequence. Where the individual decides to enhance themselves, the interference of the State, which can cause a deterioration of their own health state or the breach of his dignity, can be deemed as a violation of the Convention according to Article 8 (respect for private life) or to Article 3 (inhuman and degrading treatment), depending on the severity of the infringement. This is also true when HE is at stake. In other words, it is not the performance-enhancing nature of the treatment that is important, but the state of vulnerability of the person (the fact that the State's interference has breached their dignity or led to a deterioration in their health state).

The Price case, therefore, leads to the conclusion that in some conditions the choice of enhancing one's own body might be relevant within the ECHR. In these (limited) circumstances (i.e. the health deterioration or the breach of the personal dignity) the Convention might protect the choice to be enhanced as a right. This could occur when a performance-enhancing technology is applied to a disabled (e.g. a prosthesis) so as its removal or interruption would cause a deterioration of their health state (or the infringement of their dignity). It might also be the case when a healthy person, once enhanced by a drug, suffers consequences to their health due to a State's interference with this enhancement, for example by worsening their own health condition, due to onset insomnia, angst

---

[32] *Price v. the United Kingdom* (Appl. 33394/96) judgement of 7 July 2001 *Reports of Judgements and Decisions*, 2001-VII, par. 37.
[33] *Price v. the United Kingdom* (Appl. 33394/96) judgement of 7 July 2001 *Reports of Judgements and Decisions*, 2001-VII, par. 37.

and other disturbances. Whenever the State's interference in the private life of an enhanced person (disabled or healthy) causes a harm, relevant according to Articles 3 and 8 ECHR, the ECHR can be recalled. Clearly, the existence of a harm needs to be proven, which might not be easy. The permanent or provisional nature of the HE could also be relevant for the damage assessment. In this context we have to finally bear into mind that the societal spread of HE might affect the perception of the level of the severity of damages following the removal or interruption of a HET.

7.3.4 HE makes the individual condition of the person special. Is this altered health state of the individual relevant for the State? In the ECHR, the psychophysical condition of the individual is an essential part of one's bodily integrity.

The consideration of the psychophysical state is widely acknowledged within the ECHR framework as a premise of a State's positive obligation, which normally falls under the broad sphere of the protection of Article 8 ECHR (private and family life). As clearly stated by the Court of Strasbourg, 'the concept of private life includes a person's physical and psychological integrity [therefore] the States have a positive obligation to prevent breaches of the physical and moral integrity of an individual by other persons when the authorities knew or ought to have known of those breaches'. In this context, the health state or the dignity of enhanced persons can be covered by the positive obligation of the State.

As regards the first aspect (physical state) the *Vincent case*[34] is relevant. In the Vincent case the applicant, a four-limb deficient thalidomide victim who also suffers from kidney problems, was committed to prison for contempt of court in the course of civil proceedings. She was kept one night in a police cell, where she had to sleep in her cold wheelchair, as the bed was not specially adapted for a disabled person, in very adverse conditions. She subsequently spent two days in a normal prison, where she was dependent on the assistance of male prison guards in order to use the toilet. In this instance, the Court held that there had been a violation of Article 3 ECHR (prohibition of inhuman or degrading

---

[34]*Affaire Vincent c. France* (Requête 6253/03) arrêt de 24 octobre 2006, par. 98.

treatment). According to the Court, the fact that the applicant was forced to disassemble and reassemble her wheelchair going through doorways to get around the prison facility, can indeed be considered belittling and humiliating.[35] Also being entirely dependent on other people constituted a clear violation of the Convention. Here, the Court affirmed the principle that the mere lack of recognition of the detainee's health state without adapting the cell and other structures of the prison facility breaches the State's positive obligation of protecting the individual's health. This principle applies to each individual, independently from the fact that s/he is enhanced. One day, to detain enhanced people might require not only to let them use the devices on which they are used to count (an exoskeleton), but also to adapt prison facilities to their special needs (by construing, for example, recharge points for robotic prostheses, exoskeletons, providing technical assistance, etc.). In other words, we will have new unexpected vulnerabilities. The breach of this obligation for the State could therefore be in violation of the ECHR.

This positive obligation to recognize the health condition of the individual exists also in the face of the mental state of the individual. The lack of consideration of the psychophysical state can lead to evident abuses. For example, Fabrizio Pellegrini, a pianist, artist and painter living in Chieti (Italy), was recently sentenced for the cultivation of three plants of cannabis which he used to cure fibromyalgia, a rare disease that causes a very specific kind of pain which affects muscles and, in this case, the vertebral column.[36] The use of cannabis is an instance of CE, which has an increasing therapeutic use. The use of neurostimulators or microchips to be implanted for curing a given disease such as chronic headache can have however the same effects.

In the instance of CE, the Haas case[37] is pertinent. The Haas case concerns the administration of a lethal substance to allow

---

[35] Ivi, par. 102.

[36] See http://www.repubblica.it/salute/prevenzione/2016/08/04/news/malattie_rare_approvata_legge_che_allarga_screening_neonatale-145360112/?ref=HREC1-27.

[37] *Haas v. Switzerland* (Appl. 31322/07) Judgement of 20 January 2011 *Reports of Judgements and Decisions*, 2011.

an individual 'to commit suicide in a safe and dignified manner and without unnecessary pain and suffering' in a country where assisted suicide is legal (Switzerland). At the end, the application was dismissed. Notwithstanding the fact that physicians contacted by Mr. Haas refused to prescribe the substance because of the lack of physical illness, the Court was not persuaded that in Switzerland there were no opportunity to obtain the prescription with his mental illness. However, in deciding the case the Court affirmed the principle, that mental and physical illnesses should be equally treated. This means that, in the Haas case the Strasbourg Court recognized that Article 8 ECHR (respect for private life) protects the right to integrity of the person also with regard to one's psychological state, not only from the physical standpoint. Evidently, this principle applies to everyone, irrespective of whether or not they have been enhanced.

7.3.5 Healthcare can be revolutionized by the spread of HETs, raising standards in medicine in the future. This also raises the issue of the equitable access to healthcare. The Oviedo Convention requires that any intervention in the field of health should be carried out in accordance with professional standards and obligations (Art. 4). This does not imply a general obligation for the State to use HETs, nor to ensure anyone the equal access to HETs according to Article 3. However, where therapeutic and enhancement uses coincide (as for disability) and it is not possible to distinguish a therapeutic and enhancement treatment, the use of HETs can be considered compulsory.

As seen, the State is in charge of recognizing the peculiarities related to the health state of the individual. However, once the State recognizes the special condition of a person, it could be obliged to adopt some specific measure in accordance with his peculiar condition. According to the Court, 'the physical integrity can entail for the authorities a Convention requirement to take operational measures to prevent that risk from materialising'.[38]

---

[38] *Georgel and Georgeta Stoicescu v. Romania* (9718/03) judgement of 26 July–March 2011, par. 51.

Complaints on physical integrity that entail a positive obligation for the State normally fall within the Article 8 scope (private and family life).[39] In this context, there is not only a positive obligation to recognize the special condition of some vulnerable categories of persons, but also to take all appropriate measures (including those of therapeutic nature) necessary to deal with the peculiarities of the case, in particular for the prison system. In this regard, any lack of a measure that is able to cause a detriment of health conditions of the individual (even enhanced) or to affect one's dignity (degrading treatment, discrimination), is relevant here. This principle was also affirmed within the Council of Europe ambit as regards the mental state and prison facilities.[40] This principle appears to be applicable also for detained enhanced people in the case of temporary forms of enhancement whenever their interruption might cause a state of vulnerability due to their health deterioration or the breach of their personal dignity where the measure puts the individual in a substantially different situation vis-à-vis other individuals who are in their same conditions.

This does not exclude though that such an obligation is to be conceived also in contexts other than those of a state of detention (e.g. workplace, education system, healthcare service) in analogy to the obligation to overcome social barriers faced by persons with impairments. This is because the failure to adapt the structures and conditions of detention, workplace, education system, health service, etc. to the psychophysical state of the individual, might affect either one's health (interruption of medical treatment resulting in deterioration of their mental and physical condition), or might harm their dignity (the equal treatment of an individual requiring a 'special assistance').

In the future, when the use of a wheelchair is completely outdated and substituted by exoskeletons, for example, the adaptation of facilities by the State to accommodate the new needs they raise

---

[39] Ivi, par. 45.

[40] The Recommendation Rec(2006)2 of the Committee of Ministers to member States on the *European Prison Rules* provides that '[p]ersons who are suffering from mental illness and whose state of mental health is incompatible with detention in a prison should be detained in an establishment specially designed for the purpose' (point 12.1).

could be requested by the positive obligation of the State under the ECHR. As recalled by the Court 'Article 8 [. . .] applies only in exceptional cases where her lack of access to public buildings and buildings open to the public affects her life in such a way as to interfere with her right to personal development and her right to establish and develop relationships with other human beings and the outside world'.[41]

Accordingly, as expressed by the ECtHR jurisprudence, we need to conclude that the prompt adoption of the above-mentioned measures can relieve the State by the fulfillment of its obligation.

In the *Zarzycki case*,[42] the applicant with both his forearms amputated, was detained on suspicion of a number of offences against a minor and of coercing a person into committing perjury. He complained that his detention without adequate medical assistance for his special needs and without refunding him the cost of more advanced biomechanical prosthetic arms had been degrading (Art. 3 ECHR). In particular, he alleged that, as a result, he had been forced to rely on other inmates to help him with certain daily hygiene and dressing tasks. Notwithstanding the case appears similar to others prima facie, the Court held that there had been no violation of Article 3 ECHR (prohibition of inhuman or degrading treatment), noting the proactive attitude of the prison administration toward the applicant. What is worthy of note here is that the discharge of the State from its responsibility was due to the prompt adoption of the needed measures to cope with the applicant's special needs. In particular, the penitentiary authorities gave him a special treatment by releasing him from a prisoner's ordinary duties, such as cleaning his cell, and letting him enjoy wider privileges, such as longer family visits and a shower six times per week. Furthermore, they provided him with two basic-type mechanical, and later biomechanical, arm-prostheses. Finally, in order to find an advanced-type of biomechanical arm-prosthesis, the applicant was granted to leave from serving his sentence to seek orthopedic care outside the penitentiary system. In this regard, these measures were able to

---

[41] *Zehnalová and Zehnal v. the Czech Republic* (Appl. 38621/97) admissibility decision of 14 May 2002, p. 12.
[42] *Zarzycki v. Poland* (15351/03) judgement of 3 March 2013.

satisfy the State's positive obligation toward the applicant and to avoid the condemnation of the government.[43] When a disabled is used to resorting to a given advanced type of assistive technology (e.g. an exoskeleton), which fills or overcomes the gap with the healthy, any measure confining them to their previous condition of disability violates their health state or dignity. It is therefore possible to identify in the Strasbourg Court's case law a principle according to which the State is required to put additional effort into accommodating the particular physical needs of the individual, if the failure to do so will result in a worsening of the person's condition.

This conclusion is also confirmed when the psychological state of the person concerned, meaning that the protection of the physicality of the individual integrity does not end where the mental wellness is at stake.

In the *Dybeku case*,[44] for example, the parents and brother of a detainee suffering from schizophrenia obtained the recognition of the violation of Article 3 (inhuman and degrading treatment) because the State did not adopt special conditions of detention, as well as the appropriate medical treatment, for the person in consideration of his vulnerability due to his particular condition of mental handicap. As the Court said: 'the very nature of the *applicant's psychological condition* made him *more vulnerable* than the average detainee and that his detention may have exacerbated to a certain extent his feelings of distress, anguish and fear' (emphasis added).[45] In other words, here the Strasbourg Court affirmed the principle that the positive State's obligation also exists as regards the psycho-physical state of the individual, irrespective whether one is enhanced or not.

With regard to the State's positive obligation in the case of detention, the Court has established a set of criteria. In this type of cases, the Court must take account of three factors, in particular in assessing whether the continued detention of an applicant is compatible with his or her state of health where the latter is giving cause for concern. These are: (a) the prisoner's condition, (b) the

---

[43] Ivi, par. 125.
[44] *Dybeku v. Albania* (Appl. 41153/06) judgement of 18 December 2007 *ECHR 41153/06*.
[45] Ibid., par. 48.

quality of care provided and (c) whether or not the applicant should continue to be detained in view of his or her state of health.[46] Besides, in these instances 'a lack of resources cannot in principle justify detention conditions which are so poor as to reach the threshold of severity for Article 3 [ECHR] to apply'.[47]

In the case of HE, however, it should be shown that the detention under ordinary conditions is capable of increasing the individual's vulnerability,[48] namely, that there are 'the cumulative negative effects on his health'[49] due to the length of not adequate detention conditions.[50]

The Court reached a similar conclusion in the *Slawomir case*,[51] where the Polish government was condemned having particular regard 'to the cumulative effects of the inadequate medical care and inappropriate conditions in which the applicant was held throughout his pre-trial detention, which clearly *had a detrimental effect on his health and well-being*' (emphasis added).[52] In this case, the positive obligation, which is imposed on the State, to avoid harming the mental and physical integrity of the subject in detention can also lead to provide for special medical treatment ('special measures geared to their condition'[53]) in relation to the peculiar physical and psychological conditions of the subject.[54] In this case, there could be an obligation to act, for example to provide either specific services (e.g. delivering smart drugs, assistance) or products (e.g. prostheses), and an obligation not to interrupt a performance-enhancing treatment (e.g. a medical treatment with

---

[46] *Zarzycki v. Poland* (15351/03) judgement of 3 March 2013, par. 103.

[47] *Dybeku* (n 104) par. 50.

[48] *Z.H. v. Hungary* (28973/11) judgement of 8 November 2012.

[49] Ivi, par. 51 (emphasis added).

[50] See also *Bensaid v. the United Kingdom* (Appl. 44599/98) judgement of 6 February 2002 *Reports of Judgements and Decisions*, 2001-I, par. 37: '[d]eterioration in his already existing mental illness could involve relapse into hallucinations and psychotic delusions involving self-harm and harm to others, as well as restrictions in social functioning' (emphasis added).

[51] *Slawomir Musiał v. Poland* (Appl. 28300/06) judgement of 20 January 2009.

[52] Ivi, par. 97 (emphasis added).

[53] See *affaire Revière v. France* (Requête no.33834/07) 11 julliet 2006.

[54] On the consequences of the inadequate consideration by the State of the particular psychophysical condition of the subject see also *Raffray Taddei c. France* (App. 36435/07) 21 décembre 2010.

brain-computer interfaces). In order to call into question the ECHR, the acknowledgement of a State's positive obligation has to imply that following the interruption of any medical treatment as well as the lack of an organized assistance leads to a deterioration in health conditions of the individual, or to a worsening of conditions of detention (so as they can be deemed as degrading).

This framework is confirmed also in the case of implants. In this instance the *Grimailovs case*[55] might be pertinent.

Here the Court tackled a case where a detainee had an implant inserted in his back for supporting purposes. In June 2002 the applicant, who had a piece of metal inserted in his spine after breaking his back two years earlier, was given a five and a half years' prison sentence. He complained, *inter alia*, that the prison facilities were unsuitable for him as he was paraplegic and wheelchair-bound. Also in this case, the state of vulnerability of the applicant was addressed. The Court held that there had been a violation of Article 3 ECHR (prohibition of inhuman and degrading treatment) since the detention facility was not suitable for persons with such disability. Moreover, he had to rely on his fellow inmates to assist him with his daily routine and mobility around the prison, even though they had not been trained and did not have the necessary qualifications. Although the medical staff visited the applicant in his cell for ordinary medical check-ups, they did not provide any assistance with his special needs, for example his daily routine. According to the Court, the State's obligation to ensure adequate conditions of detention includes making provision for the special needs of prisoners with physical disabilities and the State cannot absolve itself from that obligation by shifting the responsibility to others (e.g. cellmates). These special needs could be generated by the insertion of an implant in the body of the person concerned, both in the case s/he is disabled and s/he is healthy.

7.3.6 HE falls within the space of the individual freedom, it might be the exceptional object of a right, but could it be the object of a specific obligation? The Oviedo Convention sets out some clear limits in these cases, since it protects the individual's dignity

---

[55]*Grimailovs v. Latvia* (6087/03) judgement of 25 June 2003.

and integrity in biomedical research (Art. 1) and it requests that the interest of society or science cannot prevail over the interest of the person (Art. 2). This impedes that the individual can be commoditized for any reason: scientific, commercial, military, etc. As seen, according to Article 5 no intervention in the health field may be carried out if the person concerned has not given free and informed consent to it and could continue without the patient's will. This special relevance of the individual's consent in front of the medical treatment is affirmed even when the patient has no possibility to express their consent at the time of the intervention, since the doctor must also consider wishes previously expressed (Art. 9). This principle is stressed, for example, by Article 6 of the Universal Declaration on Bioethics and Human Rights that provides: 'any preventive, diagnostic and therapeutic medical intervention is only to be carried out with the prior, free and informed consent of the person concerned, based on adequate information'. With same words the Committee on Bioethics of the Council of Europe expressed Principle 4 of the Principles of the ad hoc committee of experts on progress in the biomedical sciences (CAHBI, 1989).

This framework leads therefore to exclude that under the ECHR any obligation to be enhanced can be imposed even indirectly, as clearly stated also by the case law of the Strasbourg Court.

In this instance we can refer to the Lambert case.[56] It is related to the suspension of the decision to interrupt artificial nutrition and hydration of terminally ill. In this case the Court recognized that, under Article 8 ECHR, the individual enjoys the right to refuse and, if it has already started, to stop any medical treatments. This implies the possibility to decide whether to enhance oneself (and in some circumstances even the right to be enhanced), but also, and most of all, to reject those performance-enhancing treatments at any time. In the event that public and even private organizations decide to lay down an obligation to improve the psychophysical performance of an individual (such as a soldier), this obligation has to be deemed as illegitimate under the ECHR framework. Further, as seen above, in the case of the introduction of

---

[56]*Lambert and others v. France* (Appl. 46043/14) judgement of 5 June 2015 *Reports of Judgements and Decisions*, 2015.

performance-enhancing opportunities at the workplace, the positive obligation of the State implies that it has to ensure operating conditions for the individual's right to self-determination to be freely exerted on the basis of full information excluding any undue pressure even indirect on their free will. This obligation therefore exists even when private organizations, such as firms, are at stake, in accordance with the ESC Article 3, which requires the State to ensure safe and healthy working conditions by adopting all necessary steps including legislation to ensure this freedom to be fully guaranteed.[57]

References

Acimovic Srdjan, S., Ortega, M.A., Sanz, V., Berthelot, J., Garcia-Cordero, J.L., Renger, J., Maerkl, S.J., Kreuzer, M.P., and Quidan, R. (2014). LSPR Chip for Parallel, Rapid, and Sensitive Detection of Cancer Markers in Serum, *Nano Letters*, **14**, pp. 2636–2641.

Barr, A., and Wilsonm, R. (2014). Google's Newest Search: Cancer Cells, *The Wall Street Journal*, Oct. 29, 2014, http://www.wsj.com/articles/ google-designing-nanoparticles-to-patrol-human-body-for-disease-1414515602. Accessed 2 October 2017.

Brownsword, R. (2009). Regulating Human Enhancement, *Law Innovation & Technology*, **1**, pp. 125–152.

Brey, P. (2009). Human Enhancement and Personal Identity, in: Berg, J.K., Friis, O., Selinger E., and Riis, S. (eds.), *New Waves in Philosophy of Technology. New Waves in Philosophy*, London: Palgrave Macmillan, pp. 169–185.

Bouton, C.E., Shaikhouni, A., Annetta, N.V., Bockbrader, M.A., Friedenberg, D.A., Nielson, D.M., Sharma, G., Sederberg, P.B., Glenn, B.C., Mysiw, W.J., Morgan, A.G., Deogaonkar, M., and Rezai, A.R. (2016). Restoring Cortical Control of Functional Movement in a Human with Quadriplegia, *Nature*, **533**, pp. 247–250.

Burgess, A. (1962). *A Clockwork Orange*, London, William Heinemann Ltd.

Byk, C. (1999). Bioéthique et Convention européenne des droits de l'homme, in: Pettiti, L.E., Decaux, E., and Imbert, P.H. (eds.), *La Convention*

---

[57] *Öneryildiz v. Turkey* (App. 48939/99) judgement of 30 November 2004 *Reports of Judgement and Decisions*, 2004-XII, par. 90. On this see Xenos (2003).

européenne des droits de l'homme. Commetraire article par article, Paris: Economica, pp. 101–121.

Chen, L., Qiu, R., and Li, L. (2014). The Role of Nanotechnology in Induced Pluripotent and Embryonic Stem Cells Research, *Journal of Biomedical Nanotechnology*, **10**(12), pp. 3431–61.

Colleton, L. (2008). The Elusive Line between Enhancement and Therapy and Its Effects on Health Care in U.S., *Journal of Evolution and Technology*, **18**(1), pp. 70–78, http://jetpress.org/v18/colleton.htm. Accessed 26 October 2017.

de Salvia, C. (2000). La Convenzione del Consiglio d'Europa sui diritti dell'uomo e la biomedicina, *I diritti dell'Uomo. Cronache e battaglie*, **11**(1–2), pp. 99–109.

Foà, S. (1998). Il fondamento europeo del diritto alla salute. Competenze istituzionali e profili di tutela, in: C.E. Gallo and B. Pezzini (eds), *Profili attuali del diritto alla salute*, Milano: Giuffrè, pp. 57–93.

Greely, H., Sahakian, B., Harris, J., Kessler, R.C., Gazzaniga, M., Campbell, Ph., and Farah, M.J. (2008). Towards Responsible Use of Cognitive-Enhancing Drugs by the Healthy, *Nature*, **456**(7223), pp. 702–705.

Jurisconsult's Department (2015). *Health-related Issues in the Case-law of the European Court of Human Rights*, Council of Europe/European Court of Human Rights.

Koops, B.J., and Pirni, A. (2013). Ethical and Legal Aspects of Enhancing Human Capabilities through Robotics. Preliminary Considerations, *Law Innovation & Technologies*, **5**(2), pp. 141–146.

Larriviere, D., Williams, M.A., Rizzo, M., Bonnie, R.J., and on behalf of the AAN Ethics, Law and Humanities Committee (2009). Responding to Requests from Adult Patients for Neuroenhancements. Guidance of the Ethics, Law and Humanities Committee, *Neurology*, **73**, pp. 1406–1412.

Racine, E, and Cynthia Forlini, C. (2010). Responding to Requests from Adult Patients for Neuroenhancements: Guidance of the Ethics, Law and Humanities Committee, *Neurology*, **74**(19), pp. 1555–1556.

Roagna, I. (2012) *Protecting the Right to Respect for Private and Family Life under the European Convention on Human Rights*, Strasbourg: Council of Europe.

Ruggiu, D. (2018). Implementing a Responsible, Research and Innovation Framework for Human Enhancement According to Human Rights: The Right to Bodily Integrity and the Rise of 'Enhanced Societies', *Law, Innovation & Technology*, **10**(1), pp. 1–40.

Wu, W., Wen, X., and Wang, Z.L. (2013). Taxel-Addressable Matrix of Vertical-Nanowire Piezotronic, *Science*, **340**(6135), pp. 952–957.

Xenos, D. (2003). Asserting the Right to Life (Art. 2 ECHR) in the Context of Industry, *German Law Journal*, **8**, pp. 231–254

Chapter 8

Right to a Healthy Environment

8.1 Introduction

8.1.1 The deliberate manipulation of the earth's climate is a field of research, which promises to fix the consequences of the anthropic activity which has led to the climate changing, as is the case in solar radiation management, for example, in which aerosol gases are released into the stratosphere, emulating volcanic emissions (Robock et al., 2010, p. 530). The climate change might make geoengineering a necessity for mitigating effects at the large scale.

Some projects have been already considered at the national level triggering processes of experimentation. We can recall, for example, the SPICE project for geoengineering (Stratospheric Particle Injection for Climate Engineering project) funded by the UK Research Councils in 2010. This project tried to develop a democratic and legitimate framework for science and innovation paving the way to 'a broad anticipation, reflection and inclusive deliberation, with the aim of making policy more responsive' in order to engage society in reflecting on purposes of geoengineering while it develops (Stilgoe et al., 2013, p. 10). This made this project an interesting case of RRI.[1]

---

[1]See Chapter 3 Part I on the RRI model of governance.

Human Rights and Emerging Technologies: Analysis and Perspectives in Europe
Daniele Ruggiu
Copyright © 2018 Pan Stanford Publishing Pte. Ltd.
ISBN 978-981-4774-93-2 (Hardcover), 978-0-429-49059-0 (eBook)
www.panstanford.com

Geoengineering might solve one of the Grand Challenges such as global warming, but it also raises concerns with regard to several points, in particular from the environment.

A recently published study on Nature stressed that the side effects of CO_2-removal techniques are widely unexplored (Williamson, 2016). This cannot but have an impact on individual rights. This study notes that, so far CO_2 removal techniques lacked 'new, internationally coordinated studies to investigate the viability and relative safety of large-scale CO_2 removal' (Williamson, 2016, p. 153).

Some scholars have shown, on the basis of analogues from past volcanic eruptions and climate model experiments, that 'any high-latitude sulfate aerosol production would affect large parts of the planet', for example by 'weakening the summer monsoon over Africa and Asia and reducing precipitation' (Robock et al., 2010, p. 530). It has been noted that '[t]hey produce drought, hazy skies, much less direct solar radiation for use as solar power, and ozone depletion' (Robock, 2008, p. 1166).

This climate change directly affects anthropic activity such as agriculture, water supplies, fisheries, solar energy generation, airplanes flying in the stratosphere and breaches the population's right to food and water, the economic rights of farmers, fishermen, airlines, solar energy industries, the right to health and the right to a healthy environment for individuals.

For example, sulphur emissions in high concentrations in stratospheric geoengineering may be dangerous for health and ecosystems (Robock, 2014, pp. 165–181). Although in an asymmetric and unforeseeable manner these rights might be all affected and need a study of the effective countermeasures.

To involve population on the opportunity of adopting remedies such as those offered by engineering, whose consequences necessarily cross the limited boundaries of one country, is far from being easy. In a project of with such a planetary scope stakeholders are distributed in the global sphere. Therefore, is it acceptable for a single country to take decisions that have the potential to affect the whole planet? For this reason, some have suggested that 'any such outdoor experiments need to be evaluated by an organisation, like a United Nations commission, independent from the researchers, that evaluates an environmental impact statement from the researchers and determines that the environmental impact would be negligible, as is done now for emissions from the surface' (Robock, 2014, p. 178). An independent organization, which also has the competence of addressing aspects related to rights.

These ethical concerns accompany those over the protection of individual rights and the environment. In particular, the environment has also an individual dimension which comes to our attention in this chapter.

8.2 Human Rights and the Environmental Issue

8.2.1 Choices made by governments and private actors such as large companies, funding organizations, academic institutions, etc., that effect the environment, or that frame technological responses to environmental challenges, impact directly on the protection of human rights from the environmental standpoint. In this regard, we find the environment as the object of a specific individual right.

The individual nature of this right was however debated.

In the debate on the protection of the environment, in fact we find both who argues the environmental law, especially at the international level is able to protect this interest and who argues that despite a non-individual dimension of this interest, the environment can be and is the object of a specific human right (Rozo Acuña, 2004, pp. 151ff.). According to this latter hypothesis it would be a third generation right and precisely under this light it entered into European context.

The link between the environment and human rights has long been recognized. The *Stockholm Declaration of the United Nations Conference on the Human Environment*, held in Stockholm in 1972, and the more important *Rio Declaration on Environment and Development* (1992), show how the connection between human rights, human dignity and the environment was felt as crucial since the early stages of the attention of the United Nations on environmental problems.

In its Preamble the Stockholm Declaration states: 'Man is both creature and moulder of his environment, which gives him physical sustenance and affords him the opportunity for intellectual, moral, social and spiritual growth. In the long and tortuous evolution of the human race on this planet a stage has been reached when, through the rapid acceleration of science and technology, man has acquired the power to transform his environment in countless ways and on an unprecedented scale. Both aspects of man's environment, the natural and the man-made, are essential to his well-being and to the enjoyment of basic human rights the right to life itself'.[2] The first principle of this document affirms that 'Man has the fundamental right to freedom, equality and adequate conditions of life, in an environment of a quality that permits a life of dignity and well-being, and he bears a solemn responsibility to protect and improve the environment for present and future generations'.

This original linkage between rights of individuals and Nature is at the centre of the First principle of the 1992 Rio Declaration[3] of the UNESCO which acknowledges that '[h]uman beings are at the centre of concerns for sustainable development. They are entitled to a healthy and productive life in harmony with nature'.[4]

The finding that environmental matters transcend the mere individual sphere led to recognize that each person should be engaged in the decision-making over environmental issues. If deliberation processes must be public the society as whole should be involved when the environment is at stake.

In 1982 the World Charter for Nature[5] was adopted. In this last regard, it recognizes the right of each individual to participate to the debate and decisions on environmental matters. Its Article 23, in fact, states that '[a]ll persons, in accordance with their national legislation, shall have the opportunity to participate, individually or with others, in the formulation of decisions of direct concern to their environment, and shall have access to means of redress when their environment has suffered damage or degradation'.

---

[2] http://www.un-documents.net/unchedec.htm. Accessed 23 November 2017.
[3] The UNESCO (1992) *Convention on Biological Diversity* of Rio de Janeiro adopted on 5 June 1992.
[4] http://www.unesco.org/education/pdf/RIO_E.PDF. Accessed 23 November 2017.
[5] http://www.un.org/documents/ga/res/37/a37r007.htm. Accessed 23 November 2017.

In the context of the relation between the environment and human rights, it is worth mentioning also the work of the Commission on Human Rights of the United Nations in this ambit that addressed environmental issues through resolutions on movement and dumping of toxic and dangerous products and wastes (such as the Resolution – 1989/42) since 1989. In 1994 it adopted its first resolution with the title of Human rights and the environment. This was followed by a number of resolutions on the same subject matter in 1995 and 1996.[6]

In particular, from 2002, the Year of the World Summit on Sustainable Development, the Commission on Human Rights adopted resolutions on the environment that were entitled Human rights and the environment as part of sustainable development (Res. 2002/75; Res. 2003/71; Res. 2005/60).

The connection between the environmental protection and human rights is also addressed at the regional level. In this framework they emerge the multiple modes in which different conceptualizations of human rights operate on the basis of the different cultural context.

The African Charter,[7] for example, states that '[a]ll peoples shall have the right to a general satisfactory environment favorable to their development' (Art. 24). This document therefore acknowledges the collective dimension of all environmental issues and the environment is mainly framed as a people's right.

In the American context, the environment also found a clear expression in the Additional Protocol to the American Convention on Human Rights[8] in the Area of Economic, Social and Cultural Rights, also called 'Protocol of San Salvador' as individual right '[e]veryone shall have the right to live in a healthy environment and to have access to basic public services'.

---

[6]See Res. 1994/65; Res. 1995/14; Res. 1996/13.

[7]Organisation of African Unity, *African Charter on Human and Peoples Rights* (ACHPR) adopted in Nairobi 27 June 1981 and entered into force 21 October 1986.

[8]https://www.oas.org/dil/1988%20Additional%20Protocol%20to%20the% 20American%20Convention%20on%20Human%20Rights%20in%20the% 20Area%20of%20Economic,%20Social%20and%20Cultural%20Rights% 20(Protocol%20of%20San%20Salvador).pdf. Accessed 23 November 2017.

With regard to the Council of Europe there is also an intense activity in this field.

Since 1961 the Council of Europe has produced in this subject matter a number of documents such as the European Landscape Convention, Convention on the Conservation of European Wildlife and Natural Habitats, and the Framework Convention on the value of Cultural Heritage for Society. In particular, the Convention on the Conservation of European Wildlife and Natural Habitats, or 'Bern Convention',[9] is aimed at conserving biodiversity through a regional pan-European framework extended to include the Mediterranean regions and Africa. However, the environment emerges as a specific individual right in the framework of the Strasbourg Court jurisprudence.

8.3 The Strasbourg Court Jurisprudence Related to the Right to a Healthy Environment[10]

8.3.1 There is no right to environment established within the ECHR. As known the ECHR sets out only civil and political rights. However, like for the case of the right to health, the development of the ECtHR jurisprudence has covered also aspects of the violation of human rights that correspond to a right to the environment of individual nature. The attention of the Strasbourg Court on the economic and social implications of the civil and political rights guaranteed by the Convention has permitted the development of the right to a healthy environment with its case law.[11] The aspects of

---

[9]Council of Europe (1979) *Convention on the Conservation of European Wildlife and Natural Habitats*, or 'Bern Convention', adopted in Bern in 1979 and entered into force on 1 June 1982.

[10]This paragraph partially retakes and deepens the analysis of Ruggiu (2012, pp. 352–354).

[11]*Arrondelle v. The United Kingdom* (Appl. 7889/77) decision of the Commission of 15 July 1980 *Decision and Reports*, 19, p. 186; *Athanassoglou and others v. Switzerland* (Appl. 27644/95) judgement of 6 April 2000 *Reports of Judgements and Decisions*, 2000-IV; *Balmer-Schafroth and others v. Switzerland* (App. 22110/93) judgement of 26 August 1997 *Reports*, 1997-IV; *Baggs v. The United Kingdom* (Appl. 9310/81) decision of the Commission of 14 October 1985 *Decision and Reports*, 44; *Budayeva and others v. Russia* (Appl. 15339/02, 21116/02,11673/02 and 15343/02) judgement of 2 November 2006, selected for publication in *Reports*

the protection of health and protection of the environment are often connected. Indeed, a threat to the environment may mean a high risk to health. Therefore, the protection of the environment represents a sort of anticipated protection of the individual health. Within the ECHR framework three articles mainly come at stake when an action has environmental consequences over the life of the individual: Article 2 ECHR (right to life),[12] Article 8 ECHR (respect for private and family life)[13] and Article 1 of the Protocol 1 to

of Judgements and Decisions; Fadeyeva v. Russia (Appl. 55723/00) judgement of 9 June 2005 *Reports of Judgements and Decisions*, 2005-IV; *Fredin v. Sweden* (Appl. 12033/86) judgement of 18 February 1991 *Series A*, No. 192; *Giacomelli v. Italy* (Appl. 59909/00) judgement of 2 March 2008, selected for publication in *Reports of Judgements and Decisions; Gounaridis, Iliopoulos et Papapostoulou c. Grece* (requête n° 41207/98) décision du 21 octobre 1998; *Guerra and others v. Italy* (App. 14967/89) judgement of 19 February 1998 *Reports*, 1998-I; *Hatton and others v. The United Kingdom* (App. 36022/97) judgement of 2 October 2001 *Reports of Judgements and Decisions*, 2003-VIII; *L.C.B. v. The United Kingdom* (App. 23413/94) judgement of 9 June 1998 *Reports*, 1998-III; *Lòpez Ostra v. Spain* (Appl. 16798/90) judgement of 9 December 1994 *Series A*, No. 303-C; *MacGinley and Egan v. The United Kingdom* (App. 21825/93 and 23414/94) judgement of 26 November 1996 *Reports*, 1998-III; *MacGinley and Egan v. United Kingdom* (App. 21825/93 and 23414/94) judgement of 28 January 2000 *Reports*, 1998-III; *Powell and Rayner v. The United Kingdom* (Appl. 9310/81) judgement of 21 February 1990 *Series A*, No. 172; *Spire v. France* (Appl. 13728/88) decision of the Commission of 17 May 1990 *Decision and Reports*, 65; *Surugiu c. Romanie* (requête n° 48995/99) arrêt du 20 avril 2004; *Taşkin and others v. Turkey* (Appl. 46117/99) judgement of 10 November 2004 *Reports of Judgements and Decisions*, 2004-X; *Tauira and others v. France* (Appl. 28204/95) decision of the Commission of 4 December 1995 *Decision and Reports*, 83-B; *Zander v. Sweden* (Appl. 14282/88) judgement of 25 November 1993 *Series A*, No. 279-B; *Zimmerman and Steirner v. Switzreland* (Appl. 8737/79) judgement of 13 July 1983 *Series A*, No. 66.

[12] Article 2 – Right to life
Everyone's right to life shall be protected by law. No one shall be deprived of his life intentionally save in the execution of a sentence of a court following his conviction of a crime for which this penalty is provided by law.
Deprivation of life shall not be regarded as inflicted in contravention of this Article when it results from the use of force which is no more than absolutely necessary: (a) in defence of any person from unlawful violence; (b) in order to effect a lawful arrest or to prevent the escape of a person lawfully detained; (c) in action lawfully taken for the purpose of quelling a riot or insurrection.

[13] Article 8 – Right to respect for private and family life
1. Everyone has the right to respect for his private and family life, his home and his correspondence.
2. There shall be no interference by a public authority with the exercise of this right except such as is in accordance with the law and is necessary in a democratic society in the interests of national security, public safety or the economic well-being of

the Convention (right to property).[14] The respect for private life can be invoked when any decision of public authorities (even an authorization to a private activity) can affect the life of the individual with a deterioration of their status quo with regard health, for example, and even their goods. In this latter aspect, however Article 1 P1 can be more suitable. This article in fact protects any right of economic nature, not only property but also possession, credits, etc. Therefore, whenever an act under the State's responsibility affect the value of one's goods this article is at stake. In particular, when it has environmental consequences.

The structure of this individual right under the ECHR is peculiar.

It is worth noting that there is no right to the protection of the environment as an objective interest which everyone can defend against the State. But in certain circumstances, the subjective position of the individual may be considered under this perspective, taking into account the possible environmental damages. In this case, the protection of the environment is at the same time an individual right and a limit to the enjoyment of individual rights and freedoms (de Salvia, 1997; Del Vecchio, 2001).

First, the applicant must probe the damage to his/her own health. The absence of precaution should be included into the causal process but it should not be too remote. The *Tauira case* is well illustrative of this aspect.

8.3.2 In the *Tauira case*[15] diverse inhabitants of the French Polynesia in the South Pacific applied to the Commission of

---

the country, for the prevention of disorder or crime, for the protection of health or morals, or for the protection of the rights and freedoms of others.

[14]Article 1 – Protection of property

Every natural or legal person is entitled to the peaceful enjoyment of his possessions. No one shall be deprived of his possessions except in the public interest and subject to the conditions provided for by law and by the general principles of international law. The preceding provisions shall not, however, in any way impair the right of a State to enforce such laws as it deems necessary to control the use of property in accordance with the general interest or to secure the payment of taxes or other contributions or penalties.

[15]*Tauira and others v. France* (Appl. 28204/95) decision of the Commission of 4 December 1995 *Decision and Reports*, 83-B.

Strasbourg alleging the breach of ECHR in consequence of the French nuclear tests in that area. They specifically identified risks of fracturing on the southern flank of the atoll of Mururoa, risks of atmospheric fallout, risks of marine pollution and contamination of the food chain, alleging that these were in violation of Article 2 of the Convention (right to life) by failing to implement precautionary health measures, the violation of Article 3 (prohibition of inhuman and degrading treatment) by suffering extreme feelings of fear and anxiety, and violation of Article 8 (right to private life) by failing to comply with the State's positive obligation to protect human health and private life of the inhabitants. The Commission stressed that the right of individual petition guaranteed by the Convention implies that every applicant should have an arguable claim to be themselves a direct or indirect victim of a violation of the Convention and there must be a sufficiently direct link between the applicant and the loss which he believes he has suffered. Therefore, the ECHR does not provide any *actio popularis*.[16] In other words, the applicant must produce 'reasonable and convincing evidence of the likelihood that a violation affecting him personally will occur'.[17] Mere suspicion or conjecture is not sufficient. In that case the applicants failed to substantiate their claim that the French authorities did not take all necessary measures to prevent an accident which could have occurred at any time. Consequently the Commission rejected their application as inadmissible since the interest in the application did not appeared enough individualized.[18]

8.3.3 In the protection of the environment, another aspect that may be raised is the protection of the circumstances of one's private life (Art. 8 ECHR). In the case of *Powell and Reyner v. The United Kingdom*[19] owners of various houses situated in the neighbourhood

---

[16] Private citizens and organizations who are not themselves the victims of the crimes in the action and who proceed in the interest of a given public interest.

[17] Ibid., par. 131.

[18] See also *L.C.B. v. The United Kingdom* (App. 23413/94) judgement of 9 June 1998 *Reports*, 1998-III.

[19] *L.C.B. v. The United Kingdom* (App. 23413/94) judgement of 9 June 1998 *Reports*, 1998-III.

of Heathrow airport complained of an excessively high level of noise disturbance generated by the air traffic both day and night. The quality of the applicants' private life and the enjoyment of the amenities of their houses had been adversely affected. It is to say that Article 8 may be analysed both in terms of positive duty on the State to take reasonable and appropriate measures to secure the right to private and family life guaranteed under the first provision of that article, and in terms of a protection against any 'interference by a public authority' which needs to be justified under its second paragraph.[20] In both contexts the fair balance between competing interests (individual and public) is called into question. The Court argued that it is not for the Commission or the Court to substitute for the assessment of the national authorities where the State enjoys a wide margin of discretion in determining the steps to be taken in compliance with the positive obligation of the State,[21] yet some scrutiny does exist. In those cases the scrutiny of the Court will take into account the level of necessity in a democratic society and the degree of proportionality in relation to a legitimate interest (the economic well-being of a country) of the public interference on the private life of an.[22]

8.3.4 Whenever an individual complains of an environmental damage, the State must guarantee that their application would be reviewed within a reasonable time and by an impartial judicial body for determining a possible compensation. When a claim regards rights and duties with a civil character the protection provided by

---

[20] Ibid., par. 41.

[21] Ibid., par. 44.

[22] See *Spire v. France* (Appl. 13728/88) decision of the Commission of 17 May 1990 *Decision and Reports*, 65, par. 258. In this case the noise and other types of nuisance due to a construction of a nuclear power station in France were judged by the Commission a legitimate interference of the applicant's private life and her application dismissed because inadmissible. In that occasion the Commission acknowledged that the construction of a nuclear power station served the economic well-being of the country. See also M. de Salvia (1997).

Article 6 ECHR[23] becomes at issue.[24] In sum, Article 6 §1 provides that individuals are 'to be granted access to a court whenever they have an arguable claim that there has been an unlawful interference with the exercise of their (civil) rights as recognised under domestic law'.[25] This is relevant because in its case law, the Strasbourg Court seems to maintain that the application of Article 6 in the matter of environmental protection requires the presence of a patrimonial element (de Salvia, 1997, p. 251). Without this element Article 6 cannot be used.

In this context the jurisprudence of the Strasbourg Court[26] states that, notwithstanding the public law aspects of any decision in environmental matters, the patrimonial element is determinative

---

[23] Article 6 – Right to a fair trial
In the determination of his civil rights and obligations or of any criminal charge against him, everyone is entitled to a fair and public hearing within a reasonable time by an independent and impartial tribunal established by law. Judgement shall be pronounced publicly but the press and public may be excluded from all or part of the trial in the interests of morals, public order or national security in a democratic society, where the interests of juveniles or the protection of the private life of the parties so require, or to the extent strictly necessary in the opinion of the court in special circumstances where publicity would prejudice the interests of justice.
Everyone charged with a criminal offence shall be presumed innocent until proved guilty according to law.
Everyone charged with a criminal offence has the following minimum rights: (a) to be informed promptly, in a language which he understands and in detail, of the nature and cause of the accusation against him; (b) to have adequate time and facilities for the preparation of his defence; (c) to defend himself in person or through legal assistance of his own choosing or, if he has not sufficient means to pay for legal assistance, to be given it free when the interests of justice so require; (d) to examine or have examined witnesses against him and to obtain the attendance and examination of witnesses on his behalf under the same conditions as witnesses against him; (e) to have the free assistance of an interpreter if he cannot understand or speak the language used in court.
[24] See *Zimmerman and Steirner v. Switzreland* (Appl. 8737/79) judgement of 13 July 1983 *Series A*, No. 66.
[25] See *Athanassoglou and others v. Switzerland* (Appl. 27644/95) judgement of 6 April 2000 *Reports of Judgements and Decisions*, 2000-IV; *Balmer-Schafroth and others v. Switzerland* (App. 22110/93) judgement of 26 August 1997 *Reports*, 1997-IV.
[26] *Lòpez Ostra v. Spain* (Appl. 16798/90) judgement of 9 December 1994 *Series A*, No. 303-C.

with regard to establishing the civil nature of the complaint.[27] Even if the applicant can complain only of non-pecuniary damage, whenever the State, in environmental matter, does not guarantee to an individual judicial review of a public decision, the Court acknowledges that the public interference with the right to one's healthy life conditions, which includes enjoyment of their property right, does not make the claim lose its civil character. This is because a decrease of the quality of life cannot but have an impact on the one's property value. For this reason, the right to a fair trial (Art. 6 par. 1) can be invoked anyway.[28]

The *McGinley and Egan case*[29] is illustrative how the State's interferences can compromise the exercise of one's own rights in the environmental field constituting a breach of the Convention. The case originated by two applicants, Mr Kenneth McGinley and Mr Edward Egan, who had participated in nuclear tests conducted by the United Kingdom at Christmas Island in the Pacific Ocean in 1958. They complained, *inter alia*, that a lack of access to relevant contemporaneous records of those tests constituted a denial of their access to court in the context of their applications for service disability pensions in violation of Article 6 of the Convention and an unjustifiable interference with their private life within the meaning of Article 8. According to the applicants, in fact, the refusal of access to public records on environmental radiation on Christmas Island for security reasons, prevented the applicants from gaining access to any relevant evidence relating to the compensation claim for radiation-linked illness and cancers resulting from participation in UK's nuclear tests program, thus violating both the right to information on one's own health (according to Art. 8) and the right to fair trial (Art. 6). According to the Court whenever a State engages in hazardous activities, such as nuclear experimentations, hiding adverse consequences of tests on the health of those involved

---

[27] *Zander v. Sweden* (Appl. 14282/88) judgement of 25 November 1993 *Series A*, No. 279-B

[28] Ibid., par. 25.

[29] *MacGinley and Egan v The United Kingdom* (App. 21825/93 and 23414/94) judgement of 26 November 1996 *Reports*, 1998-III; *MacGinley and Egan v. The United Kingdom* (App. 21825/93 and 23414/94) judgement (revision) of 28 January 2000 *Reports of Judgements and Decisions*, 2000-I.

in such activities, Article 8 (respect for private and family life) requires that an effective and accessible procedure is established in view of enabling individuals to have any relevant and appropriate information in this concern.[30] The Strasbourg Court, however, dismissed the application since in the light of the circumstances of the case the failure to disclose medical records of damages applicants and military records did not amount to a denial of access to justice (Art. 6) nor breach of right of privacy (Art. 8).

8.3.5 However, as it pertains to protection of the environment, the availability of the information is another point. In essence, Article 10 ECHR (freedom of expression)[31] prohibits the government from placing restrictions on the information that a person may receive that others wish or may be willing to impart to him. Notwithstanding this freedom to receive and impart information does not include a positive obligation to collect and disseminate information of its own motion.[32] Thus, when the interference of the State with the private life of the individual consists not only in a positive action but also in a failure to act which threatens the health of public, Article 8 (respect for private and family life) may be invoked.[33] Severe environmental nuisances may affect individuals' well-being and prevent them from enjoying their homes in such way as to affect their private and family life adversely. When the lack of essential information that would enable individuals to assess the risk that they and their family will be exposed to if they continue to live in a

---

[30] Ibid., par. 20.
[31] Article 10 – Freedom of expression
Everyone has the right to freedom of expression. This right shall include freedom to hold opinions and to receive and impart information and ideas without interference by public authority and regardless of frontiers. This Article shall not prevent States from requiring the licensing of broadcasting, television or cinema enterprises.
The exercise of these freedoms, since it carries with it duties and responsibilities, may be subject to such formalities, conditions, restrictions or penalties as are prescribed by law and are necessary in a democratic society, in the interests of national security, territorial integrity or public safety, for the prevention of disorder or crime, for the protection of health or morals, for the protection of the reputation or rights of others, for preventing the disclosure of information received in confidence, or for maintaining the authority and impartiality of the judiciary.
[32] Ibid., par. 58.
[33] Ibid.

polluted zone is directly imputable to the public authorities because they have omitted to fulfil their positive obligation, the Convention and Article 8 are violated.[34] The information must regard preventive safety measures, the rules that the public must observe in case of emergency, the character of industrial activity, a risk assessment both for the workers employed in the factory and for the public as well as for the environment (de Salvia, 1997, p. 253).

8.3.6 As noted above, the protection of the environment may be very close to protection of individual health, especially when the deterioration of the health state has lethal consequences. In those cases, Article 2 may be pertinent. On 18 and 25 July 2000 a flow of mud and debris hit the Russian town of Tyranauz, flooding some of the residential quarters and killing several persons.[35] The population was aware of the risk of mudslide in that area. But the public authorities did not maintain the defences that had been constructed earlier, and did not inform the people about the preventive measures implemented to mitigate the risks posed by the regular mudslides, even though the people received a number of warnings about the increasing risks. The Court found that the authorities 'failed to discharge the positive obligation to establish the legislative and administrative framework designed to provide effective deterrence against threats to the right to life as required by Article 2 of the Convention'.[36]

8.3.7 The environmental matter involves a plurality of interests, including health and environmental interests, but also economic and societal interests relating to both the public and private sphere. In this regard it should be mentioned that the Convention also protects the right of property and the economic interests of individuals, which may clash with the public interest in the protection of the environment. The Court of Strasbourg recognized that the protection of nature is an 'increasingly important consideration' in

---

[34] Ibid., par. 60.
[35] See *Budayeva and others v. Russia* (Appl. 15339/02, 21116/02, 11673/02 and 15343/02) judgement of 2 November 2006, selected for publication in *Reports of Judgements and Decisions*.
[36] Ibid., par. 159.

the today's society.[37] It is a legitimate aim which can be pursued by the State with a wide margin of discretion, striking a fair balance between the public interest and the individual interests involved, e.g. property. In the *Fredin case* the applicants claimed that the revocation of the permit to extract the gravel from their pit violated their rights guaranteed by the Convention. After complex considerations, ECtHR argued that the measure was not disproportionate. They reasoned that the pit exploitation was only temporary without any assurance of continuing after the lapse of the permission to extract, thus the State legitimately used its wide margin of discretion in order to protect the environment.

8.3.8 The framework of the right to a healthy environment as drafted above shows that the right to a certain level of quality of life is well recognized by the Convention, but that it could be limited and restricted for the economic well-being of a country, as in the case of exploitation of nuclear power, for example, or for the production of goods as the result of the last emerging technologies like geoengineering. However, it is notable that responsibility of the State for its positive obligations may also cover the dangerous activity of industry, public or private (Xenos, 2003). As the Court stated in its jurisprudence,[38] the Convention is a living instrument and must be interpreted in the light of present-day conditions. This will guarantee a margin of change also in the face of the development of the new sciences and technologies. So, in this regard, it is arguable that the right to a healthy environment may evolve further with the norms of the Convention.

References

de Salvia, M. (1997). Ambiente e Convenzione europea dei diritti dell'uomo, *Rivista internazionale dei diritti dell'uomo*, **10**, pp. 247–257.

---

[37]See *Fredin v. Sweden* (Appl. 12033/86) judgement of 18 February 1991 *Series A*, No. 192; *Giacomelli v. Italy* (Appl. 59909/00) judgement of 2 March 2008, selected for publication in *Reports of Judgements and Decisions*, par. 48.

[38]See *Abdulaziz, Cabales and Balkandai v. the United Kingdom* (App. 9214/80, 9473/81,9474/81) judgement of 28 May 1985 *Series A*, No. 94; *Marckx v. Belgium* (App. 6833/74) judgement of 13 June 1979 *Series A*, No. 31.

Del Vecchio, A.M. (2001). Considerazioni sulla tutela dell'ambiente in dimensione internazionale ed in considerazione con la salute umana, *Rivista internazionale dei diritti dell'uomo Atti del Seminario internazionale Assistenza umanitaria e diritto internazionale umanitario (Milano, 24, maggio 2000)*, **14**, pp. 360–361.

Robock, A. (2008). Whiter Geoengineering? *Science*, **320**, pp. 1166–1167.

Robock, A. (2014). Stratospheric Aerosol Geoengineering, *Issues in Environmental Science and Technology* (special issue 'Geoengineering of the Climate System'), **38**, pp. 162–185.

Robock, A., Bunzl, M., Kravitz, B., and Stenchikov, G.L. (2010). A Test for Geoengineering? *Science*, **327**, p. 530.

Rozo Acuña, E. (2004). Lo Stato di diritto ambientale con speciale riferimento al costituzionalismo latinoamericano, in: Idem (ed.), *Profili di diritto ambientale da rio de Janeiro a Johannesburg. Saggi di diritto internazionale, pubblico comparato, penale ed amministrativo*, Torino: Giappichelli, pp. 151–171.

Ruggiu, D. (2012). Synthetic Biology and Human Rights in the European Context: Comparison of the EU and the Council of Europe Regulatory Frameworks on Health and Environment, *Biotechnology Law Report*, **31**(4), pp. 337–355.

Stilgoe, J., Owen, R., and Macnaghten, Phil. (2013). Developing a Framework for Responsible Innovation, *Research Policy*, **42**(9), pp. 1568–1580.

Williamson, P. (2016). Scrutinize CO_2 Removal Methods, *Nature*, **530**, pp. 153–155.

Xenos, D. (2003). Asserting the Right to Life (Art. 2 ECHR) in the Context of Industry, *German Law Journal*, **8**, pp. 231–254.

Chapter 9

Privacy

9.1 Introduction

9.1.1 Artificial Intelligence (AI) is spreading in almost each aspect of our daily life. Our computers and mobile phones are now using AI systems in order to make technology capable of following any move, any will, any secret desire of the user, as it almost is a second skin.

Thanks to processes of machine learning, deep learning, neural network approaches, algorithms, this technology is now able to collect, process and interpret any data, even personal, that we produce, and to take the right decision according to our personal preferences. This makes AI a sort of intimate advisor which has in its hands all secrets of our lives, although this also makes privacy one of the main aspects affected by this technology.

AI is spreading in the administration, in healthcare, in the military and in the workplace[1] and is having great consequences with regard to our rights, in particular privacy and data protection, as well as the needed guarantee that we have to devise to accompany this development.

---

[1]https://www.wsj.com/articles/how-ai-is-transforming-the-workplace-1489371060. Accessed 26 November 2017.

Human Rights and Emerging Technologies: Analysis and Perspectives in Europe
Daniele Ruggiu
Copyright © 2018 Pan Stanford Publishing Pte. Ltd.
ISBN 978-981-4774-93-2 (Hardcover), 978-0-429-49059-0 (eBook)
www.panstanford.com

The growth of the AI sector will likely revolutionize the labour field.

On the one side, AI will deeply transform our labour market by substituting some jobs, implying not only routine manufacturing tasks due to the progressive increase of computerization and automation (Frey and Osborne, 2013). The risk of computerization is now affecting occupations that before we thought could be performed only by humans (e.g. doctors, lawyers, insurance agents, financial brokers). OECD, for example, estimates that 8–10% of occupations will be automatized.[2] However, it is also possible that the growth of the sector will lead to the creation of new jobs such as designer engineers, cyber security specialists, business intelligent analysts, data scientists and data specialists, privacy experts, digital architects, vertical farmers, etc.

On the other side, however, AI will largely affect our privacy in the workplace.

Thanks to machine learning, deep learning processes, algorithms, AI systems are able to process large amounts of data which can be insignificant if taken alone (Big Data), but integrated with other information such as when a given App is used, which telephone numbers have been called, which pages or images haven been looked at, geolocation, etc. Such information can be personal, if not sensitive.[3] AI systems can be used to analyse the email traffic of employees in order to understand whether they are or not satisfied with their jobs. So their employers can control whether their work is still well performed or they can prevent non-professional activities at the workplace by exposing the firm to dangers from the outside.

The same technology with its capability of analysing enormous amounts of data can be also used to recruit the most suitable

---

[2] http://www.oecd-ilibrary.org/social-issues-migration-health/the-risk-of-automation-for-jobs-in-oecd-countries_5jlz9h56dvq7-en. Accessed 26 November 2017.

[3] According to the EU Data Protection Directive (95/46/EC), 'personal data' shall mean any information relating to an identified or identifiable natural person ('Data Subject'); an identifiable person is one who can be identified, directly or indirectly, in particular by reference to an identification number or to one or more factors specific to his physical, physiological, mental, economic, cultural or social identity. Instead, 'sensitive personal data' are data concerning the health status of an individual and judiciary data.

employee for a certain job by evaluating an endless number of possible curricula. Several companies are investing in this sector, as witnessed by a recent inquiry of the Word Street Journal (Greenwald, 2017).

The Entelo Inc.,[4] for example, has developed an App thanks to which it is possible to spot the best candidate in the Web rather than wait for them to send their curriculum. This App is already used by a digital marketing firm, HubSpot Inc.,[5] for its recruitment directorate and allows the search of the right person in a shorter time and with greater efficiency. This App takes into consideration jobs titles, employers and posts in professional forums. Potential employers can look for such criteria as gender, race and military service.

This could be relevant since how an App functions may depend on the criteria that are given to the AI system for its search. In this sense, algorithms are far from being neutral.

For example, recently a Chinese team developed a system, inspired by the Italian criminalist, Cesare Lombroso, 'using facial images of 1856 real persons controlled for race, gender, age and facial expressions, nearly half of whom were convicted criminals, for discriminating between criminals and non-criminals' (Wu and Zhang, 2016, p. 1).

Another study, more disturbing, led by Michal Kosinksi e Yilun Wang, showed that deep neural networks are able to extract features from 35,326 facial images of online dating sites in order to identify information about sexual orientation of individuals and 'correctly distinguish between gay and heterosexual men in 81% of cases' (Wang and Kosinksi, 2017). In this sense, algorithms, as well as AI, can only replicate our biases but in a more pervasive and efficient manner.

The efficiency of the enterprise can be enormously boosted by AI. For instance, AI can help monitor the work of employees in order to detect inefficiencies and promptly correct them. Several Apps and software are being developed in this field and some of them are already in use among firms, especially in U.S.

---

[4]https://www.entelo.com/. Accessed 24 November 2017.
[5]https://www.hubspot.com/. Accessed 24 November 2017.

There is the software Veriato[6] (used by Dancel Multimedia[7] of New Orleans) which recovers any activity done on a computer (including Web browsing, email, chat, used Apps, keystrokes, documents, etc.), taking periodical screenshots and storing it for the next 30 days on a customer's service for reasons of privacy. Next, the server sends the metadata with dates and times of messages for being analysed by the AI system which finds anomalies indicating a low level of productivity, situations of danger within the firm (e.g. a hacking attempt) or the wish of leaving the company by some disloyal employees. So the employer can act and efficaciously solve the problem at stake. In this case, the employees are invited to sign a special agreement where they state that they are aware that all their activities are recorded. More or less like in a store where there are surveillance cameras in action and customers are warned about this.

Veriato can optimize and reduce the margin of error for those activities which were led by the human instinct in the past. In fact, it can analyse e-mails and even statements used by employees at the workplace in order to detect those manifestations addressing a positive or a negative feeling for the company. The system can give a daily score to each employee in order to help those who are in difficulty or unhappy.

Regarding to this investigation of hidden feelings of employees, some companies such as Entelo, International Business Machines Corp. IBM[8] and Workday[9] developed software able to predict who and when they can decide to leave the enterprise so that the employer can intervene in time and avoid a damage for the enterprise.

Without touching the secrecy of feelings of employees, business can need of having an exact picture of all moves and passages within a given place in order to optimize the levels of productivity such as in airports, supermarkets, stores, etc. In these cases, firms can have the need of monitoring whereabouts of employees. Bluvision,[10] for

---

[6]https://www.veriato.com/. Accessed 24 November 2017.
[7]http://dancel.com/sitesplash/index.html. Accessed 24 November 2017.
[8]https://www.ibm.com/us-en/. Accessed 24 November 2017.
[9]https://www.workday.com/. Accessed 24 November 2017.
[10]http://bluvision.com/. Accessed 24 November 2017.

example, produces radio badges able to trace any move of people wearing them or the objects of a building which can be visualized in an easy App on the mobile. When a person attempts to entry into a protected zone without credentials the system transmit an alert. This App can be used in places where object cannot be detected easily or in order to take notice of the time passed in a place or in a given activity. Amazon, for example, uses these types of Apps in its warehouses in the world.

Wearable technologies, nanoelectronics, computer implant technology and performance-enhancing technologies in general, Internet of Things (IoT), as well as Internet of Me, the rise of smart cities, will increase the availability of data which become personal once integrated. This becomes more challenging following the growth of AI in each sector of our societies.

Recently some firms, such as Three Square Market in Wisconsin, developed microchips the size of a grain of rice that can be implanted into a hand of employees enabling them to accomplish with a swipe of the hand any task within the enterprise such as accessing the office building, opening doors, turning on their computers or cars, paying for food or a cup of coffee in the cafeteria, etc.[11] Despite the great simplification which this entails, the continuous generation of data by such a technology can pave the way to unexpected forms of control in the workplace that require adequate safeguards.

9.2 The Right to Privacy and Digital Technologies in the Workplace

9.2.1 Privacy and data protection do not conceptually coincide. In fact, privacy is a concept, covering also data protection, that basically concerns the individual and it links itself therefore to their consent. Data protection, instead, concerns the collective integrity. Therefore, elements, which are central in privacy such as consent, may not be relevant in data protection (Focarelli, 2005, p. 38).

---

[11]See, https://www.nytimes.com/2017/07/25/technology/microchips-wisconsin-company-employees.html. Accessed 25 October 2017.

This is well reflected in the EU law which is one of the more complete legislation concerning privacy at the global level thanks to the 1995 data Protection Directive[12] and, most of all, the General Data Protection Regulation[13] which already came into force. Under the European Union, indeed, the protection of individual's privacy is separated from the protection of personal data. According to Article 7 '[e]veryone has the right to respect for his or her private and family life, home and communications', while the subsequent Article 8 provides that '[e]veryone has the right to the protection of personal data concerning him or her. Such data must be processed fairly for specified purposes and on the basis of the consent of the person concerned'. This led therefore to the recognition of a right to the respect for privacy and a right to the protection of personal data in Europe.

Starting from the Universal Declaration of Human Rights,[14] privacy was included in the first wave of rights, in particular civil and political rights. Article 12 solemnly proclaims that '[n]o one shall be subjected to arbitrary interference with his privacy, family, home or correspondence, nor to attacks upon his honour and reputation. Everyone has the right to the protection of the law against such interference or attacks'. This provision is than reflected at the level

---

[12] European Union (1995) Directive 95/46/EC of the European Parliament and of the Council of 24 October 1995 *on the protection of individuals with regard to the processing of personal data and on the free movement of such data, Official Journal* L 281, 23/11/1995 P. 0031 – 0050.

[13] European Union (2016) Regulation (EU) 2016/679 of the European Parliament and of the Council of 27 April 2016 *on the protection of natural persons with regard to the processing of personal data and on the free movement of such data, and repealing Directive 95/46/EC* (General Data Protection Regulation) (Text with EEA relevance), *Official Journal* L 119, 4.5.2016, pp. 1–88; Directive (EU) 2016/680 of the European Parliament and of the Council of 27 April 2016 *on the protection of natural persons with regard to the processing of personal data by competent authorities for the purposes of the prevention, investigation, detection or prosecution of criminal offences or the execution of criminal penalties, and on the free movement of such data, and repealing Council Framework Decision 2008/977/JHA, Official Journal* L 119, 4.5.2016, pp. 89–131.

[14] ONU (1948) *The Universal Declaration on Human Rights* (UDHR), adopted in New York on 10 December 1948 by the General Assembly of the United Nations.

of Council of Europe by Article 8 ECHR[15] which covers both privacy and data protection.

The international community was soon interested in the knot of privacy and the rise of information technologies developing a set of documents which must be recalled in this context.

On 14 December 1990 the United Nations General Assembly adopted in Resolution 45/95 the Guidelines for the regulation of computerized personal data files. According to its Principle 1 information about persons should not be collected or processed in unfair or unlawful ways. Persons responsible for the compilation of files or those responsible for keeping them have an obligation to conduct regular checks on the accuracy and relevance of the data recorded (Principle 2). In particular, the controller should ensure that '(a) All the personal data collected and recorded remain relevant and adequate to the purposes so specified; (b) None of the said personal data is used or disclosed, except with the consent of the person concerned, for purposes incompatible with those specified; (c) The period for which the personal data are kept does not exceed that which would enable the achievement of the purposes so specified' (Principle 2).

The growth of digital technologies raised increasing concerns within the international community and their impacts over individuals. This led soon it to come back on the subject matter.

On 18 December 2013 the United Nations General Assembly adopted a second resolution, Resolution No. 68/167 on the right to privacy in the digital age (A/RES/68/167). According to it States have 'to respect and protect the right to privacy, including in the context of digital communication; take measures to put an end to violations of those rights and to create the conditions to prevent such violations, including by ensuring that relevant national legislation complies with their obligations under international human rights law; review their procedures, practices and legislation regarding the surveillance of communications, their interception and the collection of personal data, including mass

---

[15]*Juhnke v. Turkey* (Appl. 1620/03) judgement of 23 September 2010 *Reports of Judgements and Decisions*, 2010.

surveillance, interception and collection; establish or maintain existing independent, effective domestic oversight mechanisms capable of ensuring transparency, as appropriate, and accountability for State surveillance of communications'.

Also the International Labour Office (ILO) soon acknowledged the relevance of the digital revolution for the labour organization. This interest led to the publication of a specific code of conduct.

In 1997 ILO issued a Code of Practice on the Protection of Workers' Personal Data ('the ILO Code of Practice'), laying down a number of principles. First, '[p]ersonal data should be processed lawfully and fairly, and only for reasons directly relevant to the employment of the worker. Personal data should, in principle, be used only for the purposes for which they were originally collected. If personal data are to be processed for purposes other than those for which they were collected, the employer should ensure that they are not used in a manner incompatible with the original purpose. [...] Decisions concerning a worker should not be based solely on the automated processing of that worker's personal data. Personal data collected by electronic monitoring should not be the only factors in evaluating worker performance' (Principle 5). 'All personal data should, in principle, be obtained from the individual worker. [...] If workers are monitored they should be informed in advance of the reasons for monitoring, the time schedule, the methods and techniques used and the data to be collected, and the employer must minimize the intrusion on the privacy of workers. Secret monitoring should be permitted only: (a) if it is in conformity with national legislation; or (b) if there is suspicion on reasonable grounds of criminal activity or other serious wrongdoing. Continuous monitoring should be permitted only if required for health and safety or the protection of property' (Principle 6).

According to the ILO Code of Practice '[w]orkers should have the right to be regularly notified of the personal data held about them and the processing of that personal data [...] to have access to all their personal data, irrespective of whether the personal data are processed by automated systems or are kept in a particular manual file regarding the individual worker [...] to know about the processing of their personal data should include the right to examine and obtain a copy of any records'.

The subject matter of the data protection emerged as a sensitive theme also within the Council of Europe giving rise to a specific convention which is a reference point in this ambit: The Convention for the Protection of Individuals with regard to Automatic Processing of Personal Data[16] (Boillat and Kjaerum, 2014, p. 16).

According to it '[t]he Parties undertake to apply this Convention to automated personal data files and automatic processing of personal data in the public and private sectors' (Art. 3).

It also provides standards for processing data at the European level. 'Personal data undergoing automatic processing shall be: (a) obtained and processed fairly and lawfully; (b) stored for specified and legitimate purposes and not used in a way incompatible with those purposes; (c) adequate, relevant and not excessive in relation to the purposes for which they are stored; (d) accurate and, where necessary, kept up to date; (e) preserved in a form which permits identification of the data subjects for no longer than is required for the purpose for which those data are stored' (Art. 5).

The Convention sets safeguards for the data subject. In particular, '[a]ny person shall be enabled: (a) to establish the existence of an automated personal data file, its main purposes, as well as the identity and habitual residence or principal place of business of the controller of the file; (b) to obtain at reasonable intervals and without excessive delay or expense confirmation of whether personal data relating to him are stored in the automated data file; (c) to have a remedy if a request for confirmation or, as the case may be, communication, rectification or erasure' (Art. 8).

At the Council of Europe level also a number of recommendations in the labour field needs to be taken into account together with the Personal Data Protection Convention. This ensemble of documents

---

[16] Council of Europe (1981) *Convention for the Protection of Individuals with regard to Automatic Processing of Personal Data* (1981, ETS no. 108), adopted in Strasbourg on 25 January 1981 (entered into force on 1 October 1985).

can steer the interpretation of the ECHR provisions, especially in hard cases.

First, it is worth mentioning the Recommendation CM/Rec(2015)5 of the Committee of Ministers to member States on the processing of personal data in the context of employment, which was adopted on 1 April 2015. According to it '[e]mployers should minimise the processing of personal data to only the data necessary to the aim pursued in the individual cases concerned' (Point 4). 'Personal data collected for employment purposes should only be processed by employers for such purposes. Employers should adopt data protection policies, rules and/or other instruments on internal use of personal data in compliance with the principles of the present recommendation' (Point 6). Moreover, '[e]mployers should provide employees with the following information the categories of personal data to be processed and a description of the purposes of the processing; – the recipients, or categories of recipients of the personal data; – the means employees have of exercising the rights set out in principle 11 of the present recommendation' (Point 10).

With regard to the use of Internet and electronic communications in the workplace this recommendation also sets a group of provisions which can be enlightening when Article 8 ECHR is at stake. 'Employers should avoid unjustifiable and unreasonable interferences with employees' right to private life [. . .] The persons concerned should be properly and periodically informed [. . .] The information provided should be kept up to date and should include the purpose of the processing, the preservation or back-up period of traffic data and the archiving of professional electronic communications [. . .] in the event of processing of personal data relating to Internet or Intranet pages accessed by the employee, preference should be given to the adoption of preventive measures, such as the use of filters which prevent particular operations, and to the grading of possible monitoring on personal data, giving preference for non-individual random checks on data which are anonymous or in some way aggregated [. . .] Access by employers to the professional electronic communications of their employees who have been informed in advance of the existence of that possibility can only occur, where necessary, for security or other legitimate reasons [. . .] Access should be undertaken in the least intrusive way possible and only after having informed the employees concerned [...] The content, sending and receiving of private electronic communications at work should not be monitored under any circumstances' (Point 14).

This framework of rules stemming not only from the Council of Europe was taken into consideration by the Strasbourg Court jurisprudence.

9.3 The Strasbourg Court Jurisprudence Related to Privacy

9.3.1 The Strasbourg Court jurisprudence on privacy has developed the theme of privacy with a large series of cases related to health[17] and the work conditions. According to the Strasburg Court these aspects (privacy and data protection) fall under the Article 8 ECHR and represent the core of the meaning of this provision.

With regard to the protection of privacy in the workplace we have the *Copland case*,[18] *I v. Finland*,[19] the *Halford case*,[20] and most of all the *Bărbulescu case*.[21]

In *Copland*[22] the applicant was employed by Carmarthenshire College. In 1998 she visited another campus of the College with a male director. She subsequently became aware that the DP had contacted that campus to enquire about her visit and understood that he was suggesting an improper relationship between her and the director. During her employment, the applicant's telephone,

---

[17]For example, *I v. Finland* (Appl. 20511/03) judgement of 17 July 2008 *(2008) ECHR 623; L.H. v. Latvia* (Appl. 52019/07) judgement of 29 April 2014 *(1997) ECHR 10; Peck v. The United Kingdom* (Appl. n. 44647/98) judgement of 28 January 2003 *Reports of Judgements and Decisions*, 2003-I; *Z v. Finland* (Appl. 22009/93) judgement of 25 February 1997 *Reports*, 1997-I.

[18]*Copland v. The United Kingdom* (App. 62617/00) judgement of 3 April 2007 *Reports of Judgements and Decisions*, 2007-I.

[19]*I v. Finland* (Appl. 20511/03) judgement of 17 July 2008 *(2008) ECHR 623.*

[20]*Halford v. the United Kingdom* (Appl. 20605/92) judgement of 25 June 1997 *Reports of Judgements and Decisions*, 1997-III.

[21]*Bărbulescu v. Romania* (App. 61496/08) judgement of 5 September 2017.

[22]*Copland v. The United Kingdom* (App. 62617/00) judgement of 3 April 2007 *Reports of Judgements and Decisions*, 2007-I.

e-mail and internet usage were subjected to monitoring at the DP's instigation. In this period, she noticed that on at least one occasion the DP became aware of the name of an individual with whom she had exchanged incoming and outgoing telephone calls. According to the Government, this monitoring took place in order to ascertain whether the applicant was making excessive use of College facilities for personal purposes. This position was contested by the applicant who lodged her application before the Strasbourg Court for the violation of Article 8 ECHR.

First, according to the Court '[t]he mere fact that these data may have been legitimately obtained by the College, in the form of telephone bills, is no bar to finding an interference with rights guaranteed under Article 8'. Moreover, storing of personal data relating to the private life of an individual also falls within the application of first paragraph of Article 8 which protect private life and correspondence of persons. For considering legitimate those interferences with the rights protected by this provision, Article 8 ECHR requests: the existence of a law which must have some basic features, then that the measure pursues a legitimate aim and finally that there is proportionality between the measure at stake and the pursued aim. With regard to the first point (the existence of a law) the Court was not convinced by 'the Government's submission that the College was authorized under its statutory powers to do "anything necessary or expedient" for the purposes of providing higher and further education'.[23] Furthermore, the Government did 'not seek to argue that any provisions existed at the relevant time, either in general domestic law or in the governing instruments of the College'.[24] Accordingly, as there was no domestic law regulating monitoring at the relevant time, the interference in this case was not 'in accordance with the law' as required by the second paragraph of Article 8. For these reasons, the interference resulted in being illegitimate and it acknowledged a grave violation of the Convention.

The existence of a law protecting efficaciously the confidentiality of personal data is a crucial aspect underlined by also *I v. Finland*[25]

---

[23] Ibid., par. 47.
[24] Ibid.
[25] *I v. Finland* (Appl. 20511/03) judgement of 17 July 2008 *(2008) ECHR 623.*

is of interest in this context. 'Between 1989 and 1994 the applicant worked on fixed-term contracts as a nurse in the polyclinic for eye diseases in a public hospital. From 1987 she paid regular visits to the polyclinic for infectious diseases of the same hospital, having been diagnosed as HIV-positive.

Early in 1992 the applicant began to suspect that her colleagues were aware of her illness. At that time hospital staff had free access to the patient register which contained' sensitive 'information on patients' diagnoses and treating doctors. Having confided her suspicions to her doctor in summer 1992, the hospital's register was amended so that henceforth only the treating clinic's personnel had access to its patients' records'.[26] Following this, the applicant was registered in the patient register under a false name. In 1995 her temporary contract was not renewed. It was not possible, however, to find who had accessed her confidential patient record.

The applicant complained the violation of Article 8 since the district health authority had failed in its duties to establish a register from which her confidential patient information could not be disclosed.

The Court noticed that as noted in *Z. v. Finland*[27] the need for sufficient guarantees is particularly important when processing highly intimate and sensitive data. However, it was to be observed that 'the hospital took ad hoc measures to protect the applicant against unauthorized disclosure of her sensitive health information'[28] too late. In fact, only when the unauthorized access already occurred, in summer 1992, it amended the patient register so that only the treating personnel had access to her patient record and the applicant was registered in the system under a false name and social security number. This could avoid that in 1992 her colleagues were aware about her HIV infection. However, this could not be proven during her civil proceeding so that she lost her civil action against the hospital. For the Court, though, what was decisive was that the records system in place in the hospital was clearly not in accordance with the legal requirements contained in Section 26 of the Personal

---

[26] Ibid., par. 6–7.
[27] *Z v. Finland* (Appl. 22009/93) judgement of 25 February 1997 *Reports*, 1997-I.
[28] Ibid., par. 42.

Files Act. Consequently, the applicant's argument that her medical data were not adequately secured against unauthorized access at the material time had to be upheld and the Finland condemned for the violation of Article 8 ECHR. According to the Court 'the mere fact that the domestic legislation provided the applicant with an opportunity to claim compensation for damages caused by an alleged unlawful disclosure of personal data was not sufficient to protect her private life. What is required in this connection is practical and effective protection to exclude any possibility of unauthorised access occurring in the first place'.[29]

9.3.2 The most significant case in this field is however the *Bărbulescu case*[30] which directly deals with of the internet usage at the workplace and which criteria must be followed in the protection of personal data in this field.

Bărbulescu was a sales engineer of a Romanian private company in Bucharest between 2004 and 2007. At his employer's request, although he had already one, he created an instant messaging account using Yahoo Messenger, an online chat service offering real-time text transmission over the internet, for the purpose of responding to customers' enquiries, he created an instant messaging account using Yahoo Messenger. In quite restrictive terms, the employer's internal regulations prohibited the use of company resources by employees. As any other employee the applicant was informed of the internal regulation. However, in five occasions the applicant exchanged messages of intimate nature with his fiancée and his brother using his personal Yahoo Messenger account. Therefore, on 13 July 2007 at 4.30 p.m. the applicant was summoned by his employer to give an explanation for using the internet for personal purposes. From the notice he was informed that his Yahoo Messenger communications had been monitored. Since he denied the facts, only fifty minutes later the employer gave him forty-five pages consisted of a transcript of the messages which the applicant had exchanged with his brother and his fiancée during the period when he had been monitored.

---

[29] Ibid., par. 47.
[30] *Bărbulescu v. Romania* (App. 61496/08) judgement of 5 September 2017.

By breaching the secrecy of correspondence, according to the applicant the employer had committed a criminal offence. Following this contestation, Bărbulescu was dismissed. Domestic authorities however upheld that the applicant's right to respect for his private life and correspondence should have been balanced with his employer's right to organize and supervise work within the company.

The proceeding before the Bucharest County Court, as well as the Court of Appeal, led to the reject of his application.

Once exhausted the internal remedies, the Strasbourg Court exanimated the application under the aspect of the violation of Article 8 ECHR. In the Court's view, since the decision to dismiss the applicant had been taken by a private-law entity, the case felt within the State's positive obligations. It should therefore be determined whether the national authorities had struck a fair balance between the applicant's right to respect for his private life and correspondence and his employer's interests.

Therefore, only one right was at stake: the right to respect for private life protected under the Convention.

The judges in Strasbourg first observed that 'the kind of internet instant messaging service at issue is just one of the forms of communication enabling individuals to lead a private social life'.[31] Moreover, the sending and receiving of communications is covered by both the concept of 'private life' and the notion of 'correspondence' of Article 8, even if they are sent from an employer's computer. In the view of fostering the business of the firm, the employer adopted measures including a ban on using company resources for personal purposes. However, it was not so clear that the applicant had been informed prior to the monitoring of his communications that such a monitoring operation was to take place. This point though was not ascertained by the domestic courts.

It is worth mentioning the observations of the Government and of third parties intervened in the proceeding.

According to the Government the disciplinary proceedings had been conducted in accordance with the legislation in force as the domestic courts ascertained. Moreover, it noted that in the

---

[31] Ibid., par. 74.

proceedings before the domestic authorities the applicant himself had produced the full transcripts of his communications, without taking any precautions. Instead, they were not proved the applicant's allegations that his communications had been disclosed to his colleagues. The 'employer's decision had been necessary, since it had had to investigate the arguments raised by the applicant in the disciplinary proceedings in order to determine whether he had complied with the internal regulations'.[32] Lastly, in the Government's view, a distinction should have been made between the nature of the communications and their content since domestic courts only evaluated the first.

According the French Government, besides, 'Article 8 of the Convention was only applicable to strictly personal data, correspondence and electronic activities'.[33] The settled case-law of the French Court of Cassation hold, for instance, that 'any data processed, sent and received by means of the employer's electronic equipment were presumed to be professional in nature unless the employee designated them clearly and precisely as personal'.[34] France therefore defended the legitimate interest of public nature of monitoring data of individuals where they overflow the mere private sphere.

On the other part, the European Trade Union Confederation submitted that it was crucial to protect privacy in the working environment, taking into account the *asymmetric relation* between employers and employees characterized by a structural subordination which give rise to a state of vulnerability. In this connection, 'internet access should be regarded as a human right and that the right to respect for correspondence should be strengthened'.[35]

The legal reasoning of the judges in Strasbourg in solving the case ruled a set of criteria applicable to all digital forms of monitoring in the workplace, even by AI systems.

For the Strasbourg Court '[w]hile the essential object of Article 8 of the Convention is to protect individuals against arbitrary interference by public authorities, it may also impose on the State

---

[32] Ibid., par. 102.
[33] Ibid., par. 105.
[34] Ibid.
[35] Ibid., par. 107.

certain positive obligations to ensure effective respect for the rights protected by Article 8'.[36] In the case in question, the Court observed that the measure of the monitoring of Yahoo Messenger communications, which resulted in disciplinary proceedings against the applicant followed by his dismissal for infringing his employer's internal regulations prohibiting the personal use of company resources, was not taken by a State authority but by a private commercial company. This means that the monitoring is not an interference with the applicant's rights protected by Article 8, but the subsequent proceeding before domestic courts which failed 'on their part to secure to the applicant the enjoyment of a right enshrined in Article 8 of the Convention'.[37] Therefore, 'regard must be had in particular to the fair balance that has to be struck between the competing interests of the individual and of the community as a whole, subject in any event to the margin of appreciation enjoyed by the State'.[38]

This space of State's discretion is correctly used only whether it 'secures respect for private life in the relations between individuals by setting up a legislative framework taking into consideration the various interests to be protected in a particular context'[39] such as that of enterprises in front of the growth information technologies.

The labour law, however, has peculiar features since it involves relations of contractual nature with particular rights and obligations on either side, and are characterized by legal *subordination*.[40] Given this specificity, labour law leaves space for negotiation between the parties to the contract of employment. It is, therefore, generally for the parties themselves to regulate a significant part of the content of their relations. From the comparative-law material no European consensus emerges on this issue. For example, few member States have explicitly regulated the question of the exercise by employees of their right to respect for their private life and correspondence in the workplace.

---

[36] Ibid., par. 108.
[37] Ibid., par. 110.
[38] Ibid., par. 112.
[39] Ibid., par. 115.
[40] Ibid., par. 117.

This leads to recognize the States a great margin of appreciation, which is though not unlimited, as well as subject to the scrutiny of the Court. The State, in fact, must ensure that 'the introduction by an employer of measures to monitor correspondence and other communications, irrespective of the extent and duration of such measures, is accompanied by adequate and sufficient safeguards against abuse'.[41]

Despite rapid developments in this area, according to the Court proportionality and procedural guarantees against arbitrariness are essential. Therefore, in this context, the domestic authorities should adhere to the following criteria considering: (1) 'whether the employee has been notified of the possibility that the employer might take measures to monitor correspondence and other communications'; (2) (ii) 'the extent of the monitoring by the employer and the degree of intrusion into the employee's privacy', by making a distinction 'between monitoring of the flow of communications and of their content'; (3) 'whether the employer has provided legitimate reasons to justify monitoring the communications and accessing their actual content'; (4) 'whether it would have been possible to establish a monitoring system based on less intrusive methods and measures than directly accessing the content of the employee's communications'; (5) 'the consequences of the monitoring for the employee subjected to it', ensuring the proportionality of the interference with the exercise of freedom of expression as protected by Article 10 of the Convention; (6) 'whether the employee had been provided with adequate safeguards, especially when the employer's monitoring operations were of an intrusive nature. Such safeguards should in particular ensure that the employer cannot access the actual content of the communications concerned unless the employee has been notified in advance of that eventuality'[42]; (7) the 'access to a remedy before a judicial body with jurisdiction to determine, at least in substance, how the criteria outlined above were observed and whether the impugned measures were lawful'.[43]

---

[41] Ibid., par. 120.
[42] Ibid., par. 121.
[43] Ibid., par. 122.

These criteria should have been applied by Romanian authorities.

According to the Court, 'the employer had a legitimate interest in ensuring the smooth running of the company'. Surely, this could be done 'by establishing mechanisms for checking that its employees are performing their professional duties adequately and with the necessary diligence'.[44]

However, this should have been made by adhering to the above-mentioned criteria.

As to whether the applicant had received prior notification from his employer, the Court observed that 'he did not appear to have been informed in advance of the extent and nature of his employer's monitoring activities, or of the possibility that the employer might have access to the actual content of his messages'.[45]

As regards the scope of the monitoring and the degree of intrusion into the applicant's privacy, the Court observed that 'this question was not examined by either the County Court or the Court of Appeal, even though it appears that the employer recorded all the applicant's communications during the monitoring period in real time, accessed them and printed out their contents'.[46]

Nor does it appeared that 'the domestic courts carried out a sufficient assessment of whether there were legitimate reasons to justify monitoring the applicant's communications'.[47] The County Court had mentioned the need to avoid the company's IT systems being damaged, liability being incurred by the company in the event of illegal activities in cyberspace, and the company's trade secrets being disclosed, but did not analysed them in depth. Instead there is no evidence that the applicant had actually exposed the company to any of those risks.

'[N]either the County Court nor the Court of Appeal sufficiently examined whether the aim pursued by the employer could have been achieved by less intrusive methods'.[48]

---

[44] Ibid., par. 127.
[45] Ibid., par. 133.
[46] Ibid., par. 134.
[47] Ibid., par. 135.
[48] Ibid., par. 136.

Neither courts considered the seriousness of the consequences of the monitoring and the subsequent disciplinary proceedings. In this regard, it is worth noting that the applicant had received the most severe disciplinary sanction, namely dismissal.

Lastly, the Court observed that 'the domestic courts did not determine whether, when the employer summoned the applicant to give an explanation for his use of company resources, in particular the internet, it had in fact already accessed the contents of the communications in issue'.[49]

Therefore, the Court of Appeal's conclusion that a fair balance was struck between the interests at stake appeared to the Strasbourg Court quite questionable.

Having the domestic courts failed to determine whether the applicant had received prior notice from his employer of the possibility that his communications on Yahoo Messenger might be monitored; nor did they have regard either to the fact that he had not been informed of the nature or the extent of the monitoring, or to the degree of intrusion into his private life and correspondence, of the reasons of the measure, as well as of consequences of the breach of the internal regulation, Article 8 was violated and the State was condemned to the payment of pecuniary damages (about 60.000 euros).

Despite this conclusion, it worthwhile stressing that from this case we have a set of criteria which are to be applied for the next development of AI with regard to data protection. Although it is an open question, these criteria should likely be applied also when information is originally non-personal (Big Data) but becomes personal once it is integrated. However, the application of the criteria drawn out by the Strasbourg Court would create a case of 'privacy by design' – of paramount importance – for RRI frameworks.

References

Boillat, P., and Kjaerum, M. (2014). *Handbook on European Data Protection Law*, Luxembourg: European Union Agency for Fundamental Rights, Council of Europe.

---

[49] Ibid., par. 138.

Focarelli, C. (2015). *La privacy. Proteggere i dati personali oggi*, Bologna: Il Mulino.

Frey, C.F., and Osborne, M.A. (2013). *The Future of Employment: How Susceptible Are Jobs to Computerisation?* Oxford: Oxford University Press.

Greenwald, T. (2017). How AI Is Transforming the Workplace, *The World Street Journal*, https://www.wsj.com/articles/how-ai-is-transforming-the-workplace-1489371060 Accessed 29 November 2017.

Wu, X., and Zhang, X. (2016) Automated Inference on Criminality using Face Images, arxiv.org/abs/1611.04135:

Wang, Y., and Kosinksi, M. (16 October 2017). Deep Neural Networks Are More Accurate Than Humans at Detecting Sexual Orientation From Facial Images, *Journal of Personality and Social Psychology* (in press).

Xenos, D. (2003). Asserting the Right to Life (Art. 2 ECHR) in the Context of Industry, *German Law Journal*, **8**, pp. 231–254.

Chapter 10

Freedom of Scientific Research

10.1 Introduction

10.1.1 The spread of emerging technologies, in the healthcare for example, will pave the way to the processing of a large amount of data of personal nature. This can foster the further development of scientific research.

The convergence of emerging technologies in healthcare, in fact, can boost research in medicine such as in the case of epidemiological integrated studies of culture and population health, fostering forms of 'mass medicine' which treat at the same time the population and the individual thanks to information even more precise, if not individualized, of large samples (Stemerding, 2017).

Nanotechnologies thanks to the increased availability lab-on-a-chip, for example, will enhance the quality and the quantity of health information, allowing many to gain a better knowledge of their health status in order to promptly take the right decision (Acimovic Srdjan et al., 2014). The simplification of diagnostics, allowed by nanotechnologies, can improve the early diagnosis and control of some pathologies such as diabetes thanks to the development of all-printed temporary tattoo-based glucose sensors for noninvasive glycemic monitoring, so-called 'electronic tattoos', (Bandokar et al.,

Human Rights and Emerging Technologies: Analysis and Perspectives in Europe
Daniele Ruggiu
Copyright © 2018 Pan Stanford Publishing Pte. Ltd.
ISBN 978-981-4774-93-2 (Hardcover), 978-0-429-49059-0 (eBook)
www.panstanford.com

2015) or of bionic contact lens having the same diagnostic purpose for persons with diabetes (Lingley and Parviz, 2016).

Thanks to the popularization of emerging technologies, wearable technologies such as mobile phones, fitness and health trackers, permit forms of self-screening which meliorate the accuracy of available information for the large public (Barr and Wilson, 2014). This will lead to forms of mobile Health ('mHealth') which have the effect of directly engaging the public in the protection of their health (EGE, 2016, p. 19).

In this connection, the development of AI can gain several results thanks to the usage of Big Data in the health field supplying Apps and services of diagnostics and even, in certain instances, therapy (Annovi, 2016). In the future, the availability of a digital doctor will be a reality for many (Frey and Osborne, 2013). And the development of domotics (home automation) and automatized forms of transport, such as self-driving cars, will pave the way to the constant monitoring of our health state and prompt feeding of these health services with new updated data (e-health). This not only raises concerns related to privacy.[1]

At the same time, in fact, thanks to this new technological availability people will have the possibility of making research on their own. Therefore, forms of traditional top-down medicine will coexist with bottom-up forms, paving the way, one day, to new expressions of scientific research which will pretend to find protection by public authorities.

An illuminating example of this incoming future occurred in 2014. A group of families, parents of 4 and 5-year-old children suffering for type 1 diabetes, created for them the Nightscout Project[2] an innovative do-it-yourself mobile technology system for this type diabetes (Lee, Hirschfeld, 2016). This continuous glucose monitoring system (CGMS) was the outcome of this collective work.

First, the father of a young patient (who was a software programmer) developed 'a computer code that would enable him

---

[1] On this see Chapter 9 of Part II.
[2] https://jamanetwork.com/journals/jama/fullarticle/2512793. Accessed 29 November 2017.

to access the blood glucose readings from the CGMS receiver to the computing cloud through a smartphone. With the data in the cloud, the blood glucose levels could be viewed by the parents from anywhere to provide a continuous monitoring solution. When the father successfully transmitted his son's blood glucose data to the cloud, he sent a tweet of his achievement through the social media platform' (Lee and Hirschfeld, 2016, p. 1447).

Then, the group made the computer code open source. This included: the smartphone application for transferring data from the CGMS to the cloud, the Web application to display values stored by the CGMS, and the watch face for a wearable device that displays the values open source. The group also created a website which hosted the code and do-it-yourself written instructions and informational videos for setting up the system and a private Facebook group, called CGM, which 'rapidly increased to more than 15.000 members in the United States within 18 months, and has expanded to include more than 4.000 members in a number of other countries' (Lee and Hirschfeld, 2016, p. 1447).

This bottom-up, patient driven approach to health production represents today an important lesson for a new era of medicine powered by patient engagement, mobile technology, cloud computing, and social media (Lee and Hirschfeld, 2016, p. 1447).

Public engagement in the field of scientific research also paved the way to instances of participatory monitoring of health data in the past. These forms of participatory surveillance in the field of health, also called 'citizens' veillance', gave rise to several cases, such those occurred in Sardinia (Italy) of biobanking for genomics, namely the organized and participatory storage of human biological samples and information, where 'the Sarroch municipality, together with other stakeholders, promoted the project in 2006 as a complex set of epidemiological investigations with the purpose of using science for timely policy measures' (Tallacchini et al., 2014, p. 29).

In Sardinia, in fact, there are some communities of people who are centenary and whose genome is presumed shielding the secret to their long life. The study of their genes, therefore, is deemed of paramount importance, for example, in the pharmaceutical sector.

In the Sarroch case of citizens' veillance the biobank was physically located in the village (in the South Sardinia). The foundation was set up not as linked to a scientific institution but as an independent entity collectively owned by citizens. Moreover, in 2012 the Sarroch Bioteca Foundation was officially recognized and established as a trusted entity. All citizens were entitled to be members, but they were asked to formally give their adhesion to the project, and subsequently to freely volunteer. Therefore, they were asked to give an ad hoc informed consent before being enrolled in specific research by providing their biological samples.

This extraordinary example poses the question of the possibility of taking research back by the society, meaning that people can be entitled of a specific individual right to scientific research.[3]

10.2 The Protection for Freedom of Scientific Research in the International and European Law

10.2.1 This interesting issue was tackled by EGE in its 2015 opinion *The Ethical Implications of New Health Technologies and Citizen Participation* (EGE, 2015, pp. 45–46). This could be deemed an excellent example of the application of an approach based on human rights at the EU level. Here the legal conceptualization of citizens as 'active actors' of their lives and health was found in the human rights law both international and European.

Freedom of research appears as a cluster of rights. Within the concept of freedom of research, in fact, we can address several rights: the right to manifest one's own beliefs and ideas of scientific nature, the right to carry out research, the right to have access to scientific information, the right to participate in decisions related to science, as well as (public) health, the right to have one's own inventions, insights, and ideas protected through patent law, etc.

These aspects of freedom of scientific research were recognized at the international law level.

---

[3]On this see also Chapter 3 Part I on the socio-empirical approach to RRI.

First, freedom of research is protected by Article 27[4] of the Universal Declaration on Human Rights (UDHR)[5] which recognizes the 'right freely to participate in the cultural life of the community' as well as the right 'to share in scientific advancement and its benefits'. Moreover, it acknowledges for the inventor the protection 'of the moral and material interests resulting from any scientific'.

Next, Article 15[6] of the International Covenant on Economic, Social and Cultural Rights[7] acknowledges a specific obligation for the States to ensure not only the participation to scientific research but also to 'enjoy the benefits of scientific progress and its applications'.

At the regional level, besides, Article 22 of African Charter[8] recognizes a specific right to development having collective nature. According to it: '[a]ll peoples shall have the right to their economic, social and cultural development with due regard to their freedom and identity and in the equal enjoyment of the common heritage of mankind'. After the colonial age, African nations claimed a set

---

[4] Article 27

Everyone has the right freely to participate in the cultural life of the community, to enjoy the arts and to share in scientific advancement and its benefits.

Everyone has the right to the protection of the moral and material interests resulting from any scientific, literary or artistic production of which he is the author.

[5] ONU (1948) *The Universal Declaration on Human Rights* (UDHR), adopted in New York on 10 December 1948 by the General Assembly of the United Nations.

[6] Article 15

The States Parties to the present Covenant recognize the right of everyone:

(a) To take part in cultural life;

(b) To enjoy the benefits of scientific progress and its applications;

(c) To benefit from the protection of the moral and material interests resulting from any scientific, literary or artistic production of which he is the author.

The steps to be taken by the States Parties to the present Covenant to achieve the full realization of this right shall include those necessary for the conservation, the development and the diffusion of science and culture.

The States Parties to the present Covenant undertake to respect the freedom indispensable for scientific research and creative activity.

The States Parties to the present Covenant recognize the benefits to be derived from the encouragement and development of international contacts and co-operation in the scientific and cultural fields.

[7] ONU (1966) *International Covenant on Economic, Social and Cultural Rights* (ICESCR), adopted in New York on 16 December 1966 by the General Assembly of the United Nations (entered into force on 3 January 1976).

[8] Organisation of African Unity (1981) *African Charter on Human and Peoples Rights* (ACHPR), adopted in Nairobi 27 June 1981 and entered into force 21 October 1986.

of rights against the interference of foreign countries, *inter alia*, the right to development. This right however pones some problems (Pariotti, 2013, p. 168ff.). First as regard to the rights holder which is problematic to be identified (people, the State, individuals?). This right, therefore, represents a group right which can be not easy to be conceptualized from the legal standpoint. Moreover, it can result in conflict with other rights of individual nature. These makes difficult to create concrete mechanisms of protection.

Although disputed, it is worth noting that the right to development was subsequently proclaimed by the United Nations in 1986 in the Declaration on the Right to Development,[9] which was adopted with resolution 41/128 by the United Nations General Assembly. Then, it was reaffirmed into the 1993 Vienna Declaration[10] (Art. 10).[11] The Vienna Declaration states that '[e]veryone has the right to enjoy the benefits of scientific progress and its applications', noting 'that certain advances, notably in the biomedical and life sciences as well as in information technology, may have potentially adverse consequences for the integrity, dignity and human rights of the individual, and calls for international cooperation to ensure that human rights and dignity are fully respected in this area of universal concern'.

This potential conflict of scientific research with other human rights, in particular of environmental nature, led to the 2002 Johannesburg Declaration on sustainable development.[12]

At the European level, instead, the European Convention on Human Rights (ECHR),[13] established a freedom to participate in discussions regarding matters of public health that can be found in

---

[9]http://www.un.org/documents/ga/res/41/a41r128.htm. Accessed 30 November 2017.

[10]ONU (1993) *Vienna Declaration and Programme of Action*, adopted by the World Conference on Human Rights in Vienna on 25 June 1993.

[11]The World Conference on Human Rights reaffirms the right to development, as established in the Declaration on the Right to Development, as a universal and inalienable right and an integral part of fundamental human rights.

[12]http://www.un-documents.net/jburgdec.htm. Accessed 30 November 2017.

[13]Council of Europe (1997) *Convention for the Protection of Human Rights and Dignity of the Human Being with regard to the Application of Biology and Medicine* (Convention on Human Rights and Biomedicine or the Oviedo Convention) (CETS n. 164), adopted in Oviedo on 4 April 1997 (came into force on 1 December 1999).

Article 10 (Freedom of expression)[14] which includes in the freedom of thought the right 'to hold opinions and to receive and impart information and ideas without interference by public authority and regardless of frontiers'.

The linkage of freedom of research and freedom of thought is underlined also in other contexts. According to the Explanatory Memorandum to the EU Charter of Fundamental Rights,[15] for instance, the right to freedom of scientific research should be primarily deduced from the right to freedom of thought and expression.

However, within the Council of Europe it is again Article 8 ECHR[16] the probable main reference point of freedom of research thanks to the fact that it covers all aspects of self-determination, also those related to the development of ideas and insights of scientific nature. In this regard the general acknowledgement that the Convention is a 'living instrument' that must be interpreted according to present-day conditions by the Strasbourg Court, often linked to the application of Article 8, makes this provision the best candidate to cover these profiles (Flear et al., 2013). Therefore, when the right to carry out a research free from interferences of public

---

[14] Article 10 — the right to freedom of expression.

Everyone has the right to freedom of expression. This right shall include freedom to hold opinions and to receive and impart information and ideas without interference by public authority and regardless of frontiers. This Article shall not prevent States from requiring the licensing of broadcasting, television or cinema enterprises.

The exercise of these freedoms, since it carries with it duties and responsibilities, may be subject to such formalities, conditions, restrictions or penalties as are prescribed by law and are necessary in a democratic society, in the interests of national security, territorial integrity or public safety, for the prevention of disorder or crime, for the protection of health or morals, for the protection of the reputation or rights of others, for preventing the disclosure of information received in confidence, or for maintaining the authority and impartiality of the judiciary

[15] European Union (2000) *Charter of Fundamental Rights of the European Union*, proclaimed in Nice on 7 December 2000.

[16] Article 8 – respect for private and family life

Everyone has the right to respect for his private and family life, his home and his correspondence.

There shall be no interference by a public authority with the exercise of this right except such as is in accordance with the law and is necessary in a democratic society in the interests of national security, public safety or the economic wellbeing of the country, for the prevention of disorder or crime, for the protection of health or morals, or for the protection of the rights and freedoms of others.

powers is at stake, Article 8 is perhaps the most suitable means for this purpose.

As said, freedom of research is able to cover several other aspects involved in science. First, we need to bear in mind that this freedom of science can hide interests of public nature such those of the State in the economic development of the nation through the advances of knowledge, as well as the right to development. In these instances, the Strasbourg Court is used to acknowledging a margin of appreciation to the State subject to the scrutiny of the Court in order to determine whether the exercise of the State's discretion represents a legitimate interference with the rights protected by, for example, Article 8 ECHR.

Moreover, freedom of research can take the form of rights of economic nature (intellectual property rights) which can conflict with other rights or values of the international human rights law such as human dignity. In Chapter 5 of the Part II (human dignity) we mentioned the *Brüstle case*.[17] The *Brüstle case* is a case where the claim for the patent of their own inventions (in this instance, isolated and purified neural progenitor cells for the treatment of neural defects) may conflict with the rights and freedoms of others (such as the protection of dignity of human being). This can give rise to a tension with the rights enshrined in the Oviedo Convention[18] and its Protocols. In this case, in any event, under the ECHR system these rights can find protection by Article 1 of the Protocol 1,[19] but it can find several limitations such as those on dignity.

---

[17] *Oliver Brüstle v Greenpeace eV* (Case C-34/10) judgement of 18 October 2011 *OJ C 362*, p. 5. See also *International Stem Cell Corporation v. Comptroller General of Patents and designs and Trade Marks* (C-364/13) judgement of 18 December 2014, not published yet. On the embryonic research within the ECHR see *Parrillo v. Italia* (Appl. 46470/11) judgement of 27 August 2015.

[18] Council of Europe (1997) *Convention for the Protection of Human rights and Dignity of the Human Being with regard to the Application of Biology and Medicine* (Convention on Human rights and Biomedicine or the Oviedo Convention) (CETS n. 164), adopted in Oviedo on 4 April 1997 (entered into force on 1 December 1999).

[19] Article 1 – Protection of property Every natural or legal person is entitled to the peaceful enjoyment of his possessions. No one shall be deprived of his possessions except in the public interest and subject to the conditions provided for by law and by the general principles of international law. The preceding provisions shall not, however, in any way impair the right of a State to enforce such laws as it deems necessary to control the use of property in accordance with the general interest or to secure the payment of taxes or other contributions or penalties.

In this regard, the Council of Europe and its parliamentary Assembly treated scientific research manly as a potential threat for the human being, for example, with regard to those applications of scientific research to the human being in the field of genetic engineering, involving foetus or other vulnerable subjects. In this regard the Oviedo Convention and its Protocols are exemplary of this approach to science.

10.2.2 The right to take part actively to scientific research represents therefore only one aspect of the interests affected by scientific research. To focus only on this dimension might lead us to forget some other profiles which may result in crucial in the technology assessment. In connection to a given technology, there is often a ramification of rights which must be taken into account. Therefore, the focus on one aspect can give only a partial image of the interests involved, weakening the final framework.

The Nightscout Project, for example, did not tackle the issue of how the mobile technology system developed by patients' parents was built, and how were the work conditions of those who contributed to this technology. Nor the health issues related to CGMS are deepened but only presumed as solved. Were there health concerns related to that technology and how were thy faced? Could they be raised by the new usage which has been created for this technology? These aspects, instead, remained in the shadow. This is not a mere rhetorical argument. They further developed the CGMS as approved by the Food and Drug Administration. Therefore, the initial usage conditions of the technology mutated, especially in an open source environment. However, how a technology is made, where is built, under which conditions (labour and environmental conditions) cannot be entirely separated from the question, though important, of user participation.

Furthermore, and more importantly, nothing it is said about the safeguards for privacy and data protection of the young patients which were greatly challenged by the adoption of open source forms and by the use of social media. It is clear that a technology using sensitive data, which operates in an open environment, requests stronger guarantees with regard to the protection of data, especially because there is children's information at stake. This does not mean that the technology is not good, but that it is probably fragile. At least under these aspects.

An example of this weakness related to the privacy is another instance of citizens' veillance coming, again, from Italy. In 2000 Tiscali founded a research institution, *SharDna* aimed at studying the DNA of a long-lived population of the East Sardinia (Ogliastra). The population was involved in the project with the aim of sharing benefits of research. In 2009 the company, which substituted *SharDna*, went to bankrupt and its biobank storing more than 230.000 biological samples of this community was purchased by an English company of London, *Tiziana Life Sciences Plc.*, specialized in targeted pharmaceutical drugs, which wanted to transfer all biological samples in U.K. In the meantime, biological samples were stolen and found before they were sent abroad. In 2016 the Italian Data Protection Authority blocked the proceeding of sensitive data and ordered to gain a new consent from the population since the original usage of genetic information was mutated.[20] Furthermore, the case gave rise to a criminal proceeding for theft which is actually pending.[21]

10.3 The Strasbourg Court Jurisprudence Related to Freedom of Research

10.3.1 The difficulties of conceptualizing freedom of research within the ECHR can be well exemplified in the *Parrillo case*.[22] In this context the limitations of the right at issue, such as those concerning human dignity, can be an obstacle for its recognition at the judicial level.

In 2002 the applicant had recourse to assisted reproduction techniques, undergoing in vitro fertilization ('IVF') treatment with her partner at the Centre for reproductive medicine at the European

---

[20]http://www.quotidianosanita.it/allegati/allegato8661127.pdf.   Accessed   30 November 2017.

[21]https://www.quotidianosanita.it/sardegna/articolo.php?articolo_id= 55347. Accessed 30 November 2017.

[22]*Parrillo v. Italy* (Appl. 46470/11) judgement of 27 August 2015 *Reports of Judgements and Decisions*, 1998.

Hospital in Rome. The five embryos obtained from the IVF treatment were placed in cryopreservation. Before the embryos could be implanted the applicant's partner died, on 12 November 2003, in a bomb attack in Nasiriya (Irak) while he was reporting on the war.

After deciding not to have the embryos implanted, the applicant sought to donate them to scientific research and thus contribute to promoting advances in treatment for diseases that are difficult to cure. The applicant therefore asked the director of the centre to release the five cryopreserved embryos so that they could be used for stem-cell research. The director refused to comply with her request on the grounds that this type of research was banned and punishable as a criminal offence in Italy under Section 13 of Law No. 40/2004.

However, the arguments upheld by the applicant led to the complete dismissal of the application and to the lack of acknowledgement of her right to contribute to scientific research. According to the applicant, in fact, embryos conceived by in vitro fertilization could not be regarded as 'individuals' because if they were not implanted they were not destined to develop into foetuses and be born. She concluded that, from a legal point of view, they merely were 'possessions'.

The Strasbourg Court agreed that in certain circumstances a 'legitimate expectation' of obtaining an asset may also enjoy the protection of Article 1 of Protocol No. 1 (property). However, the Court considered that it did not apply to the present case. Having regard to the economic and pecuniary scope of that Article, human embryos cannot in principle be reduced to 'possessions' within the meaning of that provision. In this instance, human dignity was at stake, although the Court never recognized embryos as persons. Therefore, as Article 1 of Protocol No. 1 to the Convention was not applicable in the instant case, the application was rejected as incompatible *ratione materiae*.

10.3.2 A first case where the Strasbourg Court drew the freedom of research on Article 10 (freedom of thought) is the *Hertel case*.[23]

---

[23] *Hertel v Switzerland* (Appl. 59/1997/843/1049) judgement of 25 August 1998 *Reports of Judgements and Decisions*, 1998.

Freedom of scientific research, in fact, covers freedom to carry out research following personal insights.

In collaboration with a professor at the University of Lausanne and a technical adviser at the Lausanne Federal Institute of Technology, Mr. Hertel carried out a study of the effects on human beings of the consumption of food prepared in microwave ovens. Over a period of two months, the blood of eight volunteers who followed a macrobiotic diet was analysed before and after consuming eight types of food (some were cooked or defrosted in a microwave oven and the others were raw or cooked by conventional means). Then, a research paper was written in 1991.

On 1992 the Swiss Association of Manufacturers and Suppliers of Household Electrical Appliances ('the MHEA') applied to the Swiss Court for an interim order prohibiting Mr Franz Weber, on pain of the penalties, from using the image of a man's skeleton or any other image suggesting the idea of death associated with the graphic of a microwave oven, as well as from stating that microwave ovens must be abolished and their use banned, and that scientific research proves what a hazard food that has been exposed to radiation in a microwave oven is to health.

The proceeding before the Swiss Court led to a prohibition for the applicant from stating that food prepared in microwave ovens was a danger to health.

This measure was considered a breach of the Convention. According to the judges in Strasbourg, indeed, this prohibition constituted a violation of Article 10 (Freedom of expression) since '[t]he effect of the prohibition was to censor the applicant's work and substantially to reduce his ability to put forward in public views which have their place in a public debate whose existence cannot be denied. It matters little that his opinion is a minority one and may appear to be devoid of merit since, in a sphere in which it is unlikely that any certainty exists, it would be particularly unreasonable to restrict freedom of expression only to general accepted ideas'.[24]

10.3.3 Freedom of scientific research also includes the right to have access to information and to public participation in decisions

---

[24]Ibid., par. 50.

related to (public) health. In this case rather than to refer to Article 10 ECHR, which expressly imposes on the State a negative duty not to interfere with the freedom to receive and impart information, it is better to refer to Article 8 in order to claim a general right of access to information, including administrative data and documents (EGE, 2016, p. 46).

As seen in Chapter 5 of Part II (Right to health), the *Roche case*[25] is an important case on clinical trial. In this context the Strasbourg Court acknowledged the right to access to information of health nature as general principle.

Michael Roche took part to several medical experimentations during his military service between 1953 and 1968. Once tests terminated, he discovered several diseases linked to his participation in those medical experimentations. However, the Government obstructed the disclosure of health reports related to those tests. Nor it gave rise to public research studies able to overtake the legitimate concerns of the soldiers involved in experiments. Only following some proceedings, the Government decided to grant the access to a part of his medical documentation. According to the Court, notwithstanding was not proven the causal link between the participation to experiments and the disease, the State has a positive obligation to ensure the access to all information concerning the health of an individual.[26]

This right to have access to scientific data raises especially in the environmental field. As seen in Chapter 8 of Part II (Right to a healthy environment), the lack of disclosure of this information can impair the exercise of other relevant rights such as the right to fair trial (Art. 6 ECHR).[27]

---

[25] *Roche v. The United Kingdom* (App. 32555/96) judgement of 19 October 2005 *Reports of Judgements and Decisions*, 2005-IX.

[26] Ibid., par. 166.

[27] Article 6 – Right to a fair trial

In the determination of his civil rights and obligations or of any criminal charge against him, everyone is entitled to a fair and public hearing within a reasonable time by an independent and impartial tribunal established by law. Judgement shall be pronounced publicly but the press and public may be excluded from all or part of the trial in the interests of morals, public order or national security in a democratic society, where the interests of juveniles or the protection of the private life of the parties so require, or to the extent strictly necessary in the opinion of the court in

The *McGinley and Egan case*[28] gives us and idea of how issues concerning the access to information can be covered by other articles of the Convention such as Article 6. Mr Kenneth McGinley and Mr Edward Egan participated in nuclear tests conducted by the United Kingdom at Christmas Island in the Pacific Ocean in 1958. Following the discovery of a number of diseases linked to those experiments, they requested for health records in relation to the Christmas Island detonations in order to obtain a disability pension. Their request was refused on the grounds of national security by public authorities. Consequently, given the refusal from the Secretary of State, they lodged a civil proceeding for compensation, which ended with a dismissal for not having correctly applied existing laws (Rule 6 of the 1981 Rules). Although the Court acknowledged in principle the existence of a positive obligation to provide an effective and accessible procedure enabling the applicants to have access to all relevant and appropriate information, it concluded that the applicants 'had sufficiently detailed knowledge' about the existing procedures available at that time.[29] As noted, '[w]here a Government engages in hazardous activities, such as those in issue in the present case, which might have hidden adverse consequences on the health of those involved in such activities, respect for private and family life under Article 8 requires that an effective and accessible procedure be established which enables such persons to seek all

---

special circumstances where publicity would prejudice the interests of justice.

Everyone charged with a criminal offence shall be presumed innocent until proved guilty according to law.

Everyone charged with a criminal offence has the following minimum rights: (a) to be informed promptly, in a language which he understands and in detail, of the nature and cause of the accusation against him; (b) to have adequate time and facilities for the preparation of his defence; (c) to defend himself in person or through legal assistance of his own choosing or, if he has not sufficient means to pay for legal assistance, to be given it free when the interests of justice so require; (d) to examine or have examined witnesses against him and to obtain the attendance and examination of witnesses on his behalf under the same conditions as witnesses against him; (e) to have the free assistance of an interpreter if he cannot understand or speak the language used in court.

[28] *MacGinley and Egan v The United Kingdom* (App. 21825/93 and 23414/94) judgement of 26 November 1996 *Reports*, 1998-III; *MacGinley and Egan v. The United Kingdom* (App. 21825/93 and 23414/94) judgement (revision) of 28 January 2000 *Reports of Judgements and Decisions*, 2000-I.

[29] Ibid., par. 35.

relevant and appropriate information'.[30] Lastly, no infringement of the Convention occurred, but this principle constitutes a reference point for future cases concerning science.

References

Acimovic Srdjan, S., Ortega, M.A., Sanz, V., Berthelot, J., Garcia-Cordero, J.L., Renger, J., Maerkl, S.J., Kreuzer, M.P., and Quidant, R. (2014). LSPR Chip for Parallel, Rapid, and Sensitive Detection of Cancer Markers in Serum, *Nano Letters*, **14**, pp. 2636–2641.

Annovi, G. (2016). La grandezza risiede nella partecipazione, *Forward*, **4**, pp. 13–15.

Bandokar, A., Jia, W., Yardımcı, C., Wang, X., Ramirez, J., and Wang, J.(2015). Tattoo-Based Noninvasive Glucose Monitoring: A Proof-of-Concept Study, *Analytical Chemistry*, **87**, pp. 394–398.

Barr, A., and Wilson, R. (29.10.2014). Google's Newest Search: Cancer Cells. The Wall Street Journal, http://www.wsj.com/articles/google-designing-nanoparticles-to-patrol-human-body-for-disease-1414515602. Accessed 29 November 2017.

European Group on Ethics in Science and New Technologies (EGE) (2016). *The Ethical Implications of New Health Technologies and Citizen Participation*, Brussels, 13 October 2015. Luxemburg: Publications Office of the European Communities

Lee, J.M., and Hirschfeld, E. (2016). A Patient-Designed Do-It-Yourself Mobile Technology System for Diabetes Promise and Challenges for a New Era in Medicine, *Journal of American Medical Association*, **315**(14), pp. 1447–1448.

Frey, C.F., and Osborne, M.A. (2013). *The Future of Employment: How Susceptible Are Jobs to Computerisation?* Oxford: Oxford University Press.

Lingley, A., and Parviz, B. (2008). Multipurpose Integrated Active Contact Lenses, *The Neuromorphic Engineering*, http://www.ine-news.org/pdf/0056/0056.pdf. Accessed 29 November 2017.

Murphy, T., and Ó Cuinn, G. (2010). Works in Progress: New Technologies and the European Court of Human Rights, *Human Rights Law Review*, **10**(4), pp. 601–638.

---

[30]Ibid., par. 20.

Pariotti, E. (2013). *I diritti umani: concetto, teoria, evoluzione*, Padova: CEDAM.

Stemerding, D. (2017). NBIC Convergences a Challenge for Bioethics and Biopolitics, in: Caenazzo, L., Mariani, L. and Pegoraro, R. (eds.), *Convergence of New Emerging Technologies*, Padova: Piccin, pp. 23–61.

Tallacchini, M.C., Boucher, P., and Nascimento, S. (2014). *Emerging ICT for Citizens' Veillance*, European Commission, JRC Science and Policy Reports.

Acknowledgements

A book is never the outcome of one lonely individual's efforts. This too isn't. It has benefited from the results of several publications and from the workshops and conferences that I had the opportunity to take part in from 2009.

Therefore, I need to first thank Anna and my chidren, Nicolò and Sebastiano, for the endless patience they showed during the preparation of this work.

I am also grateful to Phil Macnaghten for his explanations concerning the rise of the Commission Code of Conduct for responsible nanoscience and nanotechnology research (Potsdam 14-15.04.2014) and his comments on RRI, especially with regard to the underlying reasons of its socio-empircal version (Trento 7-8.11.2015).

I would like to express my gratitude to Elena Pariotti for her suggestions on the main features of the RRI model.

I would like to express my sincere thanks also to Roger Brownsword for his clarification related to the different conceptualisations of human dignity and for his help when I worked on the part regarding the right to bodily integrity.

I am grateful to everyone at Pan Stanford Publishing who worked together with me to accomplish this result.

This work had the support of the Centre for Environmental, Ethical, Legal and Social Decisions on Emerging Technologies (CIGA) of the University of Padua, which I thank, together with my colleagues.

Lastly, demerits are all mine.

Index

Printed in the United States
by Baker & Taylor Publisher Services